Proceedings in Life Sciences

Plant Growth Regulation

Proceedings of the 9th International Conference
on Plant Growth Substances
Lausanne, August 30 - September 4, 1976

Edited by P. E. Pilet

With 128 Figures

Springer-Verlag
Berlin Heidelberg New York 1977

Dr. Paul-Emile Pilet
Professor of Plant Biology and Physiology
Director of the Institute of Plant Biology and Physiology
of the University of Lausanne, Switzerland

ISBN-13: 978-3-642-66591-2 e-ISBN-13: 978-3-642-66589-9
DOI:10.1007/ 978-3-642-66589-9

Library of Congress Cataloging in Publication Data. International Conference on Plant Growth Substances, 9th, University of Lausanne, 1976. Plant growth regulation. (Proceedings in life sciences) Includes bibliographical references and index. 1. Plant regulators--Congresses. 2. Growth (Plants)--Congresses. I. Pilet, Paul Emile. II. Title. QK745.I55 1976 581.3'1 77-2880

Softcover reprint of the hardcover 1st edition 1977

2130/3020 5 4 3 2 1 0

Preface

The 9th International Conference on Plant Growth Substances was held at the University of Lausanne, August 30-September 4, 1976. This meeting was sponsored by the National Swiss Foundation for Scientific Research (FNRS), the Fonds Herbette of the Faculty of Sciences of Lausanne, by several other official sources, and by a few Swiss industries. The conference was attended by over 420 scientists, representing 42 countries, with a total of 550 participants.

Our 9th conference was rather different from the previous ones. The usual short submitted papers followed by a too brief discussion were replaced by three kinds of scientific communications. First in the *reports*, authors gave a critical analysis of a selection of broad topics, directing their comments to new perspectives of research and methodology. Second, the *paper demonstrations*, presented by over 200 scientists, induced large and fruitful discussions and ensured that personal contacts were made. Finally, during four *workshops*, several interesting points - introduced in the reports and paper demonstrations - were developed and discussed.

In the past, it was a tradition to publish the exhaustive collection of all the short contributions presented at such conferences. However, the justification for such weighty books has grown less and less, firstly because the number of communications has constantly increased and secondly because many of the papers describe "results already published elsewhere, some with preliminary results and conclusions which may later prove to have been premature, and the rest ultimately being published in extenso in the normal journals"[1]. In any case it has never been quite clear for whom such a mammoth volume was intended. "Certainly it cannot be meant for the general plant scientist or student, since it is rather like having the papers from several monthly issues"[2] of journals in plant sciences. "The unfortunate delay in publication" - undoubtedly due to the size of such a book - "must mitigate against its usefulness to physiologists working with plant regulators"[2].

These latter comments and, as previously indicated, the changes in the structure of our conference have induced me, as editor, to present a book that contains only the formal reports read during our meeting and a list of all paper demonstrations given at the conference.

However, as a result of adopting a new way of presenting the "proceedings," the editor had to cope with other difficult technical problems. As there were thirty reports not always dealing with a single theme, it was difficult to delineate distinct chapters. The choice, as well as the order adopted, may, of course, be criticized, but some grouping was necessary to facilitate reading. This book, because it mentions the paper demonstrations that can be easily referred to, will help - I hope - all those looking for recent references and information about problems with which they have little familiarity.

---

[1] L.J. Audus, Endeavour 32, 149 (1973)
[2] J.B. Harborne, Phytochemistry 12, 1199 (1973)

I would like to thank all those colleagues who accepted the invitation to contribute their reports to the present book. I am also grateful to the staff of Springer Verlag for their help and advice during the production of this volume.

Lausanne / March 1977 P.E. Pilet

Contents

Contributors

ADDICOTT, F.T.	Department of Botany, University of California, Davies, California 95616, U S A
ALBERSHEIM, P.	Department of Chemistry, University of Colorado, Boulder, Colorado 80309, U S A
BOCION, P.F.	Biological Laboratories, Dr. R. Maag Ltd., 8157 Dielsdorf, Switzerland
BRUINSMA, J.	Department of Plant Physiology, Agricultural University, Aiboretumlaan 4, Wageningen, Netherland
CHAYLAKHYAN, M.K.	K. A. Timiryazew Institute of Plant Physiology, USSR Academy of Sciences, Botanicheskaja 35, 127273 Moskow, USSR
CHERRY, J.H.	Department of Horticulture, Purdue University, West Lafajette, Indiana 47907, U S A
COCKING, E.C.	Department of Botany University of Nottingham, University Park, Nottingham, WG72 RD, United Kingdom
CROZIER, A.	Botany Department, The University, Glasgow G 128QQ, Scotland
DAVIES, P.J.	Section of Genetics, Development and Physiology, Division of Biological Sciences, Cornell University Ithaca, New York 14853, U S A
DAVIS, W.	Department of Biochemistry, University of Cambridge, Cambridge, United Kingdom
ELLIOTT, M.C.	School of Biology, Leicester Polytechnic, Leicester LE 1 9 BH, United Kingdom
ERION, J.	Department of Biochemistry, Rothamsted Research Station, Harpenden, Herts. AL5 2JQ, United Kingdom
FOX, J.	Department of Botany, University of Kansas, Lawrence, Kansas 66045, U S A
GALSTON, A.W.	Department of Biology, Yale University, New Haven, Connecticut 06520, U S A
GIANFAGNA, T.J.	Section of Genetics, Development and Physiology, Division of Biological Sciences, Cornell University, Ithaca, New York 14853, U S A
GINKEL, U.	Botany Section, University of the Saarland, D-6600 Saarbrücken, FRG

HANSON, D.	MSU/ERDA Plant Research Laboratory, Michigan State University, East Lansing, Michigan 48824, U S A
HENSON, I.E.	Department of Botany and Microbiology, University College of Wales, Aberystwyth Dyfed SY23 3 DA, United Kingdom
HORGAN, R.	Department of Botany and Microbiology, University College of Wales, Aberystwyth Dyfed Sy23 3DA, United Kingdom
KALDEWEY, H.	Botany Section, University of the Saarland, D-6600 Saarbrücken, FRG
KARMANN, R.	Botany Section, University of the Saarland, D-6600 Saarbrücken, FRG
KEFELI, V.I.	K. A. Timiryazew Institue of Plant Physiology USSR Academy of Sciences, Botanicheskaja 35, 127273 Moscow, USSR
KENDE, H.	MSU/ERDA Plant Research Laboratory, Michigan State University, East Lansing, Michigan 48824, U S A
KLÄMBT, D.	Botanical Institute, Meckenheimer Allee 170, D-5300 Bonn, FRG
KUTACEK, M.	Institute of Experimental Botany, CSAV, Kedvoro 15 16630-Prague, CSSR
LABAVITCH, J.M.	Department of Chemistry, University of Colorado, Boulder, Colorado 80309, U S A
MACLACHLAN, G.A.	Biology Department, McGill University Montreal, Que., Canada
MACMILLAN, J.	School of Chemistry, The University, Bristol B 58 1 TS, United Kingdom
MARKER, A.	Department of Plant Physiology and Organic Chemistry, The University of Adelaide, Adelaide, S.A. 5064, Australia
MARRÈ, E.	Centro di Studio del C.N.R. per la Biologia Cellulare e Moloculare delle Piante, Instituto di Science Botaniche, Università di Milano, Milan, Italy
MASUDA, Y.	Department of Biology, Faculty of Science, Osaka City University, Sumiyoski-ku, Osaka 558, Japan
MACNIEL, M.	Department of Chemistry, University of Colorado, Boulder, Colorado 80309, U S A
MORRÈ, D.J.	Department of Medical Chemistry and Pharmacognosy and Biological Sciences, Purdue University, West Lafayette, Indiana 47907, U S A
OSBORNE, D.J.	Agricultural Research Council Unit of Developmental Botany, 181 A Huntingdon Road, Cambridge CB3 ODY, United Kingdom

PALAND, I.	Botany Section, University of the Saarland, D-6600 Saarbrücken, FRG
PALEG, L.G.	Department of Plant Physiology and Organic Chemistry, The University of Adelaide, Adelaide, S.A. 5064, Australia
PÈAUD-LENOEL, C.	Centre Universitaire de Luminy, Biochimie Fonchonnelle des Plantes ER 104, F-13288 Marseille Cedex 2, France
PILET, P.E.	Institute of Plant Biology and Physiology of the University. 1005 Lausanne, Place de la Riponne, Switzerland
PROEBSTING, W.M.	Section of Genetics, Development and Physiology, Division of Biological Sciences, Cornell University, Ithaca, New York 14853, U S A
REEVE, D.R.	Botany Department, The University, Glasgow G12 8QQ, Scotland
SIEVERS, A.	Botanical Institute, University of Bonn, Venusberg 22, D-5300 Bonn, FRG
DE SILVA, W.H.	Biological Laboratories, Dr. R. Maag Ltd., 8157 Dielsdorf, Switzerland
SPOTSWOOD, T.M.	Department of Plant Physiology and Organic Chemistry, The University of Adelaide, Adelaide, S.A. 5064, Australia
VENIS, M.A.	Shell Research Ltd., Shell Biosciences Laboratory, Sittingbourne Research Centre, Sittingbourne, Kent, ME 9 8 AG, United Kingdom
VOLKMANN, D.	Botanical Institute, University of Bonn, Venusberg 22, D-5300 Bonn, FRG
WAIN, R.L.	Department of Biochemistry, Wye College, University of London, Wye, Ashford (Kent), United Kingdom
WAREING, P.F.	Department of Botany and Microbiology, University College of Wales, Aberystwyth Dyfed Sy23 3DA, United Kingdom
WIATR, S.M.	Department of Botany, University of California, Davies, California 95616, U S A
WIGHTMAN, F.	Department of Biology, Carleton University, Ottawa, Ontario K1S 5B6, Canada
WILKINS, M.B.	Botany Department, Glasgow University, Glasgow G 12 8QQ, Scotland

I. The Wall of Growing Cells

1. The Molecular Structure of the Primary Cell Wall and Elongation Growth

P. ALBERSHEIM, M. McNEIL, and J. M. LABAVITCH

Abstract

A previously proposed molecular model for primary cell walls is briefly described, and more recent evidence that supports the model is presented. Evidence is also presented that supports the hypothesis that the primary cell walls of all dicotyledonous plants, and probably, too, the walls of all monocotyledonous plants, are constructed on the same architectural plans using the same or functionally similar polysaccharides. The results of recent experiments designed to detect chemical changes in the wall during elongation growth are summarized. We have concluded that cellular elongation results from and depends on the insertion into the wall of newly synthesized polymers.

A. The Primary Cell Wall - A Structural Model

The cell walls of growing plant tissues, primary cell walls, are able to extend to many times their initial length. The walls of cells that are no longer growing, secondary cell walls, are modified to become rigid structures that no longer can expand. The remarkable mechanical properties of the growing cell wall can be explained only if we first understand this structure. Our laboratory has investigated the nature of the constituent molecules that make up the wall, and we have presented a model (Fig.1) of how these molecules might be fitted together to produce a wall with the observed characteristics (3, 15). The model, although it remains somewhat tentative, has received considerable support since that time. During the past three years we have been trying to extend the model beyond suspension-cultured sycamore cells, from which the model was derived, and to develop new techniques to refine the model with the hope of eventually elucidating the chemistry of cell wall growth. It should be emphasized that chemical characterization of the molecules of the wall can answer only part of the problem of cell wall growth. We must know the three-dimensional relationship between the cellulose fibers; this is being elucidated in ultrastructural studies (see, e.g., 34). We must also understand how the wall is synthesized.

Ninety percent of the primary cell wall is polysaccharide, and it is this part that we have been studying. The remaining 10% of the wall is protein, and this has been studied most extensively by Lamport and his colleagues at Michigan State University (20-23).

Of the structural materials of the primary cell wall, cellulose is the best known and best understood. It consists entirely of β-1,4-glucan chains. The glucan chains are linear, and in the primary wall they aggregate to form fibers made up of about 40 chains each. The glucan chains in the cellulose fiber are held together by hydrogen bonds.

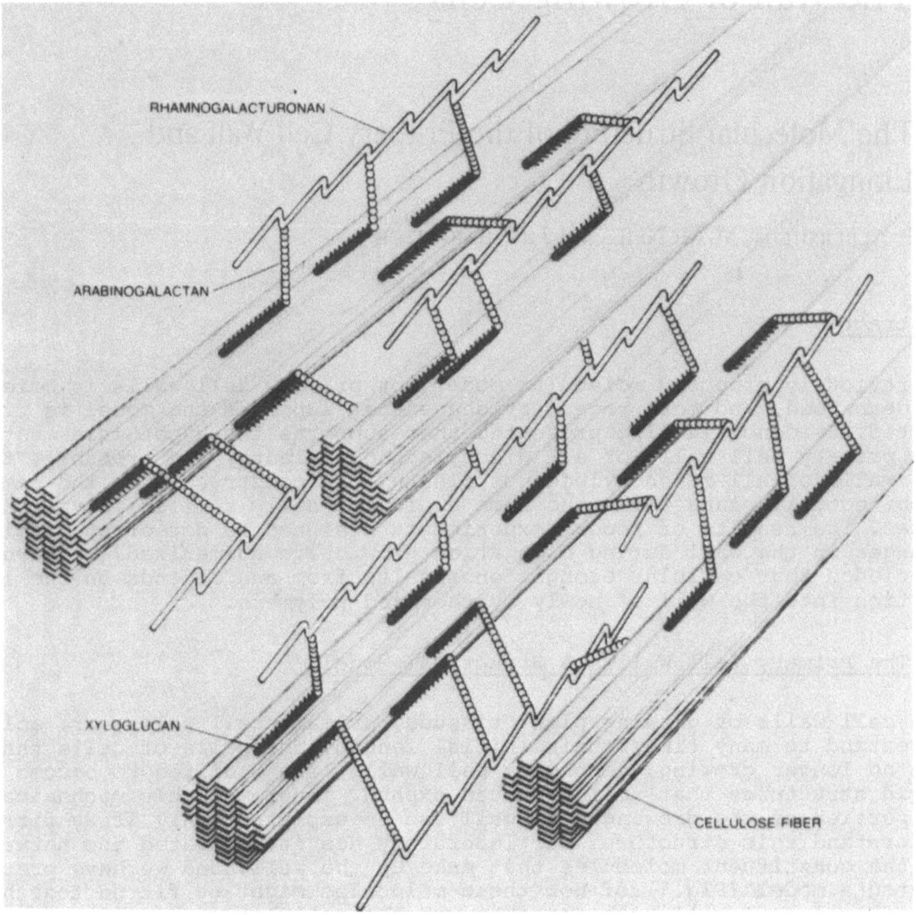

Fig. 1. Model of primary cell walls of dicots. Model is discussed in text. Courtesy of *Scientific American*

These bonds form between the hydrogen atom of a hydroxyl group on one sugar unit and an oxygen atom on another unit. Individually, hydrogen bonds are rather weak, but between the glucan chains of a cellulose fiber there are a great many hydrogen bonds. It is the cellulose fibers that are primarily responsible for the strength of the cell wall.

The large number of hydrogen bonds between glucan chains make the cellulose fiber an extremely cohesive structure. When it is attacked chemically without specific cleavage of the interchain hydrogen bonds, the covalent glycosidic bonds between glucose units frequently break before the chains can be separated. Even enzymes specific for the glycosidic bonds in the glucan chains have difficulty in degrading crystalline cellulose (4, 6, 7, 11-13, 32, 37). Because the chains are tightly packed together, β-1,4-glucanases have trouble approaching the otherwise susceptible glycosidic bonds. Of course, cellulose can eventually be completely hydrolyzed by enzymes. This process appears to involve enzymes, other than hydrolases, yet to be fully characterized.

The unraveling of the mechanisms underlying control of cellulose catabolism is a fascinating and immensely important area of research about which we now know little.

Cellulose fibers make up only a part of the primary cell wall. The fibers are embedded in a matrix of other molecules, most of which are also polysaccharides. In the cell walls that we have examined most closely, there are four other quantitatively important polysaccharides: xyloglucan, arabinan-galactan, and rhamnogalacturonan. Xyloglucan composes the classical hemicellulose fraction of the primary cell wall while the arabinan-galactan and rhamnogalacturonan compose the classical pectic fraction of the wall. The chemistry of the xyloglucan, of the arabinan-galactan, and of the rhamnogalacturonan is considerably more complex than that of cellulose. In cellulose, all of the sugar units are identical, and they are all bonded to one another in the same way. In these other polysaccharides, there are two or more kinds of sugar units, connected by several kinds of glycosidic linkages. Fortunately for us, in the molecules of the cell wall, we have encountered only about two dozen kinds of glycosidic bonds. Moreover, the complex polysaccharides in the primary cell walls appear to consist of repeating substructures. By determining how the substructures are put together, the entire molecule can then be described.

When the cell wall polymers are degraded with purified endoglycanases, some of the liberated oligosaccharides contain fragments of more than one polysaccharide. Such fragments indicate that in the cell wall the polysaccharides (excluding cellulose) are bonded covalently to one another. Such composite fragments have enabled us to determine the order in which the polysaccharides are connected within the wall, and these fragments have given us a glimpse of the geometry of the wall polymers. It is from this type of data that a model of the chemical structure of the primary cell walls of sycamore cells was formulated (5, 15, 35).

In our model each cellulose fiber is completely coated with a layer of xyloglucan one molecule thick. The multiple glucose units of the xyloglucan lie parallel to the axis of the cellulose fiber, and they are hydrogen-bonded to the glucan chains of the cellulose fibers. Hydrogen-bonding is possible because the glucan backbone of xyloglucan is structurally identical to the glucan chains of cellulose (Fig.2). Unlike cellulose, however, xyloglucan can form hydrogen bonds on only one side; bonding is impeded on the opposite side by the protruding fucose and galactose units (5, 36).

At their reducing ends, many of the xyloglucan molecules are glycosidically bonded to arabinan-galactan molecules (Fig.1). Although the exact geometric relations are not known, the arabinan-galactan chains (Fig.3) can be thought of as running radially away from the cellulose fiber like spokes radiating from a hub. The reducing ends of the arabinan-galactan chains are glycosidically linked to rhamnogalacturonans. The arabinan-galactans are believed to bond exclusively to rhamnosyl residues, the sugars that give rise to kinks in the rhamnogalacturonan chain (Fig.4). Each cellulose fiber is coated with many xyloglucan molecules. Xyloglucans are bonded to only one arabinan-galactan, and an arabinan-galactan terminates at one rhamnogalacturonan chain. The rhamnogalacturonan chains, however, receive several arabinan-galactan molecules, including chains radiating from more than one cellulose fiber. Rhamnogalacturonan thus binds the fibers into a more or less rigid matrix (Fig.1).

A protein component of the primary wall seems likely to have a structural function (20-23). The wall protein, which Lamport and his col-

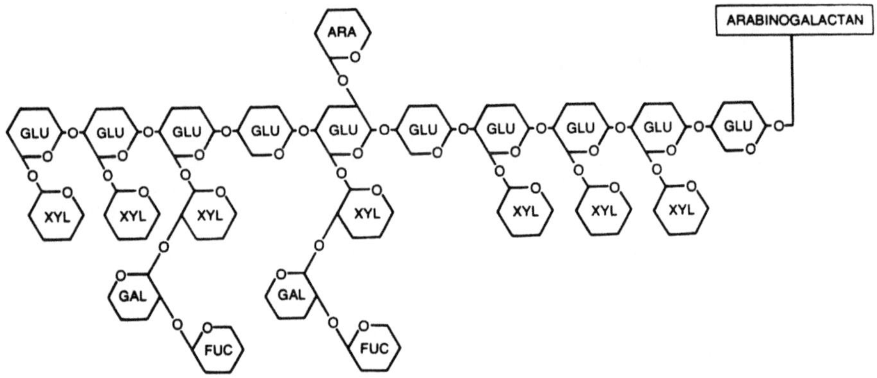

Fig. 2. Model of portion of xyloglucan, the hemicellulose of primary cell walls of dicots. *ARA*: arabinose; *FUC*: fucose; *GAL*: galactose; *GLU*: glucose; *XYL*: xylose. Courtesy of *Scientific American*

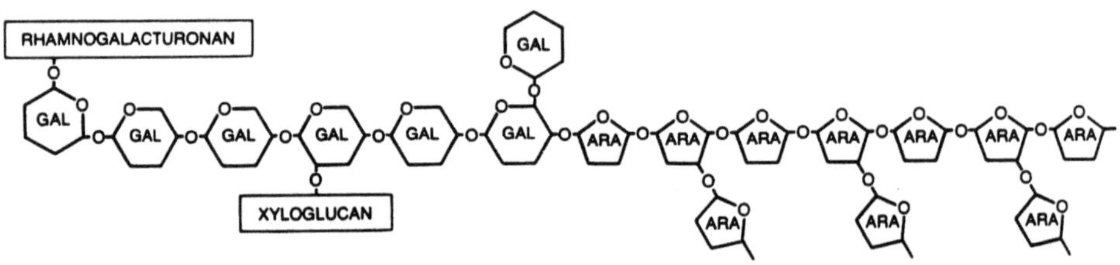

Fig. 3. Model of arabinan-galactan, neutral pectic fraction of primary cell walls of dicots. *ARA*: arabinose; *GAL*: galactose. Courtesy of *Scientific American*

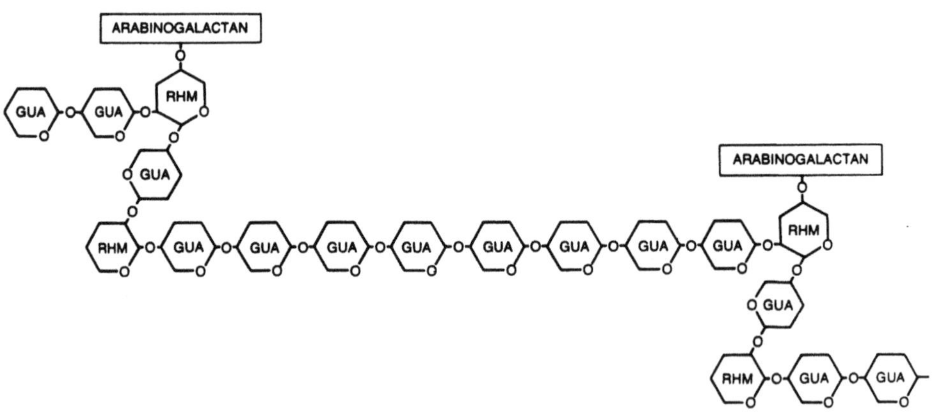

Fig. 4. Model of portion of rhamnogalacturonan, acidic pectic fraction of primary cell walls of dicots. *GUA*: galacturonic acid; *RHM*: rhamnose. Courtesy of *Scientific American*

leagues are working on, is rich in hydroxyproline, an amino acid characteristic of structural proteins in animal systems. The hydroxyproline-rich protein also has sugars attached to it. These properties suggest that this wall protein has a structural function. In our first proposal of a wall structural model (15), we indicated that a coherent model of the wall can be formulated both with and without inclusion of the hydroxyproline-rich protein. It is still true that we can draw a structural model of the wall that ignores this protein. The function of this protein within the wall awaits additional evidence.

B. More Recent Investigations into the Structure of Primary Cell Walls

Our present research emphasis is directed to the identification of the detailed structures of the polysaccharides that interconnect the cellulose fibers. The structure of the xyloglucan that coats the cellulose fibers is well understood. The structure of the rhamnogalacturonan that acts as a focal point of attachment of several cellulose fibers is fairly well understood. However, less is known about the structure of the arabinan-galactan that interconnects the xyloglucan and rhamnogalacturonan. It is not known whether the arabinan is attached to the galactan or the galactan to the arabinan. It is not known through what linkage and to what sugar xyloglucan is attached to the arabinan-galactan, nor how the arabinan-galactan is attached to the rhamnogalacturonan.

Highly purified endoglycanases are essential for structural analysis of the wall polysaccharides. The most useful enzymes to date have been a highly purified endopolygalacturonase (10, 35) and a partially purified endoglucanase (5). The latter requires additional purification, and methods are available in the literature for achieving this (4, 6, 7, 11-13, 32, 37).

No such enzymes have been available for characterizing the arabinan-galactan of the dicotyledon cell wall. We have now successfully purified an endo-β-1,4-galactanase for this purpose (16). Wild-type *Bacillus subtilis* has been induced to produce this enzyme by growing the bacterium on soybean arabinan-galactan. The enzyme from the culture fluid has been purified more than 200-fold. Affinity chromatography was the most effective step in a purification procedure that resulted in a preparation that contained only a single 40,000 mol wt protein band upon SDS-disc gel electrophoresis. The purified galactanase digests the galactan portion of the soybean arabinan-galactan as well as a β-1,4-galactan purified from citrus pectin. The purified galactanase digests partially the isolated cell walls of suspension-cultured sycamore cells. The predominant product of the enzymic degradation of the substrates tested is a 4-linked tetragalactoside. The evidence accumulated suggests very strongly that the galactanase attacks its substrates in both an exo- and endo-manner. This sort of dual activity has not, to our knowledge, been previously documented for glycanases. The polygalacturonase secreted by the pathogenic fungus, *Colletotrichum lindemuthianum*, which we have also purified (10), has been described as an endoglycanase, but product analysis suggests that this enzyme, too, can act as an exo- as well as an endoglycanase. Perhaps this is a general property of cell wall-degrading enzymes.

The *B. subtilis* galactanase has helped to elucidate the structure of soybean arabinan-galactan. A previous interpretation of the arabinan-galactan structure (29) suggested that the polymer has a backbone of 1,4-linked galactosyl residues and that the arabinosyl residues were present as mono- or disaccharide constituents on that backbone. Degradation of the polymer with the purified galactanase followed by partial characterization of the enzyme products provides a different picture. After galactanase digestion, almost all of the arabinose in the polymer, but very little of the galactose, voids Bio-Gel P-2, indicating a degree of polymerization of the arabinosides of at least 10. This indicates that the arabinan portion of the arabinan-galactan is organized primarily in rather large oligo- or polyarabinosides. This is exactly the result we get working with the cell wall arabinan-galactans of other dicots. The arabinan-galactan appears to be two homopolymers attached to each other: a highly branched arabinan and a little-branched galactan (Fig.3). It is not clear whether the arabinan is attached to the galactan or the galactan to the arabinan.

The purified endogalactanase has also been useful in studying a fragment of the wall containing the intact arabinan-galactan and containing, too, a small fragment of the xyloglucan attached to the arabinan-galactan and a small fragment of the rhamnogalacturonan to which the arabinan-galactan is attached. This piece of the wall will be called the "interconnecting fragment."

The interconnecting fragment has been obtained as follows: isolated walls were treated with the purified endopolygalacturonase and washed free of the solubilized products of this enzyme. The walls were then extracted with either dimethylsulfoxide or alkali in order to break the hydrogen bonds between the xyloglucans and the cellulose fibers. This extraction resulted in the solubilization of the xyloglucan covalently attached to the arabinan-galactan and the latter covalently attached to a small fragment of the endopolygalacturonase-degraded rhamnogalacturonan. The interconnecting fragment was further purified in a two-step procedure. The fragment was first attached to an anion exchange column utilizing the carboxyl groups of the uronosyl residues of the residual rhamnogalacturonan for binding. The interconnecting fragment was eluted from the anion exchange column in a salt gradient and then further purified by binding to cellulose. The xyloglucan portion of the interconnecting fragments binds strongly to cellulose; the fragment was eluted from the cellulose with either dimethylsulfoxide or alkali. The purified interconnecting fragment was, when desired, reduced in size by treatment with endoglucanase to remove most of the xyloglucan.

We have subjected the interconnecting fragment to the purified endogalactanase (16) in an effort to identify the linkage connecting the xyloglucan fragment to the arabinan-galactan as well as the linkage connecting the arabinan-galactan to the rhamnogalacturonan fragment. The galactanase releases as small oligosaccharides approximately 40% of the galactose of the interconnecting fragment, but does not separate any of the xyloglucan portions of the interconnecting fragment from the rhamnogalacturonan portions of the fragment, thus demonstrating that there is no exposed contiguous region of 4-linked galactosyl residues interconnecting the xyloglucan and the rhamnogalacturonan. We cannot tell whether there is a short or sterically hindered galactan participating in the interconnection of the xyloglucan and rhamnogalacturonan. The results of experiments using mild acid hydrolysis of the wall polysaccharides have suggested that a galactan is responsible, at least in part, for this interconnection (35). In order to solve this problem an enzyme must be isolated that can separate the xyloglucan from the rhamnogalacturonan or else the size of the interconnecting fragment must be substantially reduced. The most likely enzyme to provide this solution is an endo-arabinanase. An endo-arabinanase has been partially purified in our laboratory from a *Sclerotinia* species. We will either continue with the purification of this enzyme or attempt to purify the endo-arabinanase produced by a *B. subtilis* strain, the purification of which has recently been reported (14).

C. Evidence That Our Structural Model Can Be Widely Applied to Walls of Other Cells

Our structural model was derived entirely from experiments with sycamore cells cultured in suspension, but it now appears that it is valid for other kinds of cells as well. Our first solid evidence came from a comparison of the hemicellulose of bean and of sycamore (38). The walls of red kidney bean (*Phaseolus vulgaris*) suspension-cultured cells have a hemicellulose (xyloglucan) that is extraordinarily similar to that of sycamore cells. The similarity of these polymers is demonstrated by their susceptibility to the same hydrolytic enzyme, an endoglucanase, and by the fact that enzymic hydrolysis of each yields an almost identical mixture of oligosaccharide fragments. Even better evidence of this similarity has been obtained by comparing the partially methylated alditol acetates derived from the xyloglucans of each of these plants.

Not only are the xyloglucans of bean and sycamore almost identical, but the manner in which the xyloglucans are attached within these walls is also the same. This was shown by combining chemical, enzymic, and liquid chromatographic methods for isolating the xyloglucans and xyloglucan fragments from the bean and sycamore cell walls (5, 38).

An examination of the literature supplies additional evidence supporting the thesis that the primary walls of all dicotyledons are similar. Even though cell wall analyses of earlier times were generally carried out on wall fractions obtained by chemical rather that enzymatic means, it is impressive that the earlier data obtained from a variety of plants and plant tissues fit well into the structural model that we have proposed (15).

Further evidence for a generalized model of primary walls was obtained by comparing the partially methylated alditol acetates derived from sycamore and beans with those from a variety of plant cell walls. Wolfgang D. Bauer, then at Michigan State University, synthesized these derivatives from the walls of cultured tomato cells (*Lycopersicon esculentum*); Mina Fisher in our laboratory made the derivatives from the walls of suspension-cultured soybean (*Glycine max*) cells, and Bernard Nusbaum, also in our laboratory, obtained this information for both suspension-cultured red kidney bean (*P. vulgaris*) cells and for 8-day-old hypocotyls of red kidney beans. The gas chromatograms of the partially methylated alditol acetates obtained from these different plants are strikingly similar. It is evident that all the cell walls are composed of the same sugars linked in the same way. Quantitative analysis shows that the proportion of the different glycosyl linkages in the walls varies somewhat, but the major components are the same in all the walls (2). These data suggest that all of these cell walls are composed of similar polymers. This information and the fact that these various plant cell walls are degraded in the same manner by the same enzymes that degrade the sycamore walls and that this enzymatic degradation yields similar chromatographic fractions have convinced us that the suspension-cultured cells of such widely diverse plants as beans, tomatoes, and sycamore trees have, architecturally, very similar walls.

The fact that the walls of the hypocotyls of red kidney beans yield partially methylated alditol acetates similar to those obtained from suspension-cultured bean cells is evidence that the walls of suspension-cultured cells are representative of primary walls found in the intact plant. Apparently, secondary wall polysaccharides yield relatively small amounts of partially methylated alditol acetates when secondary walls are treated in the same manner that is successful in synthesizing these derivatives from the polysaccharides of primary cell walls. This finding has been confirmed by the results of the studies of the walls of pea epicotyls carried out independently by Alan Darvill and Michael Hall of the University College of Wales (Aberystwyth) and by the authors. D.H. Northcote of Cambridge University, England, told me that his group has analyzed the sugar composition of walls of excised cambial cells from sycamore trees and found the composition to be indistinguishable from that which we reported for suspension-cultured sycamore cells. T.E. Timell of the State University of New York at Syracuse has told me that he and his student, Bryan Simpson, have been investigating the cell walls of cambial cells removed from the stems of the aspen tree and have found polymers very similar to those obtained from suspension-cultured sycamore cells. Thus, we conclude that the model of the walls of sycamore cells grown in suspension-culture acts as a model for the walls of other suspension-cultured cells, and is also a valid model for the primary walls of cells in intact plants.

Burke, Kaufman, McNeil, and I surveyed the cell walls of six different suspension-cultured grasses (8) to determine whether suspension-cultured monocotyledon cell walls could be compared in structure to dicotyledon cell walls. We compared the fingerprint patterns for the partially methylated alditol acetates obtained from the cell walls of these different grasses, and found a great similarity in the glycosyl linkages in these walls. It is interesting that although the wheat and rice cultures were derived from root tissue, the oat and brome grass from embryo tissue, the sugarcane from internode tissue, and the rye grass from endosperm, the walls of all these tissue cultures are very similar. However, only the cells derived from endosperm have, in significant amounts, the partially methylated alditol acetate indicative of a β-1,3-glucan. We believe the β-1,3-glucan is a storage polysaccharide present in the endosperm walls and is not a structural component of endosperm cell walls. This conclusion was further substantiated by our findings that the walls of barley aleurone cells also completely lack 3-linked glucosyl residues (27), and that oat internode cell walls have less than 1% 3-linked glucan and that this glucan does not disappear with growth. These findings directly contradict those of Nevins and his colleagues at Iowa State University (24-26, 30, 31). These workers suggest that indoleacetic acid induces a decrease in the glucose content of a noncellulosic glucan *wall* polysaccharide and that this decrease is related to indoleacetic acid-induced stimulation of growth. Our findings suggest that Nevins et al. are looking at a loss of a storage polysaccharide that is only secondarily related to wall growth. During their purification procedure, apparently a storage polysaccharide was present in the cell walls. This is not surprising as it has been clearly shown in dicot cell walls that starch contaminates purified cell walls. The starch can be removed with amylase without altering the wall structure (35). Indeed, preliminary results from this laboratory and from Peter Kaufman's laboratory at the University of Michigan demonstrate that there are no significant changes in the composition or structure of the cell walls during the growth of internodes of *Avena* seedlings.

The proportion of the various sugar derivatives in the walls of the different suspension-cultured grass cells is not the same, but all of the walls are based on the same sugars linked in the same manner. Our evidence combined with evidence from other laboratories in which studies of cell walls of monocotyledons have been carried out indicates that the primary walls of monocots will be as structurally similar to each other as the primary cell walls of dicots have been found to be. The partially methylated alditol acetate patterns obtained from these grass cells are very different from those obtained from suspension-cultured dicotyledons, but available evidence strongly suggests that primary cell walls of monocots will have an architectural plan similar to that of dicots and that monocot cell walls will have cellulose fibers interconnected via hemicellulose and pectic polymers. The dominant hemicellulose of the primary cell walls of monocots (arabinoxylan) is not the same as the hemicellulose in the dicot (xyloglucan), but both monocots and dicots do contain, as a major primary cell wall constituent, a hemicellulose that hydrogen-bonds to cellulose (5, 27), and both monocots and dicots contain pectic polymers (8, 35).

D. Chemical Changes Accompanying Elongation Growth of Cell Walls

One possible consequence of rapid growth would be the breaking of some of the interconnections between cellulose fibers. In an elaborate series of experiments with extracts of the walls of pea stem segments, we have determined the percent of xyloglucan chains that are attached, through arabinan-galactan, to rhamnoglacturonans. In other words, we have asked how many of the interconnecting fragments that we can extract with dimethylsulfoxide are intact. We found in three sets of auxin-treated, rapidly expanding pea stem segments that about 80% of the xyloglucan chains are connected to rhamnogalacturonan chains, while in three sets of slow growing, control segments also about 80% of the xyloglucan chains are connected to rhamnogalacturonan chains. Although this experiment must be repeated, the results suggest that there are no measurable differences in the number of interconnections between cellulose fibers in growing and nongrowing primary cell walls.

Peter Kaufman and his co-workers at the University of Michigan have developed an excellent test system for studying the changes, if any, that take place in the growing cell wall. They have characterized on a physiologic level the impressive hormone-controlled elongation of the internodes of oat (*Avena*) stem segments. When such segments are incubated in the presence of gibberellic acid and sucrose, the cells elongate within two days to up to 100 times their original length (1, 28). We are cooperating with Dr. Kaufman to examine the composition and structure of the cell walls of the *Avena* internodes at various times during the elongation process. Our initial findings suggest that there are no significant differences between cell walls that are growing rapidly and cell walls that are growing slowly (Table 1). Our initial detailed investigations of the glycosyl linkage compositions of *Avena* internode cell walls, obtained at different stages of elongation, also demonstrate that there are no measurable differences in these linkages. If these results are substantiated, we will have clearly established that the bonds in the wall that are broken to permit the wall to expand must be reconstituted during growth in a manner indistinguishable from the state of those bonds before the lengthening of the wall occurred. These are important experiments, for, as our distinguished colleague, Hans Burström, said so well in a letter to me: "Nothing more is required to explain cell expansion than the synthesis of cell wall matter and its incorporation (into the wall)."

Hormone stimulation of elongation growth has been said to result from a *temporary* weakening or relaxation of the wall (9). However, during elongation growth, only a small number of the bonds within the wall are broken at any one moment. But since the wall grows throughout its length and since new material is inserted within the entire structure of the wall (33), and since there is no apparent change in wall composition or linkage pattern during growth (unpublished results), many bonds must break and be reformed. When we first proposed our cell wall model, we suggested that the bonds that might be broken are hydrogen

Table 1. Cell wall composition of oat internodes at various stages of growth in presence and absence of gibberellic acid

Growth (h)	0	12		24		48	
Hormone		-Ga	+Ga	-Ga	+Ga	-Ga	+Ga
Growth (cm)	1.0	1.1	1.6	1.3	2.7	1.5	5.9
Rhamnose	2.3	2.3	2.2	2.0	1.5	1.3	1.1
Fucose	0.5	0.6	0.5	0.4	0.5	0.3	0.2
Arabinose	27.8	27.8	28.3	28.5	28.0	28.3	25.2
Xylose	55.4	54.1	53.2	55.9	57.3	57.0	59.8
Mannose	1.4	1.6	1.4	0.9	0.7	0.6	0.7
Galactose	5.5	5.7	5.8	5.1	5.0	5.1	5.8
Glucose	7.7	7.8	8.2	7.4	7.2	7.2	7.3
Cellulose	58	54	62	60	65	44	52
Uronides	9	7	9	9	8	8	7
Proteins	7	7	8	8	7	10	10

Data (from P. Kaufman, M. McNeil, and P. Albersheim, unpublished preliminary results) are presented as % of total neutral sugars and % of wall accounted for by cellulose, uronic acids, and proteins

bonds between the xyloglucans and the cellulose fibers (15), but our more recent evidence suggests that this is not probable (36). We believe that cell wall growth is enzyme-catalyzed (9). The search for such an enzyme or enzymes requires knowledge of the substrates of the enzymes and requires that such substrates be isolated in order to assay for the enzymes.

We have also been trying to identify the specific polymers within the wall in which bonds are broken and resynthesized during elongation growth. Our efforts have been based in part on the observations of Labavitch and Ray (18, 19), who have shown that the auxin-stimulated elongation of pea stem tissues is accompanied by an enhanced rate of conversion of a glucose- and xylose-rich cell wall polysaccharide to a water-soluble form. We have demonstrated (17) that the polysaccharide that Labavitch and Ray were working with is the primary wall xyloglucan, with a structure like that described for the xyloglucans of suspension-cultured cells (5, 38). These findings provide a suggestion as to the identity of the cell wall bonds that are broken and resynthesized as cellular elongation occurs, for we found that auxin enhances the rate at which the cell wall xyloglucan, attached to a fragment of arabinan-galactan, is converted to a water-soluble form (Table 2). Thus, there is a possibility that auxin treatment promotes the cleavage of the arabinan-galactan and that growth may involve the insertion of new wall material into the cleaved crossed-connecting arabinan-galactan.

Regardless of where new wall material is inserted, the structures of the polysaccharides that interconnect the cellulose fibers are of utmost importance in achieving an understanding of elongation growth. These polymers must be cleaved in order to allow the cellulose fibers to move past one another. Further understanding of this process requires that the glycosyl sequences of the polysaccharides involved be identified; it requires that the manner in which the polymers are

Table 2. Radiolabel in component sugars of cellulose-powder-purified water-soluble pea xyloglucan from ^{14}C-labeled pea stem tissues treated for 7 h in 50 mM glucose ±17 μM indoleacetic acid

Sugar	Total radioactivity recovered in each sugar		Ratio
	+IAA	-IAA	+IAA/-IAA
	CPM		
Galactose	2650	1600	1.66
Glucose	13,450	8300	1.62
Xylose	6100	3800	1.61

Detailed description of experimental procedure has been presented elsewhere (19). Experiment was performed twice with similar results

linked together be ascertained; and it requires that sufficient amounts of the interconnecting fragments be isolated so that these polysaccharides can be used as a substrate for the assay of enzymes that can either cleave the fragment or insert additional polysaccharides into the fragment.

Acknowledgements

The authors acknowledge the financial assistance provided by a grant from the United States Energy Research Development Administration and by a University of Colorado Faculty Fellowship.

References

1. Adams, P.A., Kaufman, P.A., Ikuma, H.: Plant Physiol. 51, 1102 (1973)
2. Albersheim, P.: Tissue Culture and Plant Science 1974. H.E. Street, (ed.). London, New York: Academic Press, 1974, pp. 379
3. Albersheim, P.: Sci. Am. 232, 80 (1975)
4. Almin, K.E., Eriksson, K.E., Pettersson, B.: Europ. J. Biochem. 51, 207 (1975)
5. Bauer, W.D., Talmadge, K.W., Keegstra, K., Albersheim, P.: Plant Physiol. 51, 174 (1973)
6. Berghem, L.E.R., Pettersson, L.G.: Europ. J. Biochem. 37, 21 (1973)
7. Berghem, L.E.R., Pettersson, L.G., Axio-Fredriksson, U.B.: Europ. J. Biochem. 53, 55 (1975)
8. Burke, D., Kaufman, P., McNeil, M., Albersheim, P.: Plant Physiol. 54, 109 (1974)
9. Cleland, R.: Ann. Rev. Plant Physiol. 22, 197 (1971)
10. English, P.D., Maglothin, A., Keegstra, K., Albersheim, P.: Plant Physiol. 49, 293 (1972)
11. Eriksson, K.-E., Pettersson, B.: Europ. J. Biochem. 51, 193 (1975)
12. Eriksson, K.-E., Pettersson, B.: Europ. J. Biochem. 51, 213 (1975)
13. Eriksson, K.-E., Pettersson, B., Westermark, U.: FEBS Lett. 49, 282 (1974)
14. Kaji, A., Saheki, T.: Biochim. Biophys. Acta 410, 354 (1975)
15. Keegstra, K., Talmadge, K.W., Bauer, W.D., Albersheim, P.: Plant Physiol. 51, 188 (1973)

16. Labavitch, J., Freeman, L., Albersheim, P.: J. Biol. Chem. 251, 5904, (1976)
17. Labavitch, J.M., McNeil, M., Albersheim, P.: unpublished results
18. Labavitch, J.M., Ray, P.M.: Plant Physiol. 53, 669 (1974)
19. Labavitch, J.M., Ray, P.M.: Plant Physiol. 54, 499 (1974)
20. Lamport, D.T.A.: Biochemistry 8, 1155 (1969)
21. Lamport, D.T.A.: Ann. Rev. Plant Physiol. 21, 235 (1970)
22. Lamport, D.T.A.: Colloques Internationaux C.N.R.S. 212, 27
23. Lamport, D.T.A., Katona, L., Roerig, S.: Biochem. J. 133, 125 (1973) (1973)
24. Leoscher, W., Nevins, D.J.: Plant Physiol. 50, 556 (1972)
25. Leoscher, W.H., Nevins, D.J.: Plant Physiol. 52, 248 (1973)
26. Leoscher, W., Nevins, D.J.: Mech. Reg. Plant Growth 12, 729 (1974)
27. McNeil, M., Albersheim, P., Taiz, L., Jones, R.L.: Plant Physiol. 55, 64 (1975)
28. Montague, M.J., Ikuma, H., Kaufman, P.B.: Plant Physiol. 51, 1026 (1973)
29. Morita, M.: Agr. Biol. Chem. (Tokyo) 29, 626 (1965)
30. Nevins, D.J.: Plant Cell Physiol. (Tokyo) 16, 495 (1975)
31. Nevins, D.J., Loescher, W.: Plant Growth Substances 828 (1973). Proc. 8th Intern. Conf. Plant Growth Substances. Tokyo (1974)
32. Pettersson, G.: Arch. Biochem. Biophys. 130, 286 (1969)
33. Ray, P.M.: J. Cell Biol. 35, 659 (1967)
34. Roland, J.C., Vian, B., Reis, D.: J. Cell Sci. 19, 239 (1975)
35. Talmadge, K.W., Keegstra, K., Bauer, W.D., Albersheim, P.: Plant Physiol. 51, 158 (1973)
36. Valent, B.S., Albersheim, P.: Plant Physiol. 54, 105 (1974)
37. Westermark, U., Eriksson, K.-E.: Acta Chem. Scand. 29, 419 (1975)
38. Wilder, B.M., Albersheim, P.: Plant Physiol. 51, 889 (1973)

Paper Demonstration References (see Section IX):

56 - 73 - 76 - 116 - 117 - 137 - 159 - 162 - 167 - 186

2. Cellulose Metabolism and Cell Growth

G. A. MACLACHLAN

Abstract

The cellulose framework of the primary cell wall extends greatly in mass and area during auxin-evoked cell expansion. Data from studies with auxin-treated pea epicotyls show that β-glucan synthetase activity at the cell surface (tissue slices) can generate alkali-insoluble products from uridine diphosphoglucose (UDPG), part of which contain β-1,4-linkages and possess a high molecular weight (DP > 2000). Surface synthetase is particularly susceptible to inactivation by soluble heat-labile intracellular factor(s) from peas, which may explain its lability during homogenization. Preincubation of tissue slices with pure pea cellulases enhances subsequent glucan synthesis in this system. Thus, availability of acceptor chain ends in the existing wall may be a limiting factor for synthesis. These cellulases are induced by auxin without mediation by ethylene. It is tentatively proposed that cellulose is metabolised in expanding primary walls by tightly regulated enzymic complexes that act at the cell surface to insert new carbohydrate units into the continuous network of wall cellulose, thus extending the restrictive framework surrounding young cells.

A. Introduction

Close examination of the many hundreds of published electron micrographs of growing primary walls does not reveal discontinuities in the microfibrillar cellulose framework at any stage of growth, yet radioautographs show clearly that new cellulose is inserted into the inner surface of the wall at multiple loci during growth (8). This raises serious questions, not only about how the expanding cell loosens or relaxes the circumscribing network of partially crystalline polymer, but also about how new cellulose can be deposited in such a way that fibril integrity is maintained and no discrete new fibril ends appear. The problem is to identify the key reactions required for microfibril extension: they are obviously prime candidates for the elusive target events through which hormones exert control over cell expansion.

It would seem necessary to postulate the existence of a coordinated series of lytic, synthetic, and perhaps even repairing reactions in surface enzyme complexes, the sum of which results in intercalation of new sugar or cellodextrin units into existing microfibrils and extension of the cellulose framework. Such a complex would bristle, of course, with components that growth hormones could potentially regulate, and this brief report can only survey some of the possibilities. It deals primarily with enzyme reactions that catalyse hydrolysis and synthesis of the β-1,4-glucosidic bond as they have been studied in hormone-treated peas. Of necessity, with the present exploratory state of technical expertise in this field, results are often incomplete and inconclusive and must be interpreted critically and with great caution. This is a frontier.

B. Materials and Methods

Growing regions of decapitated pea epicotyls (apical 10 mm segment) were treated with auxin (indole-3-acetic acid (IAA) or 2,4-D) as described earlier (2, 6). Extraction and assay procedures have been published for endogenous pea alkali-insoluble glucan (13, 15), mean degree of polymerization (16), β-glucosidases (5), and cellulases (2). β-glucan synthetase activity was assayed by providing particulate fractions (13) or slices of segments (14, 15) with ^{14}C-labelled glucose nucleotide, 20 mM MOPS buffer, pH 7.25, 10 mM $MgCl_2$, 5 mM dithioerythritol in a total volume of 250 µl. Incubations were for 10 min at 35°C, and incorporation of ^{14}C was measured in material insoluble in hot water, chloroform-methanol and hot (85°C) 1 M NaOH-0.1% $NaBH_4$. All enzyme assays were conducted with limiting enzyme concentrations for a period during which activity was linear. In the case of the viscometric assay used for cellulase activity, a conversion factor of 300 pmol reducing groups produced per unit of viscosity loss was employed to obtain values recorded in the tables.

Analyses of proportions of β-1,4- and β-1,3-linkages in glucan were carried out by assaying hydrolysis products formed following hydrolase treatments or methylation (Hakamori) using standard techniques. Molecular weight distribution of synthetase products was studied following acetylation (Tanghe method) adapted to a microscale by fractionating products soluble in dichloromethane:methanol (9:1) on columns (1 × 100 cm) of glass beads (CPG-500).

C. Results and Conclusions

I. Metabolism in vivo

Table 1 contains a summary of effects of auxin treatment on levels of various components potentially related to metabolism of cellulose in the growing region of the decapitated pea epicotyl. In this system, auxin evokes a 2- to 3-fold increase in fresh weight, protein, and RNA levels in 2 to 3 days, due mainly to lateral expansion and synthesis in cortical cells (6, 17). Concurrently, there is a massive increase in the amount of glucan that is insoluble in lipid solvents or alkali (hot 1*N* and 4*N* at 25°C) (see also 13, 14).

This alkali-insoluble glucan is mainly cellulose. It can be seen in shadowed electron-microscopic preparations to be composed of clean microfibrils (8). It contains glucose only, combined in polymers, which, when nitrated, dissolve completely in acetone and yield a very high viscosity-average molecular weight (degree of polymerization (DP) > 8000, Table 1). Recent assays of its linkages in this laboratory by methylation analysis show a preponderance (> 90%) of β-1,4-glucan.

Alkali-insoluble glucan may not represent all of the cellulose in these growing cells. Alkali-soluble glucan has a substantial content of β-1,4-linkages, and it is possible that endogenous cellulase activity leads to formation of alkali-soluble cellodextrins (6, 8, 18). Glycoprotein (7) and glycolipid (3) containing short chains of β-1,4-linked dextrin are probably also present. Accordingly, values for high-DP cellulose deposition, as assayed in Table 1, should be regarded as reflecting minimal rates of β-1,4-glucan synthesis.

Similar increments occur in alkali-insoluble glucan when the pea segment is allowed to elongate in the intact epicotyl, i.e., in the pre-

Table 1. Effect of auxin treatment of decapitated pea epicotyls on levels of cellulose and enzymes related to cellulose metabolism

Component per 10 mm apical segment	Zero time	Treated 3 days	
		Control	+Auxin
Alkali-insoluble glucan (nmol glu)	510	790	2190
Degree of polymerisation (DP × 10^{-3})	8.4	11.0	9.2
Enzyme activities (nmol red. groups formed or transferred/h)			
β-glucan synthetase (alkali-insoluble glucan from 5 mM UDPG)			
Particulate (ER + Golgi)	1.9	1.1	2.0
Cell surface (10 seg. slices)	19.6	45.8	79.0
β-glucosidase (vs. 10 mM cellobiose)			
Buffer-soluble + insoluble	16.7	16.0	63.9
β-glucan endo-hydrolase (vs. 0.75% CMC)			
Buffer-soluble (BS)	0.4	0.4	12.3
Buffer-insoluble (BI)	0.2	0.3	15.5

See methods for details of extraction and estimation procedures. Note that auxin increased average rate of deposition of alkali-insoluble glucan in vivo from ca. 4 to 33 nmol glucose equiv/h over the 3 days of this test

sence of natural hormone derived from attached plumule and hook (15). When the decapitated epicotyl is treated with gibberellin, the apical segment also elongates with increases in segment fresh weight and glucan comparable to those evoked in swelling auxin-treated tissue (Y. Raymond, unpublished). Ethylene gas inhibits elongation but causes some swelling of the epicotyl, during which there are corresponding increases in the level of alkali-insoluble glucan (15). Cytokinins have little effect on growth in peas when added alone but, when added together with auxin, they synergistically enhance swelling, and glucan deposition is even greater than with auxin alone (unpublished data). In all of these instances, the levels of alkali-insoluble glucan per unit fresh weight of tissue are maintained or increased during hormone-evoked cell expansion, whatever the direction or magnitude of growth.

The question thus arises of whether cellulose deposition is an indirect consequence of cell expansion, albeit closely correlated with it, or whether all plant hormones are able to regulate one or more of the many reactions potentially capable of modulating cellulose metabolism. Progress in answering this question, as it relates to known auxin effects in peas, is summarized below.

II. Substrate Metabolism

With respect to the source of substrate for cellulose synthesis, UDPG (the major sugar nucleotide present in peas) or guanosine diphosphoglucose (GDPG) may be generated from glucose-1-phosphate by the appropriate pyrophosphorylase (10), or UDPG may arise by transglycosylation from sucrose by reversal of sucrose synthetase activity (12). The ability of sucrose to donate its glucose moiety to alkali-insoluble glucan in a sequence of UDP-dependent reactions has been demonstrated in vitro. Under optimal conditions, pea sucrose synthetase acts to cleave sucrose and form UDPG much more rapidly than it acts to generate sucrose (12). The rate (approx. 120 nmol glucose transferred/h/seg) is more than sufficient to account for the observed maximum rates of deposition of alkali-insoluble glucan in vivo (cf. Table 1). Since free sugars in the growing region of peas often reach levels comparable to the K_m value for UDPG synthesis from sucrose, the amounts of these components are not likely to be limiting factors for cellulose synthesis.

Hormonal regulation of sugar nucleotide synthesis is still possible via mechanisms that do not involve changes in levels of essential enzymes or precursors. Little is known about intracellular distribution of sucrose synthetase or pyrophosphorylase in relation to glucan synthetases. Hormonal effects on compartmentation or on pH could exert dramatic control on magnitude and direction of glucosyl transfer in vivo.

III. β-Glucan Synthetase Activities

Alkali-insoluble glucan synthetase has been detected and assayed in particulate fractions from peas using low (< 10 μM) or high (> 500 μM) concentrations of UDPG or GDPG as substrate. Following differential centrifugation of particles in density gradients, all of these activities concentrate in regions rich in Golgi dictyosomes (11, 14). If segments are pre-treated with auxin (or ethylene), or assayed after growth has ceased, synthetase activity also appears in lighter particles rich in endoplasmic reticulum (ER) vesicles (14, 15). The yields of total particulate activity, however assayed, decline in segments that are allowed to age in the absence of auxin. In some instances particulate synthetase activity may disappear altogether from controls, although it reappears upon subsequent addition of auxin (unless inhibitors of protein synthesis are also added (13)).

Such results may indicate that auxin is required for the generation of glucan synthetase. They may mean instead/also that auxin influences intracellular transport and secretion of synthetase to loci where they are labile or not isolated with the particulate fraction.

Synthetase activity can also be assayed by supplying UDPG directly to living tissue slices (14, 15). Initial rates of incorporation of glucose into alkali-insoluble glucan are much greater (10-40-fold) than those recovered in total particulate membranes from the same segments, especially in slices from auxin-treated tissue (Table 1). GDPG is less effective under the same conditions (Table 2). Such synthesis has been referred to as "cell surface" activity on the assumption that living cells would not be permeable to sugar nucleotides in the brief period (10 min) of these assays.

The notion that cell surface synthetase activity is peculiarly labile when tissues are homogenized (14) is supported by the observation (Table 2) that this activity is severely inhibited by the addition of soluble extracts derived from the same tissue. The suppression of surface activity is progressive with time (H.Y. Chung et al., unpublished) and appears to be due to an inactivating factor that is heat-labile. Boiled epicotyl extracts activate the system. Particulate UDPG-dependent activity is less affected by addition of these extracts, and GDPG-dependent activity is only stimulated by them. Changes in levels and distribution of such modifying factors presumably could regulate glu-

Table 2. Factors in pea extracts that can activate or inactivate pea β-glucan synthetase activities

Reaction mixture	nmol alkali-insoluble glucan formed from 0.6 mM substrate/h/seg			
	Particulate		Cell surface	
	UDPG	GDPG	UDPG	GDPG
Standard	1.7	0.4	21.4	1.0
+ soluble extract	0.9	1.0	0.8	1.3
+ boiled extract	3.7	0.4	37.5	1.1

Assays were performed in quadruplicate as described in methods (average S.E. = ±8% of values given) using total particulate or 10 slices from one apical 10 mm segment ± crude soluble supernatant extracted by 1 mM MOPS (pH 7.25) from one segment in a total vol of 250 μl

can synthesis in vivo, and certainly could drastically affect the values obtained in attempts to assay synthetase in vitro.

There are many other qualifications that must be made before current estimates of glucan synthetase activity can be extrapolated to conclusions about regulation of cellulose synthesis. Alkali insolubility, including the classical "scouring" technique for ridding cotton or wood cellulose of other polysaccharides, is not a sufficient criterion for identifying small amounts of newly synthesized glucan as cellulose. There are many reports in the literature of β-1,3-linkages in alkali-insoluble products, and certain heteropolysaccharides (e.g., some glucomannans) are also insoluble. Nor is hydrolysis by fungal cellulase preparations to glucose and cellobiose adequate proof that trace amounts of labelled glucan are entirely cellulose. In our experience, such "cellulase" preparations are often contaminated with β-1,3-glucanase and a particularly active laminaribiase, which convert any β-1,3-glucan in the product to glucose. Methylation analysis of the alkali-insoluble glucan synthesized in peas (by Dr. Debbie Delmer at Michigan and Dr. Geoffrey Fincher in this laboratory) has shown that substantial amounts of β-1,3-linkages are present. In our tests, the percentage of β-1,4-linkage in glucan increases as substrate UDPG concentration decreases, which may indicate co-existence of at least one β-1,4-glucan synthetase with a low K_m value for UDPG and a β-1,3-glucan synthetase with a comparatively high K_m value.

We have also recently attempted to determine the DP distribution of the glucan formed from high UDPG concentrations in peas by acetylating it and fractionating on columns of glass beads (see Methods). The product contains a major component with a DP between 2000 and 3000, but it also contains components with low DP (below 250) and other products that do not have the stability or solubility characteristic of cellulose acetates.

It will be necessary to separate, solubilize, and purify the glucan synthetases, and probably to re-assemble glycolipid, glycoprotein, enzyme, activator, and acceptor components into fully active systems that can form authentic high-DP β-1,4-linked glucan, before it will be possible to assess the roles that hormones may have in regulating cellulose synthetase.

IV. Cellulase Activities

In the presence of auxin, pea epicotyls generate two distinct endo-β-1,4-glucan hydrolases (Tables 1 and 2), both meriting designation as true cellulases (E.C. 3.2.1.4). One is buffer-soluble (BS) with a mol wt of approx. 20,000, the other buffer-insoluble (BI) but salt-soluble, mol wt 70,000 (2). They have been purified to homogeneity by several standard criteria. Marked differences in their amino acid compositions and in their physical and immunologic properties indicate that one is not derived from the other.

The pure enzymes have now been used (18) in extensive kinetic studies of substrate specificities and affinities. Both preferentially hydrolyse higher DP cellodextrins in a random manner at rates comparable to the rates at which they attack soluble cellulose derivatives (e.g., carboxymethyl cellulose (CMC)). They also hydrolyse certain mixed-linkage glucans (e.g., barley glucan), but only if these contain contiguous β-1,4-linkages. Cellobiose, laminaridextrins, glucan with alternating β-1,3-/1,4-linkages, and other β-1,4-linked polysaccharides (xylan, chitin) are not hydrolysed.

Cellobiose is the major final product produced by purified pea cellulases from defined cellodextrins, Whatman cellulose, or the alkali-insoluble glucan synthesized from UDPG by pea preparations (18). Crude epicotyl extracts also contain β-glucosidases that hydrolyse cellobiose (5). V_{max} values for these hydrolases relative to those of the pea cellulases (Table 1) are more than sufficient to ensure that any disaccharide generated by the latter would never accumulate in vitro. Presumably this explains why cellobiose is never found in extracts of pea tissue.

The very large increase in BS and BI cellulase activities that are evoked by auxin treatment of the epicotyl (Tables 1 and 3) are not accompanied by similar increases in levels of other glucanases, pectinases, or glycosidases (4, 5), and can be completely prevented by adding inhibitors of RNA or protein synthesis (6, 9). Messenger RNA for BS cellulase has been estimated in pea polysomes by extracting total mRNA, translating it, and measuring the amount of product that precipitates with monospecific antibodies prepared to the pure enzyme (17). By this means, pea polysomes have been shown to acquire cellulase mRNA following auxin treatment of the epicotyl before levels of cellulase activity begin to increase, as would be expected of true enzyme induction.

Other plant hormones have no such effect on pea cellulase. In particular, treatment with ethylene gas, although it causes some epicotyl swelling and minor increments in cellulase levels, severely inhibits the generation of high cellulase levels when it is added together with auxin (Table 3). The auxin/cellulase relationship appears to be highly specific, and not mediated by any effects auxin may have on ethylene production or on other hormones.

Antibodies to pure pea cellulases have also been labelled with ferritin and used to examine the intracellular distribution of cellulase in thin sections of auxin-treated epicotyls (1). Most BS and some BI cellulase is located within ER vesicles. This is consistent with observations that pea microsomes can generate cellulase activity in vitro (9) and rough ER is particularly rich in mRNA for this enzyme (17). Most BI cellulase is found in close association with microfibrils on the inner wall surfaces, not attached to plasma membranes. After several days of auxin treatment, when cellulase levels reach their maximum, cell walls of many swollen cortical cells in the epicotyl begin to disintegrate

Table 3. Inhibiting effect of ethylene on the induction of BS and BI cellulase activities by auxin

Treatment	Swelling (mg/mm)	Cellulase activity (nmol reducing groups releas- from 1.0 ml 1% CMC/h/seg)	
		BS	BI
Zero time	2.5	0.9	0.3
3 days, + ethylene	4.0	3.0	0.9
3 days, + IAA	6.7	32.7	54.0
3 days, + IAA + ethylene	6.6	15.0	27.6

8-day-old decapitated etiolated pea seedlings grown in jars were treated with lanolin paste ±0.5% IAA and flushed with gentle stream of air ± 100 ppm ethylene gas. Segments (apical 10 mm) were extracted and assayed for BS and BI cellulase activity as described in Methods

visibly (6), and cytochemical tests for cellulase (1) show very high activity concentrated at such loci. There can be no doubt that auxin-induced cellulase is secreted from cells and attacks native cellulose, on occasion to the point of fragmenting the fibril framework. It is hardly surprising that the formation and distribution of pea cellulases, which are as active as any fungal cellulase per unit protein, are closely regulated by the plant.

Following pea segment homogenization, part of the auxin-induced cellulase activity remains particle-bound (9) and, after isopycnic density-gradient centrifugation, bands at a density (1.11 g/cc) coincident with ER vesicles and a major part of β-glucan synthetase activity (8). There may be a period during the generation of cellulase and synthetase when both are present in the same transport organelle. Whether they continue to be associated in an active complex when excreted to the cell surface remains to be established.

Results and arguments to suggest that modest levels of cellulase activity may, paradoxically, play a constructive role in cellulose deposition during growth have been presented and discussed in detail earlier (8). The most direct evidence consists of observations that incorporation of ^{14}C-sucrose into alkali-insoluble glucan by excised pea segments can be enhanced by concurrent provision of small amounts of fungal cellulase in the outside medium. Similar experiments have now been conducted (18) using tissue slices pretreated briefly with pure BS or BI cellulase, followed by incubation in labelled UDPG. Both cellulases markedly enhance the rate of glycosyl transfer to alkali-insoluble glucan, provided they are used at a moderate concentration (comparable in the outside medium to levels per unit fresh weight attained in auxin-treated peas). Very high levels of supplied fungal or pea cellulases reduce the net yields of new glucan synthesized. These results imply that one of the limiting factors for cellulose deposition in peas is the availability of endogenous β-1,4-glucan chain ends to act as acceptor molecules for transglycosylation, and that partial endohydrolysis by cellulase can overcome this limitation.

Acknowledgements

Studies reported from this laboratory have been financed by grants from the National Research Council of Canada and the Quebec Department of Education. Part of the work was carried out in the laboratory of

Prof. Bruce Stone, La Trobe University, Melbourne, while the author was Visiting Commonwealth Fellow. Grateful thanks are extended to Prof. Stone and my collaborators in these studies at McGill.

References

1. Bal, A.K., Verma, D.P.S., Byrne, H., Maclachlan, G.A.: J. Cell Biol. 69, 97 (1976)
2. Byrne, H., Christou, N.V., Verma, D.P.S., Maclachlan, G.A.: J. Biol. Chem. 250, 1012 (1975)
3. Colvin, R.: Applied Polymer Symp. 28, 623 (1976)
4. Datko, A.H., Maclachlan, G.A.: Plant Physiol. 43, 735 (1968)
5. Datko, A.H., Maclachlan, G.A.: Can. J. Botany 48, 1165 (1970)
6. Fan, D.F., Maclachlan, G.A.: Plant Physiol. 42, 1114 (1967)
7. Franz, G.: Applied Polymer Symp. 28, 611 (1976)
8. Maclachlan, G.A.: Applied Polymer Symp. 28, 645 (1976)
9. Maclachlan, G.A., Davies, E., Fan, D.F.: In: Biochem. Physiol. Plant Growth Substances. Ottawa: Runge Press, 1968, p. 443
10. Peaud-Lenoel, C., Axelos, M.: Europ. J. Biochem. 4, 561 (1968)
11. Ray, P.M., Shininger, T.L., Ray, M.M.: Proc. Natl. Acad. Sci. U.S. 64, 605 (1969)
12. Rollit, J., Maclachlan, G.A.: Phytochemistry 13, 367 (1974)
13. Shore, G., Maclachlan, G.A.: Biochim. Biophys. Acta 329, 271 (1973)
14. Shore, G., Maclachlan, G.A.: J. Cell Biol. 64, 557 (1975)
15. Shore, G., Raymond, Y., Maclachlan, G.A.: Plant Physiol. 56, 34 (1975)
16. Spencer, F.S., Maclachlan, G.A.: Plant Physiol. 49, 58 (1972)
17. Verma, D.P.S., Maclachlan, G.A., Byrne, H., Ewings, D.: J. Biol. Chem. 250, 1019 (1975)
18. Wong, Y-S.: M.Sc. thesis, McGill Univ., Montreal (1976)

Paper Demonstration References (see Section IX):

21 - 38 - 39 - 50 - 53 - 98 - 116 - 119 - 167 - 191

3. Wall Extensibility in Relation to Auxin Effects

Y. MASUDA

Abstract

Decrease in the minimum stress-relaxation time, To, is well correlated with the capacity of the cell wall to extend, but the increase in extensibility appears to be the expression of an extension that has already occurred. Decrease in To of a cell wall specimen results following treatment with a fungal exo-type β-1,3-glucanase or 0.1*N* KOH, but only to a limited degree with cellulase or 0.1% ammonium oxalate. Auxin causes a decrease in the hemicellulosic glucose level in the cell wall after a lag period of 2 h when it induces cell extension. Hydrogen ions cause cell extension without decreasing the glucose level in the wall. Auxin effects on extension and glucose level are inhibited by nojirimycin, a potent inhibitor of glucanases, which does not, however, inhibit hydrogen ion-induced cell extension and To decrease.

A. Introduction

Cell wall extension occurs as a continuous series of independent extension steps, and the auxin-induced extension step is initiated by a biochemical modification in the cell wall components, resulting in cell wall loosening. The significance of cell wall loosening in auxin-induced cell extension has long been recognized, and many papers on this subject have been published in the last 40 to 50 years (3). Basic information on cell wall loosening in relation to auxin-induced cell extension has been obtained from these publications. They cover three major problems (1) measurement of mechanical properties of the cell wall, and physical and chemical implications of the parameters, (2) relationship between mechanical properties of the cell wall and cell extension, (3) biochemical modifications of the cell wall components leading to cell wall loosening. We know, however, very little about the relationship between physical parameters and biochemical modifications. The present paper discusses this point using the oat coleoptile system.

B. Material and Methods

Oat (*Avena sativa* L, cv. Victory) seedlings were grown as reported previously (8). A 5-mm segment was excised from the upper region of the coleoptile with the first leaf removed. Segments were pooled in distilled water and incubated for various periods of time in test solutions at 25°C. Changes in the segment length were traced under a binocular microscope (×10) under dim green light. In another series of experiments for sugar determination of the cell wall, segments, ca. 15 mm long, were starved in water for 6 h and 10-mm segments excised from them were then incubated under different conditions (8).

Stress-relaxation analysis was done as reported previously (7, 13). Segments were methanol-killed and then pronase-treated. A rehydrated specimen was fixed between two clamps of an Instron tensile tester (TM-M) connected to a HITAC-10 mini computer, and the parameter To, the minimum relaxation time, was automatically calculated using the equation:

$$S = b \cdot \log \frac{t + Tm}{t + To}$$

where S is stress, t is time, and b is a constant (Tm: maximum relaxation time). Extensibility, mm/g, was also measured.
Cell walls for sugar analysis were prepared by crushing the coleoptile segments between two glass plates followed by washing (×3) with water, acetone, and a 1:1(vol/vol) methanol-chloroform mixture. The noncellulosic polysaccharides of the dried cell walls were hydrolyzed in 2*N* trifluoroacetic acid, and the liberated sugars were reduced with sodium borohydride, acetylated in the presence of acetic anhydride, and analyzed by gas-liquid chromatography (Hitachi 063) (1).

C. Results and Discussion

A variety of techniques have been developed to measure the mechanical properties of the cell wall, and some parameters obtained by these techniques have been widely used as representing auxin-induced cell wall loosening. However, we are often confronted with the difficulty of not being certain which of the extension steps are represented by these parameters. They do not always represent cell wall loosening, often being a different expression of cell wall extension, i.e., growth.

Among the parameters, the following have been widely used or recently developed for use: extensibility, compliance, and stress-relaxation time. Extensibility is the increase in length in response to a specific load applied to a tissue segment, usually by stretching it using an Instron tensile tester. Compliance is also a measure of strain/stress, i.e., a modulus of extensibility, determined also using a tensile tester (2). If a cell wall specimen is quickly stretched with a tensile tester and stretching is stopped, the produced stress decays with a fixed strain. The minimum relaxation time, To, is the time when stress starts to decay (13).

In order to separate the change in mechanical properties of the cell wall representing wall loosening from the subsequent wall extension, tissue segments are incubated in a mannitol solution in which the former occurs without the latter. Thus, the direct effect of auxin on cell wall loosening can be seen. Although indole-3-acetic (IAA) has no effect on the length of segments incubated in mannitol solutions at concentrations greater than 0.2 M (Fig.1A-D), subsequent extension in the absence of mannitol and IAA (Fig.1B) is much larger in the segments preincubated in IAA-containing mannitol solution than in those without IAA. The To values of the cell walls of segments incubated in mannitol solutions are smaller for the segments treated with IAA than those for the control segments (Fig.1C), when they are incubated in mannitol solutions of less than 0.3 M, which is equivalent to the diffusion pressure deficit of the tissue. The auxin effect of decreasing To seems even larger if the segments are treated in a mannitol solution of about 0.2 M, in which IAA causes no extension. A good correlation thus is observed between the IAA-induced decrease in the To value of the cell wall and the capacity of the segments to extend (extension after mannitol preincubation). The IAA-induced To decrease is, however, not correlated with the length of the segments in mannitol solutions.

Auxin-induced increase in extensibility measured after incubation in mannitol solutions, unlike the To value, has some correlation with the

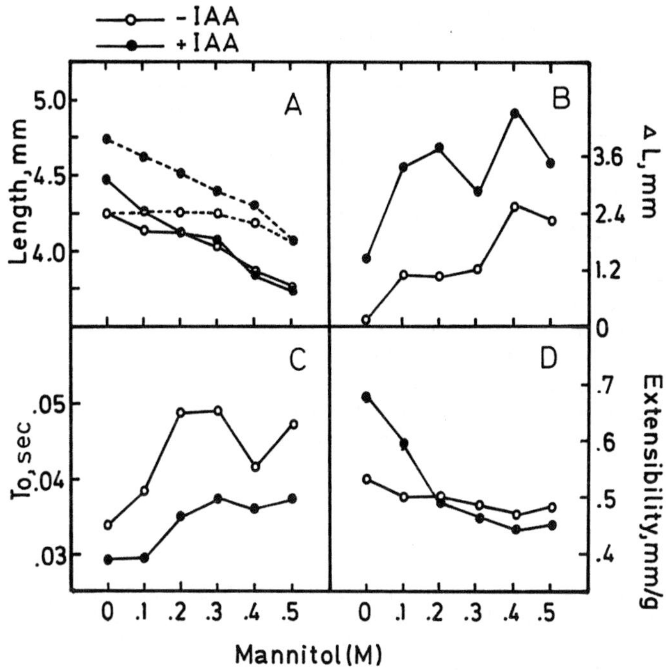

Fig. 1 A-D. Effect of IAA (10^{-5} M) on extension (A), To (C), and extensibility (D) of oat coleoptile segments incubated in mannitol solutions of various concentrations and residual extension (B) in absence of auxin and mannitol. Segments were incubated in mannitol solutions for 90 min. Segment length was then measured and segments were methanol-killed, deproteinized, and subjected to To and extensibility determinations. In separate series of experiments, lengths of segments after mannitol incubation (——— in A) and after extension for 90 min in buffer in absence of IAA and mannitol (----- in A) were measured. *Curves* in (B) differences between data represented by ——— and -----. For determinations of cell wall parameters, see Material and Methods

length of the segments in mannitol solution, but not with their capacity to extend, as shown in Figure 1D. These data suggest that cell wall loosening is well represented by the minimum stress-relaxation time, To, but that extensibility is an expression of extension that has already occurred. We conclude that the initial step of extension is represented by To and that extensibility is an expression of the later step.

The stress-relaxation process can be basically simulated by a viscoelastic model consisting of a spring and a dash-pot connected in series, with relaxation time being represented by η/G, where η is the viscosity of the dash-pot and G is the elasticity of the spring. The decrease in stress-relaxation time is caused either by the decrease in η and/or increase in G. In our cell wall system, the former rather than the latter, most likely occurs. The η is assumed to be represented by the noncellulosic polysaccharide matrix of the cell wall components and G, by the crystalline component, i.e., cellulose microfibrils. It has been reported that auxin stimulates metabolic turnover of hemicellulose when it induces cell extension (12). Table 1 contains a summary of data of experiments in which oat coleoptile segments, after being fixed and

Table 1. Effect of cell-wall-degrading enzymes, ammonia oxalate, and alkali solution on To of oat coleoptile cell wall

Treatment	To ($\times 10^{-3}$ sec)
Water	45 ± 3
Cellulase (200 μg/ml), 6 h[a]	39 ± 3
β-1,3-glucanase (0.3 U/ml), 6 h[a]	25 ± 3
Ammonia oxalate (0.25%), 10 min[b]	33 ± 4
KOH (0.1*N*), 10 min[c]	23 ± 2

Coleoptile segments were methanol-killed and deproteinized before being subjected to in vitro treatment and then stress relaxation analysis.

[a] Cellulase: an enzyme from Worthington Biochem. Co., β-1,3-glucanase: a purified fungal exo-type enzyme from culture of *Sclerotinia libertiana*. Enzyme activity: 40.0 U/mg protein.

[b] Incubated at 95°C

[c] Incubated at 25°C

deproteinized, were extracted with fungal β-1,3-glucanase, cellulase, ammonia oxalate, or KOH, then subjected to stress-relaxation analysis. Clearly, the To decrease is caused by glucanase and KOH, but not by cellulase or ammonia oxalate.

Loescher and Nevins (5, 6) have reported that auxin causes a decrease in glucose of noncellulosic components of oat coleoptile cell wall when it initiates elongation. Nevins (9) found that β-glucanase, when administered to coleoptile cell wall, causes release of glucose from the wall. Fig.2 illustrates changes in the levels of different sugars, i.e., xylose, arabinose, glucose, and galactose, in the noncellulosic components of the oat cell wall when the segments are treated with auxin. The majority of the noncellulosic components in oat coleoptile obtained by the present technique is hemicellulose (5). IAA causes a significant decrease in the level of only glucose, as already reported by Loescher and Nevins (5, 6), and not in those of the other sugars. The decrease in glucose caused by auxin is observed after a lag of 2 h.

Auxin-induced cell extension has recently been interpreted as being mediated by hydrogen ions (11). Low-pH buffer solution in fact causes extension of tissue segments and changes in mechanical properties of the cell wall (7, 11). Table 2 contains data indicating the effects of auxin, low-pH buffer solution, and nojirimycin (5-amino-5-deoxyglucopyranose), a potent inhibitor of β-glucanase (10), on the level of glucose in oat coleoptile cell wall, with elongation and To values. Auxin again causes a decrease in the glucose level, but hydrogen ions have essentially no effect on it, even though they cause extension and a decrease in To. This suggests that auxin and hydrogen ions have different effects on biochemical modifications of the cell wall, as judged by the change in glucose level. Nojirimycin inhibits the auxin-induced decrease in the glucose level as well as cell extension and the decrease in To, but not hydrogen ion-induced cell extension and To decrease. These data indicate that hemicellulosic glucan degradation is involved in auxin-induced cell wall loosening but not in that induced by hydrogen ions. However, note that a 2-h lag period is observed for the auxin-induced decrease in the glucose level in the cell wall. Since auxin is known to cause quick hydrogen ion secretion from oat

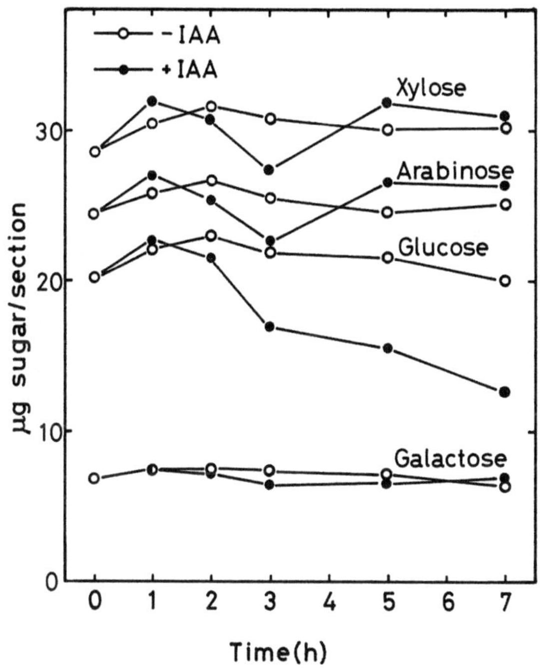

Fig. 2. Effect of IAA (10^{-5} M) on time-course changes of different sugars in noncellulosic cell wall components of oat coleoptile segments. Segments, 10-mm long, were starved for 6 h in water and then incubated in buffer and in IAA solution. For sugar determinations, see Material and Methods

Table 2. Effect of auxin, low-pH buffer solution, and nojirimycin on cell extension, To, and hemicellulosic glucose level of oat coleoptile cell wall

Treatment	Extension, mm	To ($\times 10^{-3}$ sec)	Glucose (μg/seg)
Buffer, pH 6.5	0.6 ± 0.10	27.0 ± 1.1	16.9
Nojirimycin, 10^{-3} M	0.7 ± 0.11	23.7 ± 1.5	16.9
IAA, 10^{-5} M (pH 6.5)	3.6 ± 0.49	18.5 ± 0.8	14.8
IAA + nojirimycin	1.5 ± 0.31	21.7 ± 1.2	16.2
Buffer, pH 4.0	2.5 ± 0.27	18.7 ± 1.2	17.6
Buffer (pH 4.0) + nojirimycin	2.1 ± 0.19	16.3 ± 1.0	17.5

Coleoptile segments were starved for 6 h in water and incubated for 4 h in test solutions. Data are means of 20 segments. Buffer: 0.01 M K-citrate

coleoptile segments (4), hydrogen ion secretion might participate in the auxin effect, particularly in its early phase. The decrease in To is caused by both auxin and hydrogen ions; but only the former causes a decrease in the glucose level in the cell wall, which is observed 2 h after auxin application. This is because (1) the action of auxin on hemicellulosic glucan needs a 2-h lag, or (2) the glucose decrease occurs very rapidly but can be observed only after a certain period of time, i.e., about 2 h. This problem should be studied further.

D. Conclusions

Mechanical properties representing cell wall loosening, such as the initial step of auxin-induced cell extension, are well elicited by a decrease in the parameter To, the minimum stress-relaxation time. The decrease in To is caused by both hydrogen ions and auxin, but the decrease in the glucose level in cell wall hemicellulose is caused only by the latter and not by the former, which might cause the To decrease through a different mechanism. The cell wall loosening process appears to consist of at least two phases, both affected by auxin: the initial one, which continues for about 2 h in oat coleoptile segments, is caused by hydrogen ions, and the later, caused by hemicellulosic glucan degradation, which is observable only as a decrease in the glucose level in the cell wall after a lag period. Auxin-induced hydrogen ion secretion may thus play a crucial role in the whole cell wall loosening process.

References

1. Albersheim, P., Nevins, D.J., English, P.D., Karr, A.: Carbohydrate Res. 5, 340 (1967)
2. Cleland, R.: Planta 74, 197 (1967)
3. Cleland, R.: Cell wall extension. Ann. Rev. Plant Physiol. 22, 197 (1971)
4. Cleland, R.: Planta 127, 233 (1975)
5. Loescher, W.H., Nevins, D.J.: Plant Physiol. 50, 556 (1972)
6. Loescher, W.H., Nevins, D.J.: Plant Physiol. 52, 248 (1973)
7. Masuda, Y., Yamamoto, R., Maki, K., Yamagata, Y.: In: Plant Growth Substances, 1973. Tokyo: Hirokawa Publ. Co., 1974, p. 806
8. Masuda, Y., Yamamoto, R., Kawamura, H., Yamagata, Y.: Plant Cell Physiol. (Kyoto) 15, 1083 (1974)
9. Nevins, D.J.: Plant Cell Physiol. (Kyoto) 16, 495 (1975)
10. Nevins, D.J.: Plant Cell Physiol. (Kyoto) 16, 347 (1975)
11. Rayle, D.L., Cleland, R.: Plant Physiol. 46, 250 (1970)
12. Wada, S., Tanimoto, E., Masuda, Y.: Plant Cell Physiol. (Kyoto) 9, 369 (1968)
13. Yamamoto, R., Fujihara, S., Masuda, Y.: In: Plant Growth Substances, 1973. Tokyo: Hirokawa Publ. Co., 1974, p. 798

Paper Demonstration References (see Section IX):

50 - 64 - 73 - 76 - 116 - 137 - 140

II. Hormones – Membranes

4. Membrane-Bound Auxin Receptors: Solubilization and Characterization

M. A. Venis

Abstract

Two compounds have been tested as affinity labels for auxin binding sites in corn coleoptile membranes. Reacting the membranes at pH 8-9 with the diazonium salt of CAPA (2-chloro-4-aminophenoxyacetic acid) reduces their subsequent ability to bind NAA (1-naphthylacetic acid). Diazo-Chloramben (2,5-dichloro-3-aminobenzoic acid) is also effective, and inhibition is largely independent of reaction pH over the range pH 6-9. Prior addition of NAA protects the binding sites against diazo-Chloramben inhibition when the azo-coupling reaction is carried out at pH 6.

From these results and others with sulphydryl reagents, from the pH dependence of auxin binding, and from consideration of structure-activity rules, several of the amino acid residues in the binding site environment have been tentatively assigned.

Auxin-binding proteins have been solubilized from the membranes without detergent, and purified by ion exchange and gel permeation chromatography. Gel electrophoresis indicates the presence of two main proteins in the purified fraction, of 50,000 and 54,000 daltons. Both proteins appear to be dimers.

A. Introduction

Saturable binding of the auxins IAA (indole-3-acetic acid) and NAA (1-naphthylacetic acid) to membranous preparations from corn coleoptiles was first demonstrated by Hertel and co-workers (3). Subsequently we investigated the binding kinetics and specificity of this system in greater detail and demonstrated the presence of two distinct classes of auxin binding sites with differing affinities and specificities (1, 2). Site 1 (K_D for NAA = 0.18 μM) was able to bind both auxins and physiologically inactive analogues, whereas site 2 (K_D for NAA = 1.6 μM) exhibited binding specificity compatible with that expected of an auxin receptor, and was located in a heavier membrane population, enriched in plasma membrane.

To characterize further the binding sites, we have examined the effects of group-modifying reagents on binding activity. In particular, we have explored the potential of two diazonium auxin analogues as affinity labels, or active site-directed irreversible inhibitors, i.e., compounds that bear sufficient structural resemblance to an auxin to have preferential affinity for the receptor sites over other potential binding sites, and that contain a chemically reactive function capable of attaching the molecule covalently to a suitable amino acid residue in the active site environment. The technique is one that can provide valuable information about the amino acids of the binding site. In addition, the use of a radioactive site-directed inhibitor results in a receptor labelled covalently with a radioisotope, which can then be used to facilitate purification and degradation studies on the receptor. Before proceeding to a radiolabelled compound, it is advisable to obtain some evidence that the molecule can in fact couple to the receptor site. Such experiments are described here, where we have investigated the effects of reacting membranes with the non-radioactive diazonium salts on the subsequent ability of the membranes to bind NAA-^{14}C.

We also report on the removal of auxin-binding proteins from the membranes in a soluble form without detergent treatment, their purification, and preliminary electrophoretic characterization.

B. Materials and Methods

I. Membrane Preparation

Corn coleoptile membranes were prepared by differential centrifugation between 4000 g and 80,000 g with the grinding and wash buffers described previously (2) except that the pellets were washed with buffer of pH 8.0-8.1 rather than pH 6.0.

II. Diazotization Conditions

To 5 µmol of CAPA (2-chloro-4-aminophenoxyacetic acid) or Chloramben (2,5-dichloro-3-aminobenzoic acid) in 250 µl of 0.5*N* HCl at 0-5°C, 50 µl of freshly prepared sodium nitrite solution (8 mg/ml) were added. After 10-15 min, the diazonium salt was diluted with 1.2 ml of 50 mM borate pH 8.5, and aliquots of the mixture were used immediately for coupling to the membranes. In all experiments a control mixture without any amino compound was prepared.

III. Reaction with Membranes

Membranes were resuspended at the required pH at a concentration equivalent to 0.5 g tissue fresh wt/ml. Resuspension buffers were of grinding medium composition (2) except that Tris-acetate was replaced either by potassium phosphate (pH 6-8) or by sodium borate (pH 8-9). Suspensions were incubated at 25°C (generally for 15 min) with diazonium salts or sulphydryl reagents, or with solution blanks as controls. The reaction mixtures were then diluted into 2-3 vol of cold pH 6 wash medium and centrifuged at 80,000 g for 30 min. The pelleted membranes were suspended in binding buffer, pH 5.5 (2), and saturable binding of NAA-^{14}C was determined as described previously (2). Total binding (site 1 + site 2) was assayed as Δ cpm 0.2-1.0 µM NAA, or alternatively, site 1 was estimated from Δ cpm 0.2-0.4 µM NAA and site 2, from Δ cpm 0.4-1.0 µM NAA.

IV. Solubilization and Purification of Auxin-Binding Proteins

Membranes were suspended in grinding buffer and added to 20 vol of ice-cold acetone, with stirring. The precipitate was collected by centrifugation or filtration, washed with cold acetone, and suspended in a fivefold dilution of binding buffer (DBB). Insoluble material was removed by centrifugation (38,000 g, 15 min) and the solution applied to a column of DEAE-Bio-Gel. The column was eluted successively with DBB, binding buffer, and 0.5 M NaCl in binding buffer. Fractions with auxin-binding activity (assayed by equilibrium dialysis) were pooled, concentrated, and fractionated on a column of Sephadex G100 (Pharmacia) in binding buffer. The column was calibrated with proteins of known molecular weight.

V. Polyacrylamide Gel Electrophoresis

Auxin-binding fractions from the gel filtration column were concentrated and run on 7.5% Biophore (Bio-Rad) gels in tris-glycine buffer. Alternatively, they were treated with 1% sodium dodecyl sulphate (SDS) and 5% 2-mercaptoethanol at 90°C for 2 min and run on 12% Biophore gels in tris-acetate-SDS buffer. The latter gels were calibrated with oligomeric cross-linked protein standards (British Drug Houses), monomer size 14,300 daltons.

C. Results and Discussion

The diazonium salt of CAPA was examined first as an affinity label. (The synthesis and biological activity of the corresponding 4-azido analogue, a potential photo-affinity label, have recently been described (4), but without any affinity labelling data.) When the compound is reacted with the membranes at different pH values from pH 6-9, it is found that diazo-CAPA does inhibit subsequent binding of NAA-^{14}C, but only if the coupling reaction is carried out at pH 8-9 (Fig. 1). If the compound is acting as a genuine affinity label, i.e., coupling covalently with a reactive residue in the neighbourhood of the receptor site, then prior addition of an auxin should impede the coupling of diazo-CAPA to the receptor. After separation from the reaction mixture by centrifugation and resuspension in buffer, the 'protected' membranes should retain greater auxin-binding activity than unprotected membranes. Unfortunately, the conditions required for coupling of diazo-CAPA (pH 8 or above, Fig.1) are exceedingly unfavourable for auxin binding (pH optimum 5.5, ref. 1). It was therefore not surprising to find that auxin pretreatment at pH 8 affords no significant protection. The compound may indeed be acting as an affinity label, but it is not possible to obtain supporting evidence by ligand protection.

Investigation of other amino auxin analogues revealed that reaction of membranes with the diazonium salt of the auxin Chloramben at pH values as low as 6 (25°C, 15 min) resulted in substantial inhibition of their auxin-binding capacity (Fig.1). Inhibition is largely independent of reaction pH over the range pH 6-9, in contrast to the behaviour of diazo-CAPA. Figure 2 illustrates the concentration dependence of diazo-Chloramben inhibition at pH 7 and 25°C; reaction at 0°C reduces inhibition about fourfold. Experiments of this type are difficult to perform at pH < 6 because incubation of the membranes at 25°C under these conditions results in a marked reduction in control binding activity. Nevertheless, pH 6 is sufficiently close to the pH optimum for auxin binding (5.5) to permit investigation of auxin protection against diazo-Chloramben reaction. Addition of 25 μM NAA immediately prior to the diazonium salt (150 μM) does indeed appear to reduce substantially the extent of coupling at both site 1 and site 2, as judged by subsequent ability to bind NAA-^{14}C (Table 1). Benzoic acid (BA), on the other hand, protects only site 1, which is in accordance with the site binding specificity observed for this inactive compound (2).

From the preceding results, from the known variation of auxin binding with pH and from consideration of structure-activity rules, several

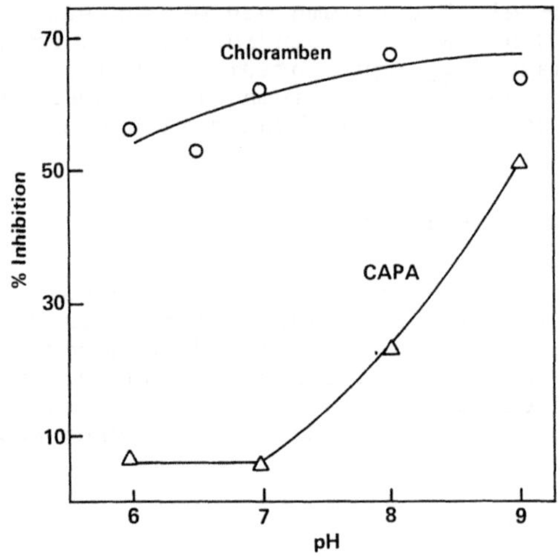

Fig. 1. Effect of reaction pH on inhibition, by diazonium salts of CAPA or Chloramben (150 μM), of NAA-^{14}C binding (site 1 + site 2) by corn coleoptile membranes. See Materials and Methods for experimental details

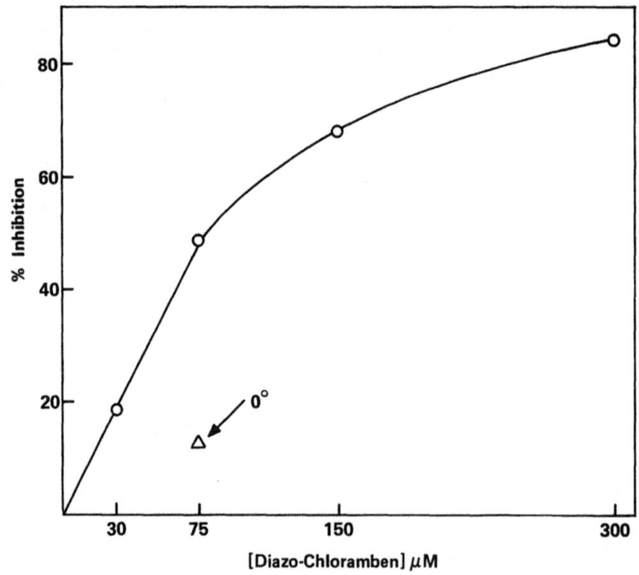

Fig. 2. Inhibition of NAA-^{14}C binding to corn coleoptile membranes (site 1 + site 2) after treatment with different concentrations of diazo-Chloramben at pH 7 for 15 min. Reaction temp 25°C except as noted. Control binding after 25°C incubation = 1027 cpm

Table 1. Affinity labelling with diazo-Chloramben in presence or absence of NAA or BA

	Inhibition of subsequent auxin binding, %			
	-NAA	+NAA	-BA	+BA
Site 1	72.1	25.4	54.5	29.0
Site 2	50.6	13.4	39.1	35.7

Membranes were resuspended at pH 6 and reacted, in presence or absence of 25 μM NAA or BA, with 150 μM diazo-Chloramben (25°C, 15 min). Control membranes received diazotisation mixture without Chloramben ±NAA or BA. Inhibition values were calculated relative to appropriate controls (±NAA or BA during the 25°C pre-incubation). Protective effects of NAA and BA were determined in separate experiments. See Materials and Methods for further details

tentative conclusions can be drawn regarding the environment of the receptor sites:

1. The presence of a free anionic function, normally carboxyl, is one of the essential structural requirements for auxin activity, suggesting binding of the carboxyl anion to a positively charged amino acid group in the receptor site. Binding of NAA is maximal at pH 5.5 and falls to very low values at pH 5 or pH 7 (1). The imidazolyl $\equiv NH^+$ of histidine (pK_a = 6.0, ref. 5) is the only candidate carboxyl-binding group whose ionization properties are compatible with loss of binding activity over the range pH 6-7. Reduction of binding on the acid side of the pH optimum can be attributed to protonation of the carboxyl anion (pK_aNAA = 4.3).

2. A fractional positive charge 0.55 nm distant from the $-COO^-$ is also regarded as an essential feature of active auxin molecules (6, and references therein). This requirement probably implies electrostatic bonding to an anionic residue in the binding site located at the appropriate distance from the carboxyl-binding imidazolyl function. At pH 5.5 the choice of negatively charged non-terminal polypeptide groups is restricted to the β- or γ-carboxyl of aspartate or glutamate, respectively (pK_a ca. 4.0, ref. 5). Protonation of this carboxyl would also contribute towards the fall-off in auxin binding at low pH.

3. The pH dependence of diazo-CAPA inhibition (Fig.1) is consistent with an azo coupling controlled by a functional group of an amino acid with a high pK, i.e., tyrosine or lysine.

4. The effectiveness of diazo-Chloramben at pH 6 (Fig.1) suggests that under these conditions, where tyrosine and lysine would not be expected to react, coupling to a histidine residue is involved. From Dreiding models it appears improbable that a Chloramben molecule can both azo couple via the 3-amino group and bind through its carboxyl group to the same histidine residue. This would imply either the presence of a second histidine near the receptor site, or azo coupling to the carboxyl-binding histidine, in which case the Chloramben carboxyl would not be suitably positioned for electrostatic bonding.

The use of general group-modifying reagents is also being examined, initially with sulphydryl reagents. When several -SH reagents were tested at pH 7 under conditions similar to those used for reaction of the membranes with diazo salts, only PHMB effectively inhibited the

Table 2. Effects of sulphydryl reagents on auxin binding

Treatment	Auxin binding, cpm (site 1 + site 2)	Inhibition, %
Control	919	-
Iodacetic 1 mM	947	-3.0
Iodoacetamide 1 mM	867	5.7
N-ethyl maleimide 1 mM	886	3.6
DTNB 1 mM	895	2.6
PHMB 0.1 mM	382	58.4
Control	818	-
PHMB 0.1 mM at +5'	424	48.2
PHMB 0.1 mM at 0' + GSH 1 mM at +10'	786	3.9
GSH 1 mM at +10'	793	3.1

DTNB = 5,5'-dithiobis (2-nitrobenzoic acid); PHMB = p-hydroxy-mercuribenzoate; GSH = glutathione
Membranes were resuspended at pH 7 and reacted at 25°C for 15 min. Additions were made at zero time unless noted otherwise. Membranes were re-isolated and auxin binding determined as described in Materials and Methods

ability of the re-isolated membranes to bind NAA-^{14}C (Table 2). Iodoacetic acid, iodoacetamide, N-ethyl maleimide, and DTNB were largely ineffective, even at ten times the PHMB concentration used (Table 2). Although PHMB can sometimes produce inhibitory effects through reaction with non-thiol groups (7), the ability of glutathione (Table 2) and dithiothreitol (not shown) to reverse PHMB inhibition renders this possibility less likely. In addition, the effectiveness of PHMB is enhanced as the pH is reduced, which is also characteristic of its reaction with -SH groups (7). Treatment of membranes with 0.1 mM PHMB at pH 6 leads to almost complete loss of auxin-binding activity (Fig.3). The superiority of PHMB to other thiol reagents may be due to its general similarity to an auxin structure, permitting it to act as a pseudo-affinity label. The suggestion that PHMB may be combining with a cysteine residue at the auxin binding site is supported by the ability of NAA to protect the membranes against modification by PHMB (Table 3). Table 4 contains a summary of current speculation on some of the amino acid residues that may be present at or near the auxin binding sites in corn coleoptile membranes.

Auxin-binding activity in the membranes can be obtained in a buffer-soluble form by extraction of acetone-delipidated membranes. When such extracts are fractionated on anion-exchange gels (Materials and Methods), binding activity is found only in the minor protein peak eluted with binding buffer. Further fractionation of this material on Sephadex G100 yields a single auxin-binding peak of apparent molecular weight 43,000 daltons. Overall purification on a protein basis is about 100-fold, with retention of 65-70% of the binding activity in the crude extract. Gel electrophoresis of the fraction with maximum binding activity reveals two protein bands in non-denaturing gels. In SDS gels, stained bands are visible at positions corresponding to molecular weights of 50,000 (main band) and 54,000, together with the probable monomers at 25,000 and 27,000 daltons. In addition, faint bands at 41,000 and 46,000 daltons are detectable.

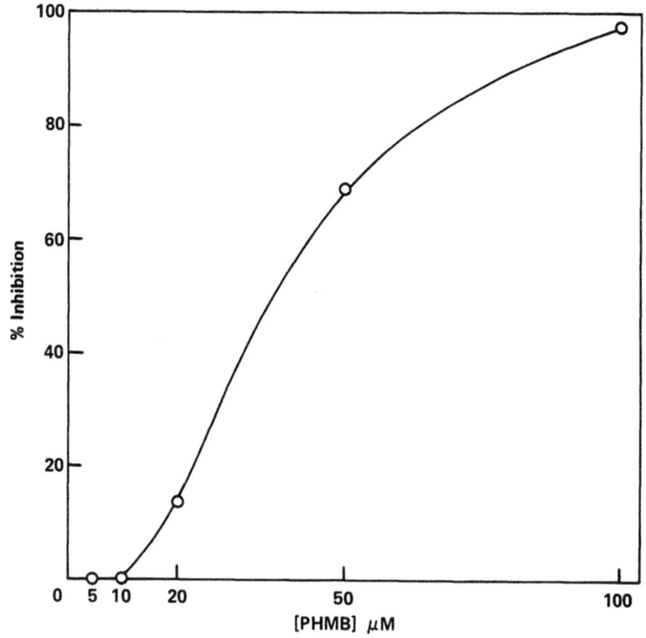

Fig. 3. Inhibition of NAA-^{14}C binding (site 1 + site 2) following incubation of membranes with varying concentrations of PHMB at pH 6, 25°C for 15 min. Control binding = 717 cpm

Table 3. Protection by NAA against PHMB inhibition of auxin binding

Addition	Auxin binding, cpm (site 1 + site 2)		PHMB inhibition, %
	Control	+50 μM PHMB	
None	541	133	75.4
5 μM NAA	503	292	42.0
10 μM NAA	487	401	17.7

Membranes were resuspended at pH 6 and incubated at 25°C for 15 min ±PHMB. Additions of NAA were made immediately prior to PHMB. Auxin binding was determined in re-isolated membranes

Table 4. Possible amino acid environment of auxin receptor sites

Observation	Inference
1. pH dependence of auxin binding	Histidine (-COO'-binding) Aspartate/glutamate (δ^+-binding)
2. Diazo-Chloramben inhibition at pH 6	Histidine
3. Diazo-CAPA inhibition at pH 8-9	Tyrosine/lysine
4. PHMB inhibition	Cysteine

D. Conclusions

Since we anticipate that receptor modification by diazo-Chloramben at pH 6 may be more specific than modification by diazo-CAPA at higher pH, and because protection by auxin can be demonstrated at pH 6 (Table 1), Chloramben has been selected for tritiation in preference to CAPA. With the labelled compound we hope to determine the specificity of receptor modification in both membrane-bound and solubilized states and to use the compound to facilitate further investigations on the binding site environment.

The auxin-binding material solubilized from the membranes is heat labile and binding saturates with NAA concentration. We have yet to carry out full kinetic analyses to determine whether site 1 and site 2 are both present. It is tempting to speculate that the SDS gel bands at 54,000 and 50,000 daltons may represent site 1 and site 2, respectively. It has been suggested that site 1, the non-specific binding site, could represent a 'pro-receptor' that undergoes a maturation process to a final auxin-specific form, i.e., site 2 (8). The molecular size relationship between the two bands is compatible with this suggestion, and their relative abundance appears to be in approximate correspondence with the ratio of site 1:site 2 deduced previously (2) from binding experiments (ca. 1:3). With larger quantities of purified material, we intend to investigate these questions and pursue further the properties of the auxin-binding proteins of corn coleoptile membranes.

Acknowledgements

I wish to thank Richard Bumpus for able technical assistance and George Earlam for synthesis of CAPA.

References

1. Batt,S., Venis, M.A.: Planta 130, 15 (1976)
2. Batt, S., Wilkins, M.B., Venis, M.A.: Planta 130, 7 (1976)
3. Hertel, R., Thomson, K.-St., Russo, V.E.A.: Planta 107, 325 (1972)
4. Leonard, N.J., Greenfield, J.C., Schmitz, R.Y., Skoog, F.: Plant Physiol. 55, 1057 (1975)
5. Morriss, J.G.: A Biologist's Physical Chemistry. 2nd ed. London: Edward Arnold (1974)
6. Porter, W.L., Thimann, K.V.: Phytochemistry 4, 229 (1965)
7. Riordan, J.F., Vallee, B.L.: In: Methods in Enzymology. Hirs, C.W., Timasheff, S.N. (eds.) London, New York: Academic Press, 1972, Vol. 25, p. 451
8. Venis, M.A.: In: Advances in Botanical Research. Woolhouse, H.W. (ed.) London, New York: Academic Press, Vol. 5, in press

Paper Demonstration References (see Section IX):

1 - 29 - 63 - 70 - 82 - 118 - 125 - 144 - 193

5. Auxin Hormone – Plasma Membrane Interactions

D. J. MORRÉ and J. H. CHERRY

Abstract

Evidence for interaction of auxin hormones with the plasma membrane of plant cells includes: (1) a thinning of the membrane observed by electron microscopy in situ and in vitro following auxin treatment as well as the following observations from isolated plasma membrane vesicles treated with auxins such as (2) an increase in fluorescence polarization of a membrane-associated probe, (3) binding of auxin and an auxin antagonist (NPA), and, (4) release of a transcription factor that stimulates RNA polymerase activity in vitro. At least an active component of the latter may be of low molecular weight and soluble in water and ethanol. These responses to auxin are rapid ($T_{1/2}$ of about 1 min), appear auxin specific, and are overcome by or not elicited by auxin antagonists (antiauxins and/or calcium ions). Our findings are compatible with a single master reaction of auxin action in which the initial receptor resides on the plasma membrane.

A. Introduction

Indications of auxin-membrane interactions in plant cells have come from a variety of sources. Brauner and Deimer (5) found that unilateral application of auxin to the hypocotyls of *Helianthus* and to coleoptiles of *Zea mays* produced a potential difference across the hypocotyl or coleoptile. Studies by Etherton (17) with microelectrodes have revealed that auxin induces a more negative membrane potential of oat coleoptile segments and that the magnitude of this change parallels the effect of auxin on elongation of the coleoptile segments. An auxin-membrane interaction was inferred also from early studies of structure-function correlations of Bungenberg de Jong and colleagues (6, 50) and those of other auxin pioneers (49, 50) and from the ability of organic solvents to mimic the action of auxin on growth of roots and stem segments (35). Weigl (52) showed that auxin binds strongly to lecithin, a common membrane phospholipid, while Stowe and colleagues (e.g., 45) have approached the problem by showing that auxin-induced elongation of pea stem segments is stimulated by lipids.

The first in vitro evidence for an auxin-plasma membrane interaction came from the studies of Lembi et al. (27) in which the auxin antagonist, NPA (N-1-naphthylphthalamic acid) was shown to bind specifically to plasma membranes of fractions from maize coleoptiles. Later, VanDerWoude et al. (48) reported an in vitro stimulation of a plasma membrane-associated glucan synthetase by 2,4-D that may or may not be related to the in vivo response of this enzyme to the hormone (1, 39). The ability of H^+ ions to mimic auxin in stimulation of growth is well established (40) as is the ability of auxin to stimulate production or secretion of H^+ ions (8, 9). The presence of a hydrogen pump has been demonstrated for animal and bacterial plasma membranes (36); the auxin-stimulated hydrogen pump of plant cells may also be plasma membrane associated, but the evidence is still indirect (18).

Direct evidence for an effect of auxin on plant plasma membranes first came from in vitro studies of Hardin et al. (21) in which a transcription factor, a substance that stimulates RNA polymerase in vitro, was found to be released from purified isolates of plasma membrane in the presence of low concentrations of auxin (1 μM 2,4-D or indole-3-acetic acid (IAA)). Several researchers have presented evidence from mammalian systems to show that one of the primary sites of hormone activity or interaction is the plasma membrane (10, 11, 14, 25, 42). In this report, recent evidence from our laboratories bearing on auxin hormone-plasma membrane interactions is summarized.

B. Materials and Methods

Soybean seeds (*Glycine max* L. var. Wayne) were soaked in water overnight and sown in moist vermiculite. Seeds were germinated in darkness at 29°C in a constant humidity chamber. After 4 days of germination, hypocotyls were excised and used for either chromatin or plasma membrane isolations, or for experiments where 1-cm sections were floated on various test solutions.

Plasma membranes were isolated according to the method of Lembi et al. (27) as detailed by Hardin et al. (21) except that dithiotheritol was not routinely included in the medium. RNA polymerase was solubilized from isolated chromatin and assayed as described (19). Under the conditions used (19, 21), the polymerase reaction goes to completion. Fractions and tissues were prepared for electron microscopy as described by Roland et al. (43). Sections were stained with a periodate-chromate-phosphotungstate (PACP) procedure that accentuates and identifies plasma membranes. Electron micrographs were obtained with a Philips EM200, calibrated and corrected for reproducible size measurements (32). Fluorescence polarization measurements were as described previously (22). Auxin binding assays were performed according to Hertel et al. (23), detailed by Williamson et al. (54). The methodology for determination of cyclic AMP (cAMP) and adenylate cyclase activity is presented in a forthcoming publication (55).

C. Results and Discussion

Two major concepts of the site of action of auxin have evolved in the nearly three decades since the discovery of this class of hormones (12). One of these has centered on the cell wall as the site of action (8, 33, 44), while the other focuses on nucleic acid metabolism (26). A role for the plasma membrane in regulating cell wall properties is implicit from the physical closeness of the two structures. There is also a potential role for the plasma membrane in regulating nucleic acid metabolism by release of a transcription factor as part of the overall auxin-plasma membrane interaction (21). The latter findings have formed a basis for integration of the seemingly disparate fast reactions of auxin action (2, 16, 40, 47, see 15, for review) and the delayed transcriptional responses (see 12 and 26, for references) into a postulated mechanism involving a single master reaction (e.g., 46, Fig.2).

Auxin stimulation of cell enlargement in soybean hypocotyls is accompanied by an increase in chromatin-bound RNA polymerase activity (38). Hardin et al. (20) isolated a factor that stimulates chromatin-bound RNA polymerase from control tissue but produces no corresponding stimulation from auxin-treated tissue. Matthysse and Phillips (28, 29) isolated a similar fraction, which in the presence of auxin, caused an increase in RNA synthesis in a chromatin system fortified with *E. coli*

RNA polymerase. Using 2,4-D-Sepharose affinity columns, Venis (51) isolated a factor that stimulated RNA synthesis in systems combining various isolated DNAs and *E. coli* RNA polymerase. Thus, numerous lines of evidence point to the presence of an auxin-responsive cytoplasmic factor, presumably a protein, capable of interacting with chromatin to enhance transcriptional activity. However, repeated attempts to purify a protein fraction with factor activity have been unsuccessful.
More recent studies suggest that the factor activity of water or buffer-soluble extracts of plasma membrane is also soluble in ethanol and that the factor or an active component of the factor may be nonproteinaceous. Clark et al. (7) used preparative thin-layer chromatography to resolve two active zones from crude ethanol extracts. The combined activity of the two active zones was approximately equal to that of the starting ethanol extract (e.g., Table 1). Both active zones coincided with materials that were reactive with α-naphthol- or phosphotungstic acid-detecting reagents and that may represent glycolipid materials. The properties of the active fraction from the ethanol extracts of plasma membranes are summarized in Table 2.

The initial experiments of Hardin et al. (21) employed a crude soybean RNA preparation having low specific activity. In comparison to the control, a relatively small amount of stimulation in activity by the transcription factor gave a large precentage increase in activity. Routinely, with polymerase preparations of greater purity and specific activity (41), we find that the plasma membrane associated and/or released factor stimulates RNA polymerase activity by about 10-15% over the control values (Table 1). This increase in total activity may be considered too small to be significant. Yet, the possibility remains that the plasma membrane factor does change the specificity of the RNA polymerase to (1) enhance activity or (2) cause a more significant change in DNA transcription with only a small overall enhancement of total activity. However, in terms of total transcribable DNA sequences, a 10% stimulation by an auxin-released factor is perhaps more realistic than a 100% stimulation in view of the known in vivo transcription effects (24).

We assume that the active material of the ethanol extracts of plasma membrane is related to the material detected by Hardin et al. (21). It might represent, for example, the prosthetic group of a larger proteinaceous complex. Whether or not such a transcription factor operates in vivo must remain speculative until suitable methodology for its detection in vivo is developed. Studies by Yunghans and Morré (55) show that the factor is not cAMP. We cannot detect either cAMP, the cyclase responsible for cAMP production, or the phosphodiesterase responsible for cAMP degradation in hypocotyl tissue of soybean, or in purified plasma membrane fractions from this tissue.

Meanwhile, additional experimental results have accumulated that demonstrate by other methods an interaction of auxins with the plasma membrane. In studies by Morré and Bracker (32), plasma membranes in isolated preparations and in whole tissues stained with the PACP procedure (43) exhibited changes in thickness after treatment with IAA (Table 3, 32). The response to IAA is rapid with a $T_{1/2}$ of about 1 min and is maximal within 20 min after the IAA is added. Membrane thickness is little affected by pH over the range of pH 5 to 8, and the dimensional changes are reversible at 26° but completely eliminated or much reduced at 0°. Benzoic acid and growth inactive analogs of 2,4-D (2,3-D and 2,5-D) have no effect on membrane thickness.

The dose-response relationships of membrane thinning by auxin show the biphasic function characteristic of the growth response of stem tis-

Table 1. Effect of plasma membrane and plasma membrane subfractions from soybean hypocotyl on soybean RNA polymerase activity

Addition	(3H) UMP incorporated[a]
RNA polymerase	5560[b]
+ freshly isolated plasma membrane + 1 μM 2,4-D	6180[c,e]
+ 85% ethanol extract of plasma membrane	6570[c,f]
+ 85% ethanol extract of plasma membrane + 1 μM 2,4-D	6560[c,f]
+ residue from ethanol extract of plasma membrane	5370[b,d]

[a] pmol/15/min mg polymerase protein from 3-5 determinations. From Clark et al. (7)

[b,c] Numbers not followed by same letter were significantly different at 1% confidence level

[d-f] Numbers not followed by same letter were significantly different at 5% confidence level

Table 2. Summary of properties of transcription factor of soybean plasma membran

1. Soluble in water, buffer and 85% ethanol but not in chloroform: not a pure protein
2. Migrates on thin-layer chromatographly, glucose > X < neutral glycolipid: low molecular weight but may be originally complexed with protein
3. Reacts with α-naphthol- and phosphotungstic acid-detecting reagents on thin-layer plates: may contain sugars, e.g., a glycolipid?
4. At least two components on thin-layer chromatography: II may be a breakdown product of I and less active
5. Released from membranes in greater amounts by either IAA or 2,4-D

Table 3. Thickness of plasma membranes of tissue explants of soybean hypocotyl in situ[a]

Treatment	Membrane thickness (Å) after staining	
	Lead citrate	PACP
H_2O	87 ± 5	101 ± 3
1 μM IAA	88 ± 6	87 ± 2
0.5 M CaCl	97 ± 5	115 ± 5

[a] Treatments were with intact tissue segments floated on solutions indicated for 1 h after which tissues were fixed and prepared for electron microscopy (43). From Morré and Bracker (32)

sues to auxins (32). There is an optimum at about 1 μM with an ascending limb (descending for growth) in the supraoptimal dose range and a descending limb (ascending for growth) in the suboptimal dose range. To our knowledge, the membrane response is one of the few auxin responses that shows the biphasic function clearly (the change in sulfhydryl content and attendant changes in heat coagulability of cytoplasmic proteins, being another, see ref. 31) in support of the traditional two-point attachment theory of auxin action (30). Plant membrane fractions do bind auxins (3, 4, 23, 54). At least a major auxin-binding fraction is consistently enriched in morphologically identified plasma membrane vesicles (54).

The nature of the auxin-induced change remains problematic. Data of Helgerson et al. (22) on fluorescence polarization and lifetime measurements of a membrane-bound probe, N-phenyl-1-naphthylamine, imply that the dimensional changes are accompanied by an increase in microviscosity of local regions around the dye. These microviscosity changes could be a cause or a consequence of a more compact or condensed membrane resulting from auxin treatment (22). Preliminary observations (and data of Table 3) suggest that some component of the membrane, i.e., that portion that stains most intensely with the PACP procedure, is lost from the membrane. These observations are consistent with the findings of Hardin et al. (21) and of Clark et al. (7) that some component having transcription factor activity is released from isolated plasma membrane fragments treated with auxins.

In other experiments, the PACP (phosphotungstic acid at low pH) staining material of plant plasma membranes appears to be extracted from the tissue or membranes with ethanol (W.N. Yunghans, J.E. Clark, D.J. Morré and E.D. Clegg, in preparation). Perhaps by coincidence, a fraction with transcription factor (TF) activity is also extracted from plasma membranes with ethanol and contains components with TF activity that coincide with components reactive with α-naphthol- or phosphotungstic acid-detecting reagents on thin-layer plates (7). It is no longer probable that all of the PACP staining material and the material with TF activity are identical. However, it does seem probable that the materials with TF activity are at least constituents of the total complex that stains with PACP.

In contrast to IAA, concentrations of calcium ions, in the range 5 to 100 mM, which inhibit auxin-induced cell elongation and wall extensibility of plant stems (Figs.1 and 2) cause an opposite effect on isolated plasma membranes, i.e., a thickening of the membranes (32, 34; Fig.1). Inhibition of wall extensibility by calcium ions requires that the tissue be living (33). Thus, growth inhibition by calcium ions cannot be explained on the basis of cross-linking of carboxyl groups of cell wall uronides by calcium ions. In electron-microscopic studies reported previously (34), the increase in membrane thickness was accompanied by lateral fusion of the membranes so that large aggregates with complex architecture were formed. Findings of J.E. Clark (unpublished) suggest that plasma membranes isolated in the presence of 0.5 M $CaCl_2$ fail to release active TF when subsequently incubated in the presence of 1μM 2,4-D or IAA plus $CaCl_2$. The effect of calcium on membrane thickness was also observed in tissue sections (Table 3), was reversible at both 26° and 0°, and was accompanied by a formation of vesicle aggregates that was monitored spectrophotometrically (53).

The effect of calcium on membrane thickness is different in important respects from that of IAA. A major difference is that the IAA effect is temperature dependent whereas the calcium effect is demonstrable even at ice bath temperatures. A second important difference is that plasma membranes isolated in media containing EDTA do not respond to calcium ions whereas the response to IAA is unaffected (32, 34 Fig.1). This latter finding suggests that the calcium response is mediated by some surface component of the membrane that is extracted by EDTA while the IAA effect is with some other membrane component.

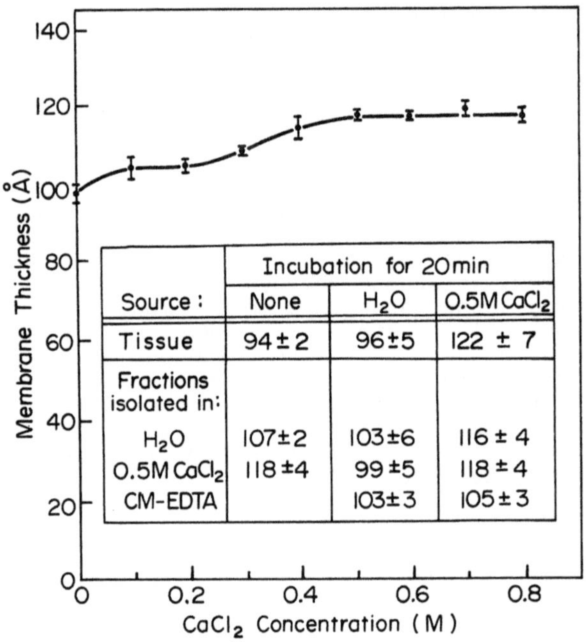

Source:	Incubation for 20min None	H_2O	$0.5M\ CaCl_2$
Tissue	94±2	96±5	122 ± 7
Fractions isolated in:			
H_2O	107±2	103±6	116 ± 4
$0.5M\ CaCl_2$	118 ±4	99 ±5	118 ± 4
CM-EDTA		103±3	105±3

Fig. 1. Thicknesses of isolated plasma membranes (PACP stained, see ref. 43, for procedure) as function of $CaCl_2$ concentration (20-min incubation). Inset shows tabular data. Calcium effect on membrane thickness is observed with both lead citrate- and PACP-stained sections, but PACP-stained plasma membranes are about 30% thicker (see also Table 3). Data from three experiments with 50 measurements each from at least 15 electron micrographs. From Morré et al. (34). Procedural details given by Morré and Bracker (32)

Morré et al. (34) attempted to relate the calcium-responsive component of membrane vesicles to the presence of a flocculant (or filamentous) layer, the latter subject to removal by EDTA or low salt. The identity of the flocculant layer is unknown. Cytochalasin B partially inhibits auxin-induced growth of soybean hypocotyls (34), but experiments to determine if the layer contains microfilamentous material of the cortial cytoplasm or an "actin-like" protein have been inconclusive. However, the findings, taken together, suggest an interaction of auxin, calcium ions, and related regulatory agents with both extrinsic and intrinsic components of the plant plasma membrane that may be important to the overall mechanism of cell expansion (Fig.2).

D. Conclusions

Ultrastructural changes were observed in isolated and in situ plasma membranes of etiolated hypocotyls of soybean (*Glycine max*) as the result of treatment of the membranes or intact tissues with the auxin hormone, indole-3-acetic acid (IAA = auxin). The response was rapid, auxin-specific, temperature-dependent, overcome by auxin antagonists (anti auxins and calcium ions), and reversible. An increase in fluorescence polarization of the membrane probe, N-phenyl-1-naphthylamine, was also recorded for auxin-treated membranes. The findings show a direct response of plasma membranes to growth regulating agents and provide

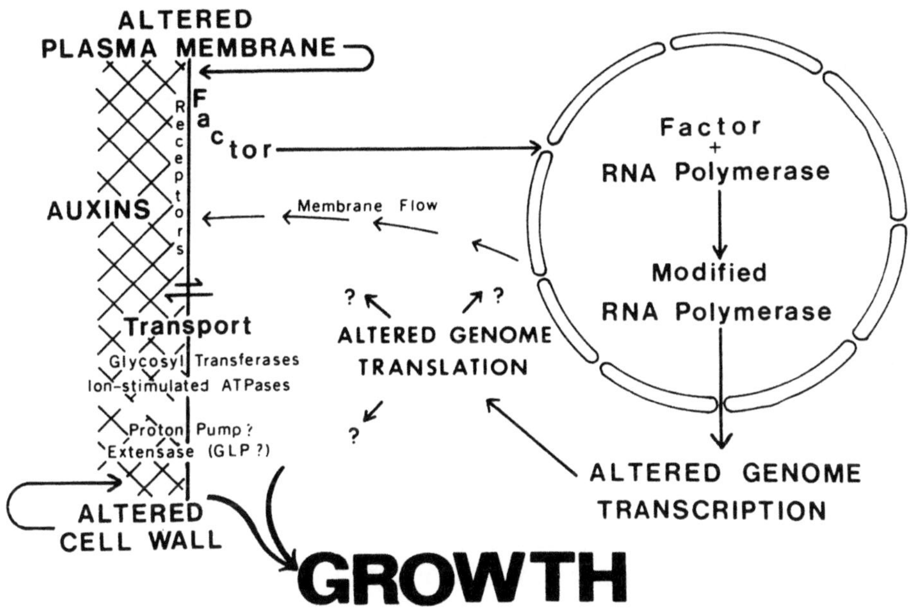

Fig. 2. Hypothetical scheme for auxin regulation through interaction with specific receptor molecules of plasma membrane. Amplification of initial "master reaction" might be achieved through conformational change in membrane affecting multiple plasma membrane-associated activities and/or release of secondary effectors or "factors." Concomitant transport of auxin molecules across membrane may or may not be involved

evidence for a cell-free response of isolated plasma membranes to a hormone. Plasma membranes bind auxin, although auxin-binding activity may not be totally restricted to plasma membrane-rich fractions.

Previous work suggesting that a transcription factor (stimulator of exogenous RNA polymerase) was released by treatment of isolated vesicles of plasma membrane of soybean hypocotyls with low concentrations (1 μM) of either 2,4-D or IAA was confirmed. The factor is not cAMP nor does cAMP appear to be involved in the mechanism of auxin action in this tissue. Efforts to isolate and identify the factor have lead to the possibility that the factor (or its prosthetic group) is nonproteinaceous and both ethanol- and water-soluble. When fractionated by thin-layer chromatography, stimulatory components were obtained from the ethanol extracts of isolated plasma membranes. Membranes treated with auxins released materials with mobilities corresponding to the stimulatory components. The chemical characteristics of the factor(s), the mode of release from the membrane, and the basis for stimulation of RNA polymerase by the factor(s) are yet to be elucidated. Suitable methodology for the isolation of the factor(s) in useful yield and purity must first be developed.

References

1. Abdul-Baki, A.A., Ray, P.M.: Plant Physiol. 47, 537 (1971)
2. Barkley, G.M., Evans, M.L.: Plant Physiol. 45, 143 (1970)
3. Batt, S., Venis, M.A.: Planta 130, 15 (1976)

4. Batt, S., Wilkins, M.B., Venis, M.A.: Planta 130, 7 (1976)
5. Brauner, L., Deimer, R.: Planta 77, 1 (1967)
6. Bungenberg de Jong, H.G., Bonner, J.: Protoplasma 24, 198 (1935)
7. Clark, J.E., Morré, D.J., Cherry, J.H., Yunghans, W.N.: Plant Sci. Lett. 7, 233 (1976)
8. Cleland, R.: Ann. Rev. Plant Physiol. 22, 197 (1971)
9. Cleland, R.: Planta 127, 233 (1975)
10. Cuatrecasas, P.: Proc. Natl. Acad. Sci. U.S. 63, 450 (1969)
11. Cuatrecasas, P.: Proc. Natl. Acad. Sci. U.S. 69, 318 (1972)
12. Davies, P.J.: Botan. Rev. 39, 139 (1973)
13. DeLaFuente, R.K., Leopold, A.C.: Plant Physiol. 25, 195 (1974)
14. Dunnick, J.K., Marienetti, G.V.: Biochim. Biophys. Acta 249, 122 (1971)
15. Evans, M.L.: Ann. Rev. Plant Physiol. 25, 195 (1974)
16. Evans, M.L., Ray, P.M.: J. Gen. Physiol. 53, 1 (1969)
17. Etherton, B.: Plant Physiol. 45, 527 (1970)
18. Hager, A., Manzel, H., Krauss, A.: Planta 100, 47 (1971)
19. Hardin, J.W., Cherry, J.H.: Biochem. Biophys. Res. Commun. 48, 299 (1972)
20. Hardin, J.W., O'Brien, T.J., Cherry, J.H.: Biochim. Biophys. Acta 224, 667 (1970)
21. Hardin, J.W., Cherry, J.H., Morré, D.J., Lembi, C.A.: Proc. Natl. Acad. Sci. U.S. 69, 3146 (1972)
22. Helgerson, S.L., Cramer, W.A., Morré, D.J.: Plant Physiol. 58, 544 (1976)
23. Hertel, R., Thomson, K.S., Russo, V.E.A.: Planta 107, 325 (1972)
24. Holm, R.E., O'Brien, T.J., Key, J.L., Cherry, J.H.: Plant Physiol. 548 (1976)
25. Kasai, M., Changeux, J.P.: J. Membrane Biol. 6, 1 (1971)
26. Key, J.L.: Ann. Rev. Plant Physiol. 20, 449 (1969)
27. Lembi, C.A., Morré, D.J., Thomson, K.S., Hertel, R.: Planta 99, 147 (1971)
28. Matthysse, A.G.: Biochim. Biophys. Acta 199, 519 (1970)
29. Matthysse, A.G., Phillips, C.: Proc. Natl. Acad. Sci. U.S. 63, 897 (1969)
30. McRae, D.H., Bonner, J.: Physiol. Plant 6, 485 (1953)
31. Morré, D.J.: Physiol. Plantarum 23, 38 (1970)
32. Morré, D.J., Bracker, C.E.: Plant Physiol. 58, 544 (1976)
33. Morré, D.J., Eisinger, W.R.: In: Wightman, F., Setterfield, G. (eds.), Biochemistry and Physiology of Plant Growth Substances, Ottawa: Runge Press, 1968, p. 625
34. Morré, D.J., Bracker, C.E., VanDerWoude, W.J.: St. Louis, Mo.: Proc. Elect. Microscop. Soc. Am. , 32nd, 32, 154 (1974)
35. Morré, D.J., Rogers, B.J., Gamble, R.: Phyton 22, 7 (1965)
36. Nakazawa, T., Asami, K., Shoger, R., Fujiwara, A., Yasumasu, I.: Exptl. Cell Res. 63, 143 (1970)
37. Nissl, D., Zink, M.H.: Planta 89, 323 (1969)
38. O'Brien, T.J., Jarvis, B.C., Cherry, J.H., Hanson, J.B.: Biochim. Biophys. Acta 169, 35 (1968)
39. Ray, P.M.: Plant Physiol. 51, 601; 609 (1973)
40. Rayle, D.L., Haughton, P.M., Cleland, R.: Proc. Natl. Acad. Sci. U.S. 67, 1814 (1970)
41. Rizzo, P.J., Cherry, J.H.: Phytochemistry 15, 637 (1976)
42. Rodbell, M., Kraus, H.M.J., Pohl, S.L., Birnbaumes, L.: J. Biol. Chem. 246, 1861 (1971)
43. Roland, J.C., Lembi, C.A., Morré, D.J.: Stain Technol. 47, 195 (1972)
44. Roland, J.C., Pilet, P.E.: Experientia 30, 441 (1974)
45. Stowe, B., Dotts, M.: Plant Physiol. 48, 559 (1971)
46. Thimann, K.V.: In: Pincus, G., Thimann, K.V. (eds.). The Hormones. New York: Academic Press, 1948, p. 5

47. Thimann, K.V., Sweeny, B.: J. Gen. Physiol. 21, 123 (1937)
48. VanDerWoude, W.J., Lembi, C.A., Morré, D.J.: Biochem. Biophys. Res. Commun. 46, 245 (1972)
49. VanOverbeek, J.: In: Klein, R. (ed.). Plant Growth Substances, Ames, Iowa: Iowa State Univ. Press, 1961, p. 449
50. Veldstra, H.: Ann. Rev. Plant Physiol. 4, 151 (1953)
51. Venis, H.A.: Proc. Natl. Acad. Sci. U.S. 68, 1824 (1971)
52. Weigl, J.: Z. Naturforsch. 24b, 365 (1969)
53. Williamson, F.A., Morré, D.J.: Proc. Indiana Acad. Sci. 82, 142 (1973)
54. Williamson, F.A., Morré, D.J., Hess, K.: In preparation
55. Yunghans, W.N., Morré, D.J.: Plant Physiol. (in press)

Paper Demonstration References (see Section IX):

25 - 29 - 39 - 63 - 70 - 82 - 90 - 118 - 134 - 144

6. Interactions Between Indole-3-Acetic Acid and Phospholipids in Deuterochloroform as Determined by ^1H, ^{13}C and ^{31}P NMR Spectrometry

A. MARKER, L. G. PALEG, and T. M. SPOTSWOOD

Abstract

Interactions between indole-3-acetic acid (IAA) and phosphatidyl choline (PC), dipalmityl phosphatidyl choline (DPPC), and phosphatidyl ethanolamine (PE) in deuterochloroform were examined with ^1H, ^{13}C, and ^{31}P nuclear magnetic resonance (NMR) spectroscopy. IAA caused large shifts in various resonances of each of the phospholipids. The chemical shifts measured with a single probe on the three lipids were significantly different from one another, and, like the dissociation constants (K_D), indicative of a 1:1 IAA:phospholipid interaction. K_D values were determined for each of the IAA-lipid interactions and, for the most part, were about 10^{-2} molal. K_D values could also be calculated if stoichiometries of 1(dimer):1 and/or 1(dimer):2 IAA:phospholipid were assumed. No meaningful K_D values could be obtained from the data if a 2:1 stoichiometry was assumed.

The fact that large chemical shifts are induced by IAA in the resonances of the $-\overset{+}{N}(CH_3)_3$ and $-\overset{+}{N}H_3$ groups as well as in those of the PO_4^- group suggests that IAA is bound at two points to the phospholipid; the carboxyl group of IAA may hydrogen bond to the lipid PO_4^- and the aromatic ring of the indole probably shields the $-\overset{+}{N}(CH_3)_3$ or $-\overset{+}{N}H_3$ group of the lipids. The results suggest that changing the fatty acid components and the polar head group influences the response of the phospholipid to IAA, at least in the interactions occurring in deuterochloroform. It seems possible, therefore, that molecular variation between phospholipids may provide the basic elements necessary for the elaboration of a concept of target-tissue specificity.

A. Introduction

Nuclear magnetic resonance (NMR) spectroscopy is now firmly established as a major investigative technique for determining molecular structure and stereochemistry and investigating interactions between molecules in solution. The technique depends on the reorientation of nuclear spins in a magnetic field induced by radiofrequency radiation of the appropriate frequency. The nuclei that have been the subject of major investigations are ^1H, ^{19}F, ^{13}C, ^{31}P and ^{11}B, all of which have nuclear spin quantum numbers of 1/2.

The five nuclei absorb radiofrequency radiation in quite distinct and non-overlapping spectral regions at constant magnetic field strength, and the precise frequency at which a nucleus absorbs energy depends both on its internal molecular environment and on its external environment (solvent, temperature, etc.). It is this latter feature that makes NMR spectroscopy particularly valuable for studying interactions in solution. ^{13}C NMR spectroscopy suffers from a lack of sensitivity relative to the other nuclei, ^{13}C being only 1.1% abundant, and effective study of this nucleus has become possible only in recent years through the use of computer averaging and Fourier transform techniques.

In two earlier publications the interactions of GA_3 (13) and IAA (5) with phosphatidyl choline (PC) in deuterochloroform were explored briefly with NMR spectroscopy. The results were obtained with a Varian DA-60-IL spectrometer, using a proton probe. As a consequence, assessment of the interaction was essentially restricted to the determination of the influence of the hormones on the characteristics of the PC trimethylamino protons. The present report describes preliminary results obtained with a Bruker HX-90E NMR spectrometer, with Fourier transformation capabilities, using three (^1H, ^{13}C, and ^{31}P) probes and three phospholipids, i.e., PC, dipalmityl phosphatidyl choline (DPPC), and phosphatidyl ethanolamine (PE) to examine the nature of complexes formed with IAA in deuterochloroform.

B. Materials and Methods

Commercial plant PE and PC were obtained from P.L. Biochemicals (Lots 471-4 and 471-1, respectively) and DPPC was from Calbiochem (Lot 300319); these were not further purified. Soya PE was prepared from a modified Singleton procedure (13) from Sigma crude soy lecithin (Lot 12C3010). IAA (BDH Laboratory Reagent) was recrystallized from $C_2H_4Cl_2$ several times and the $CDCl_3$ was obtained from C.E.A., France.

Molecular weights of 760 and 800 were assumed for PE and PC, respectively. Generally, lipids and IAA were weighed directly into NMR tubes before adding $CDCl_3$. Lipid concentrations were usually ca. 0.03-0.04 molal, which was also around the upper IAA concentration.

^{13}C and ^{31}P NMR spectra were obtained with a Bruker HX-90E spectrometer at frequencies of 22.625 and 36.43 MHz, respectively, in the pulsed Fourier transform mode with broad-band proton decoupling, at a temperature of ca. 30°C. For ^{13}C, the spectral width was 5000 Hz with an uncertainty of ±0.05 ppm (for shifts relative to $CDCl_3$). The spectral width of ^{31}P spectra was 1000 Hz with ±0.03 ppm instrumental uncertainty (relative to trimethylorthophosphate (TMP)). ^1H spectra were obtained on a Varian T-60 spectrometer at ca. 40°C in unlocked mode with spectral widths of 250 or 500 Hz and sweep times of 50 sec. Chemical shift reproducibility was within ±0.03 ppm relative to the internal reference (residual $CHCl_3$, proton resonances of TMP, or the $-CH_2$-fatty acid resonance).

The method of Nicholson and Spotswood (4) and Sykes (6) was used to calculate K_D and chemical shift values for 1:1 stoichiometry. This procedure was modified for the calculations pertaining to other stoichiometries.

C. Results

The practical operation of NMR theory is partially illustrated in Figure 1, which is a ^{13}C NMR spectrum of DPPC in deuterochloroform. The precise location of each carbon atom or group of equivalent carbon atoms can be ascertained, as can other characteristics like peak height, peak width, whether signal is as a single peak, doublet, or triplet (as $CDCl_3$), etc. Although there is no example of a proton or phosphorus scan, the principles and content are essentially similar to the scan shown here.

Since the precise resonance condition (in particular, the location of the resonance peak) is a function of its chemical environment, the introduction of a second component into the solution may or may not al-

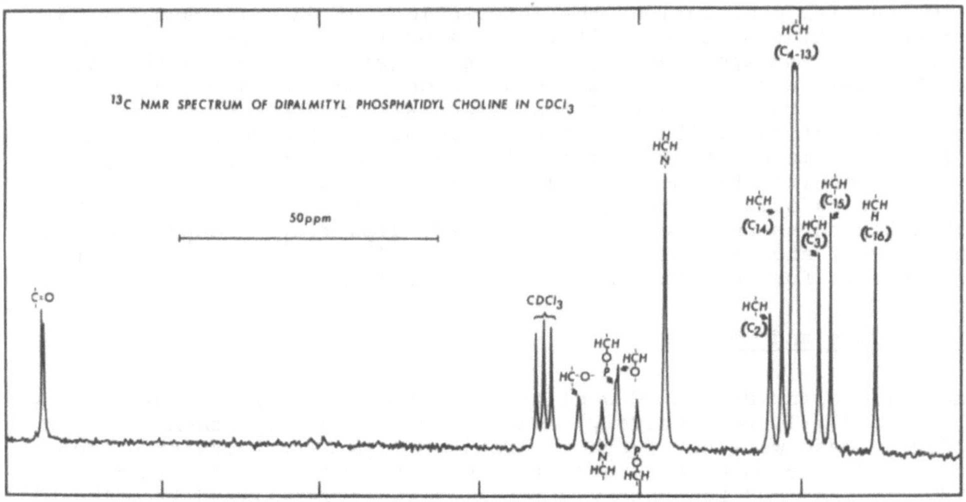

Fig. 1. ^{13}C (natural abundance) NMR spectrum of DPPC in $CDCl_3$

ter the location of any specific resonance. For example, when solid IAA was added directly to a $CDCl_3$ solution of DPPC ((IAA) / (DPPC) = 1:1), and the change in location of the discernible ^{13}C peaks of the DPPC was measured, the results shown on the left-hand side of Figure 2 were obtained. The values indicated as shift differences pertain to the change in location of the resonance peak associated with each of the specific carbon atoms. Some of the shifts are upfield (negative) and others downfield (positive), depending on the nature of the energy change induced. The plus and minus values are resolution factors rather than standard errors. Since the magnitude of the shift is indicative of the magnitude of the change in the environment of a particular atom, we can conclude that IAA exerts a reasonably strong effect on the resonance characteristics of the carbon atoms of the choline head group. The interaction with IAA influences, though less strongly, the resonance behaviour of the carbon atoms of the glycerol moiety also, and, interestingly, an effect is even exerted on both of the carbonyl carbons (C_1) of the fatty acid chains. No shift (influence of IAA) was observed for any of the other carbons of the fatty acids.

On the right side of Figure 2 the outline of PC, commercially derived from soybeans, is illustrated. The fatty acid composition is uncertain but since similar samples contained over 70% linoleic acid (13), the fatty acid indicated is linoleic acid with its two ethylenic bonds. (The two phospholipid molecules differ only in respect to their fatty acids.) Addition of IAA at the same mole ratio (1:1) as was added to the DPPC solution, caused significant shifts for each of the carbons of both the choline and the glycerol moieties of the PC, and, in contrast to the DPPC, a small shift was found with only one of the carbonyl carbons of the fatty acids. The differences in chemical shifts induced by IAA in the two phospholipids suggest that the effect of the hormone on the two lipids is different. Thus, a change in the nature of the fatty acid alters the effect of the hormone on the phospholipid, as determined by ^{13}C probe analysis.

The values shown in Figure 2 were derived from 1:1 mole ratio of IAA and phospholipid. If, on the other hand, the effect on PC of increas-

DIFFERENCE IN CHEMICAL SHIFTS OF ^{13}C ATOMS OF PHOSPHOLIPIDS IN THE PRESENCE AND ABSENCE OF INDOLEACETIC ACID

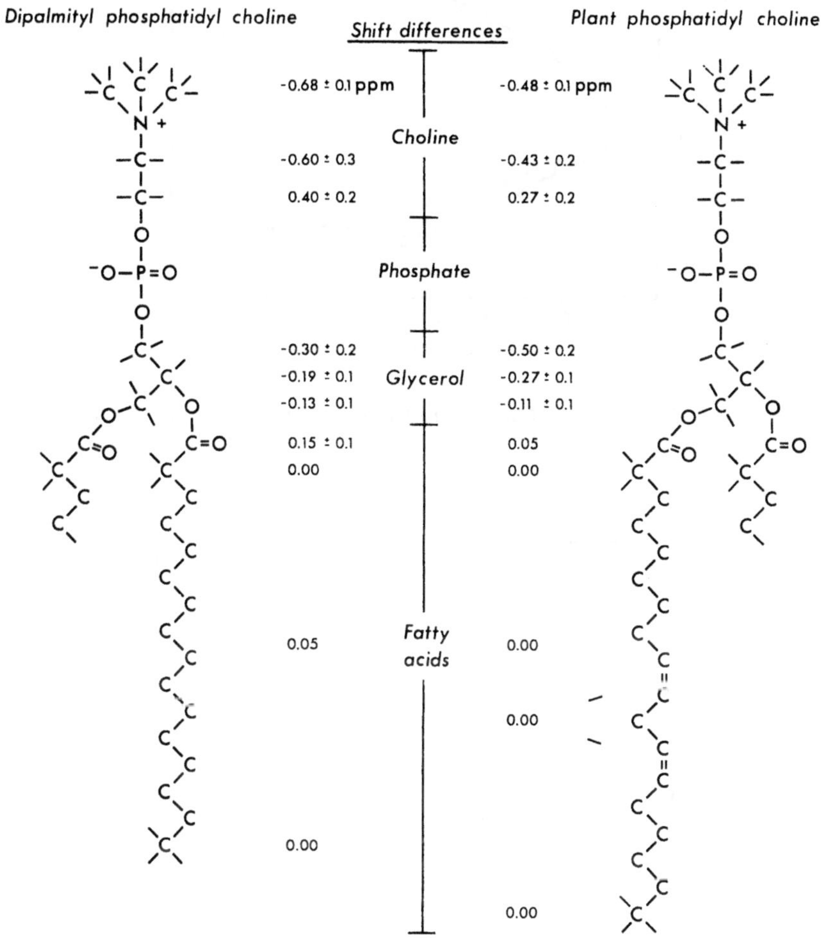

Fig. 2. Comparison of IAA-induced changes in chemical shifts of resolved carbon atoms of DPPC and PC. Shifts calculated from $CDCl_3$ solutions of 1:1 {IAA}/ {lipid}

ing the IAA concentration is determined with ^{13}C, ^{31}P, and ^{1}H probes, the curves in Figure 3 are obtained. (The ^{13}C and the ^{1}H values are derived from shifts induced in resonance characteristics of the carbon and hydrogen atoms of the quaternary nitrogen methyl groups.) The magnitude of the individual shifts increased as the IAA concentration increased, and, as indicated in Figure 4, there is a straight line relation between the shifts (1) as measured by the different probes, and (2) induced in different parts (atoms) of the molecule. A plot of change in ^{13}C chemical shifts against either of the other two (^{1}H or ^{31}P) functions would also result in a proportional relationship. Such proportionalities, of course, will only be obtained if the interaction influences the resonance characteristics of the particular nuclei observed, and if the nature of the interaction does not alter as the concentration ratio changes.

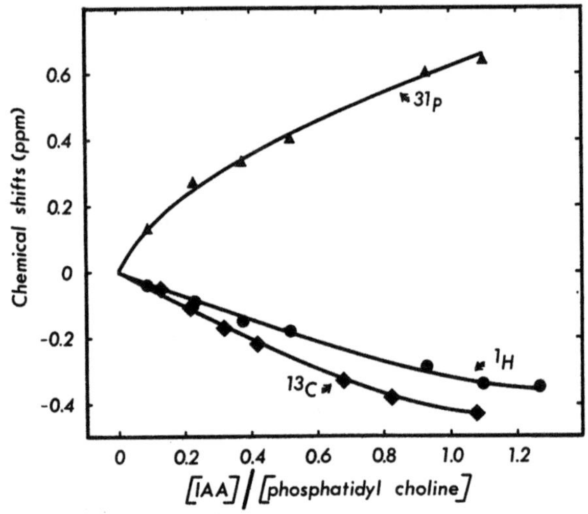

Fig. 3. Effect of different IAA/PC ratios on changes in chemical shifts observed with ^{1}H, ^{13}C, and ^{31}P probes

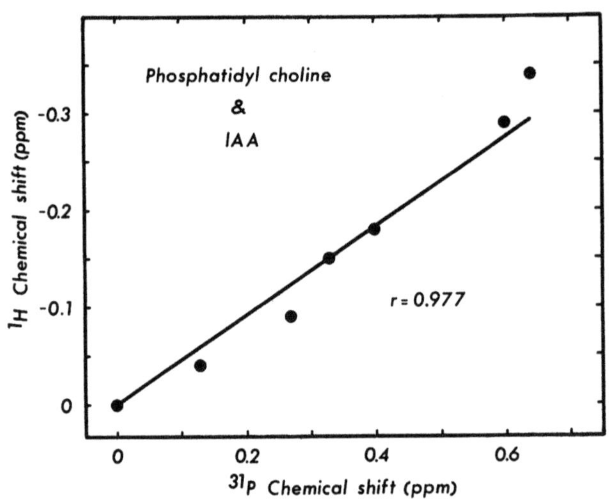

Fig. 4. Correlation between IAA-induced changes in chemical shifts measured with ^{1}H and ^{31}P probes at different IAA:PC ratios

When, instead of examining a single interaction with different probes as was done in Figure 3, one probe (^{31}P) was used to assess the interaction between IAA and three different phospholipids (i.e., PE, PC, and DPPC), the results illustrated in Figure 5 were obtained. As the concentration of IAA relative to phospholipid increased, the induced shifts also increased. The difference between PC and DPPC resides solely in the fatty acid components. However, PE has an ethanolamine group in place of the choline of the other two lipids, and the fatty acids may also differ. Thus, although the nature of the interaction

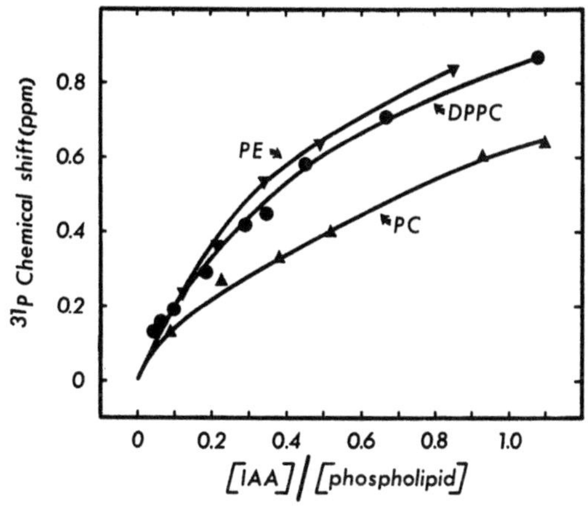

Fig. 5. Effect of different IAA/lipid ratios on changes in ^{31}P chemical shifts observed in PE, DPPC, and PC

Table 1. Chemical shift values for various IAA-phospholipid interactions

System (IAA + ...)	Probe	Shift (ppm)
Commercial plant PE	^{1}H	- 1.06 ± 0.13
Commercial plant PE	^{31}P	1.83 ± 0.13
Soya plant PE	^{1}H	- 0.64 ± 0.02
Soya plant PE	^{31}P	0.79 ± 0.07
DPPC	^{31}P	1.34 ± 0.07
DPPC	^{13}C	- 0.89 ± 0.02
Plant PC	^{1}H	- 0.61 ± 0.01
Plant PC	^{31}P	1.01 ± 0.05
Plant PC	^{13}C	- 0.58 ± 0.01

between IAA and the individual phospholipids may be similar, the specific effects induced by IAA on different parts of the different phospholipid molecules appear to differ (Figs.3 and 5).

This interpretation is highlighted when two other functions of the various interactions are examined. On the basis of the curves presented in Figures 3 and 5, a calculated value can be obtained that represents the chemical shift that might be anticipated if all of the phospholipid present was in the form of a complex with IAA. These derived values are shown in Table 1, and have been formulated for PC, DPPC, and PE, using the data obtained with all three (^{1}H, ^{13}C, and ^{31}P) NMR probes. Contrary to the chemical shift values indicated in Figure 2, the derived values in Table 1 are shown with standard errors of the means. All of the values are calculated on the assumption of a 1:1 IAA / phospholipid complex and are significantly different from one another. In other words, complex formation with IAA influenced the magnetic resonance characteristics of the phosphorus nucleus of PE differently than it influenced the phosphorus of PC and/or DPPC. The same will be true for the resonance characteristics of the C and H atoms of the quaternary nitrogen methyl groups, the observed shifts of which

Table 2. Dissociation constants (K_D) for various IAA-phospholipid interactions assuming different stoichiometric relationships

System (IAA + ...)	Probe	K_D values at IAA:lipid stoichiometrics of			
		1:1[a]	2:1	1(dimer):1[a]	1(dimer):2[b]
Commercial plant PE	^{1}H	11 ± 5	X[c]	21 ± 17	210 ± 12
Commercial plant PE	^{31}P	14 ± 7	X	22 ± 11	330 ± 130
Soya plant PE	^{1}H	2.3±1.7	X	21 ± 5	X
Soya plant PE	^{31}P	11 ± 7	X	26 ± 15	320 ± 120
DPPC	^{31}P	11 ± 5	X	27 ± 10	320 ± 90
DPPC	^{13}C	6 ± 2	X	47 ± 9	40 ± 25
Plant PC	^{1}H	14 ± 2	X	17 ± 2	X
Plant PC	^{31}P	10 ± 4	X	19 ± 7	370 ± 110
Plant PC	^{13}C	4 ± 1	X	19 ± 4	50 ± 2

[a] $K_D \times 10^3$ molal

[b] $K_D \times 10^6$ molal^2

[c] No K_D value could be calculated for this stoichiometry

were used for the calculations of ^{1}H and ^{13}C shifts within the table. (Two values obtained from a second PE sample are included in both Tables 1 and 2.)

The second type of derived values that can be obtained from the data in Figures 3 and 5 are the dissociation constants for the various complexes, and they are shown in Table 2. If calculated on the assumption of 1:1 mole ratio of IAA:lipid, the K_D values indicate a strong similarity in the strength of the interaction between IAA and the various phospholipids, and, in general, suggest the formation of complexes all of which have dissociation constants of about 10^{-2} molal, about the strength of weak enzyme:inhibitor complexes (4).

The preceding results were obtained by assuming a 1:1 stoichiometry of IAA/phospholipid. It is also possible, however, to assume different stoichiometries, and the computer programme employed (3,4) is able to determine whether any particular assumption can produce meaningful values. The calculations indicated that a 2:1 stoichiometry of IAA/phospholipid was unreasonable for any of the interactions, whereas meaningful K_D values were obtained if a 1(dimer):1 stoichiometry of IAA/phospholipid or a 1(dimer):2 were assumed (Table 2). At the moment, our state of ignorance does not enable us to choose between the alternatives, although there seems no good reason to assume a more complicated stoichiometry than 1:1.

D. Discussion

The major concept embodied in this work, of course, is not new. The possibility of some sort of physiologically important, though uncertain, interaction between auxin-type growth substances and a lipid of one kind or another, is found at least as far back as the hydrophilic/lipophilic balance theory of Veldstra and Booij (9) in 1949, and probably even earlier (7, 8). The concept recurs during the auxin structure-activity studies up to the early 1950s, after which, like old generals, it faded away. There have been occasional proponents of auxin-lipid interactions, such as Brian (2), Weigl (11), Kennedy (3) and Wood and Paleg (12, 13), but the idea has not won a large following.

The present work, in which new techniques have been used to examine hormone-lipid interactions, was envisaged as only a preliminary look at the question of target-tissue (or receptor-site) specificity of lipids. If membrane lipids are really a potential site of hormone action, some mechanism must exist to determine which lipids will react with and respond to the hormone. On the other hand, an equally good explanation of receptor-site specificity can be developed from a situation in which many lipids react with the hormone, but the hormone induces different responses in the different lipids.

The results presented here suggest that NMR spectroscopy is an excellent technique to explore the question of hormone-lipid interactions in precise and definitive ways. The use of three (^{1}H, ^{13}C, ^{31}P) probes, for example, provides a means for determining whether an interaction takes place and for evaluating the influence of the interaction (if it occurs) on almost every atom or group of atoms in either or both molecules. The data obtained with the ^{13}C probe (and indicated in Fig.2), for example, makes it clear that IAA influences electronic factors associated with every carbon atom of the choline and glycerol portions of dipalmityl phosphatidyl choline, and the effect is even observed on the carbonyl carbons (C_1) of the fatty acid chains.

The sample of PC used to compare with DPPC had an as yet undefined fatty acid composition, although it was probably between 50 and 70% linoleic acid, and certainly different than the DPPC. Thus, the differences in the responses induced in the two molecules by the presence of IAA are due to changes in the fatty acid composition of the phospholipids. Interestingly, this is one of the most (if not the most) variable characteristics of membrane phospholipids.

The second aspect examined, the nature of the head group, also suggests a measure of specificity in response. Figure 5 and Table 1 indicate that PE may respond differently than PC. In this case, however, the fatty acid composition may also differ, making a difference due to head group changes less certain. The results with both groups of samples suggest that (1) changes in the fatty acids of a phospholipid will alter the effect of an interaction of the lipid with IAA; (2) changes in the polar head group of a phospholipid will probably alter the effect of an interaction of the lipid with IAA; and (3) the specificity features essential for IAA to affect different cells in different ways (or not at all), through an interaction with phospholipids, seem to be present at least to a limited extent, in interactions occurring in deuterochloroform.

Previous results with IAA communicated three years ago (5), described a strong effect of IAA on the protons of the PC quaternary nitrogen methyl groups. At that time it was not possible to determine accurately whether other carbon atoms or nuclei were influenced by the complex. However, the strength and electronic consequences of the interaction prompted us to suggest that the aromatic ring of the indole moiety complexes with the quaternary nitrogen. It was not certain whether the carboxyl group projected upwards or downwards, although since K_D calculations suggested a 2:1 IAA/phospholipid complex, it seemed likely that IAA might project upwards and possibly even dimerize through the association of the carboxylic acid groups. The present results indicate similar strengths and electronic consequences of complex formation, and also describe effects down through the glycerol carbons. These chemical shifts suggest again that the indole ring is associated with the quaternary nitrogen, but also that the carboxyl may project downwards and possibly become associated (hydrogen bond) with an oxygen of the phosphate group. This orientation would help explain the extent of IAA effect down into the glycerol region of the lipid. Contrary to the results described previously, it was not possible to derive satisfactory K_D values in the present work, assuming a 2:1 ratio of IAA:lipid. No explanation can be given for the difference, although it is obvious that the 1:1 ratio suggested by the extensive range of samples, and the different probes, as well as the chemical shift data, makes a 2:1 ratio most unlikely.

Weigl (10) also investigated IAA-phospholipid interactions, using solubility changes, and postulated a proton binding between the proton on the indole nitrogen and the PC phosphate. In addition, he pictured a second binding between the IAA carboxyl group and the PC quaternary nitrogen. Weigl's interactions took place in aqueous media and thus are not directly comparable with the results obtained in the present work. It is interesting, however, that he also observed differences in the consequences of IAA-lipid interaction depending on the nature of the phospholipid. He ascribed the differences to differences in head groups and gave no attention to the possible effect of differences in fatty acid composition.

With only one or two exceptions, all of the K_D values (with different lipids and with different probes) (Table 2) are about 10^{-2} molal. These values are determined by characteristics such as charge interactions, stereochemistry, etc., and provide a measure of the strength of the interaction between the hormone and the lipid. The magnitude of these K_D values does not suggest a particularly strong interaction, though about what has been found with some enzyme-inhibitor complexes (4). The chemical shifts, on the other hand, may be thought of as characteristic of and dependent on the interaction, and a reflection of the specific molecular orientation within the complex. The chemical shifts induced by IAA-phospholipid complex formation are rather large; however, the magnitude of the chemical shifts are not necessarily related to the magnitude of the interaction (K_D).

Finally, it is interesting to note that findings similar to those presented here were also obtained by Bray et al. (1) who examined the interaction of indole, skatole, and 1-methylindole with hen egg yolk lecithin. Some of their work was done in chloroform, and K_D values of 54×10^{-3} molal for skatole-lecithin and 12.5×10^{-3} molal for indole-lecithin were obtained. They concluded that the chemical shifts (which were of a magnitude similar to those recorded here) provided evidence of a strong interaction between indole and lecithin and skatole and lecithin. In addition, however, they found no interaction between 1-methylindole and lecithin, and suggested that the amine proton of the indole ring must be free to enter into a hydrogen bonding association with the lecithin phosphate, which then oriented the benzene ring portion next to the quaternary nitrogen.

It is obvious from this work that there are still many unanswered questions about the interaction between indoles and phospholipids in chloroform solutions. However, since those answers are unlikely to provide better indications of interactions in living cells, perhaps it is sufficient to conclude from the present work that there is reason to explore the interaction of indoles and phospholipid membrane components in aqueous systems.

Acknowledgements

The authors would like to gratefully acknowledge the assistance provided by the Australian Research Grants Committee.

References

1. Bray, T.M., Magnuson, J.A., Carlson, J.R.: J. Biol. Chem. 249, 914 (1974)
2. Brian, R.C.: Plant Physiol. 33, 431 (1958)
3. Kennedy, C.D.: Pestic. Sci. 2, 69 (1971)
4. Nicholson, B.C., Spotswood, T.M.: Australian J. Chem. 26, 135 (1973)
5. Paleg, L.G., Wood, A., Spotswood, T.M.: In: Plant Growth Substances, 1973. Tokyo: Hirokawa Publ. Co., 1975, P. 732
6. Sykes, B.D.: J. Am. Chem. Soc. 91, 949 (1969)
7. Veldstra, H.: Enzymologia 11, 97 (1944)
8. Veldstra, H.: Biochim. Biophys. Acta 1, 364 (1947)
9. Veldstra, H., Booij, H.L.: Biochim. Biophys. Acta 3, 278 (1949)
10. Weigl, J.: Z. Naturforsch. 24b, 1046 (1969)
11. Weigl, J.: Z. Naturforsch. 24b, 367 (1969)
12. Wood, A., Paleg, L.G.: Plant Physiol. 50, 103 (1972)
13. Wood, A., Paleg, L.G., Spotswood, T.M.: Australian J. Plant Physiol. 1, 167 (1974)

Paper-Demonstration References (see Section IX):

18 - 29 - 45 - 63 - 70 - 82 - 97 - 118 - 125 - 144

7. Physiologic Implications of the Hormonal Control of Ion Transport in Plants

E. MARRÈ

Abstract

Fusicoccin (FC) markedly increases proton secretion and K^+ uptake in a number of plant tissues. The present evidence concerning the mechanism of this action suggests that FC directly activates at the cell membrane level an ATPase mediating the electrogenic H^+/monovalent cation (preferentially K^+) exchange. Most of the physiologic and biochemical effects of FC (promotion of cell enlargement, seed germination, opening of stomata, increase of CO_2 fixation and of malate and pyruvate levels) can be interpreted as consequences of the primary activation by FC of the H^+/K^+ exchange system. The finding that FC induces substantially similar responses in practically all higher plant materials so far investigated suggests the general occurrence and thus the fundamental physiologic role of the FC-sensitive system.

Activation or inhibition of the FC-sensitive system seems to mediate various important physiologic effects of the natural plant hormones, such as indole-3-acetic acid (IAA)- and cytokinin-induced stimulation of cell enlargement, abscisic acid (ABA)-induced inhibition of stomata opening and of seed germination, and also, probably, some aspects of GA_3-induced seed germination. Other hormone-regulated processes - particularly those involving specific protein synthesis - appear to be independent of the regulation of the H^+/K^+ exchange rate.

A. Introduction

In recent years important progress has been achieved in understanding the mechanism of action of plant hormones. In particular, evidence has accumulated indicating that the capacity of auxin to influence energy-dependent proton transport at the cell membrane level plays a major role in auxin-induced cell enlargement (13, 14, 20, 32, 41, 46). On the other hand, the study of this effect of auxin (and of other growth promoters) shows that important changes of the fluxes of K^+ and other ions (7, 9, 19, 26, 33) as well as of the transmembrane electric potential (PD) (9, 11, 32, 37) are intimately associated with the hormone-induced changes of H^+ transport.

It seems, therefore, that some fundamental membrane activities can be regulated by hormones, and the problem arises of how many of the physiologic effects of the various hormones can be causally related to changes at the membrane level. The recognition of the mechanism of action of the hormones on such structures as plasmalemma, tonoplast, endoplasmic reticulum, and Golgi membranes becomes a necessary step in order to understand hormone action. Also, the possibility of influencing ion transport by hormone treatment opens new, interesting possibilities for investigating membrane functions.

In the present paper, evidence pointing to the following conclusions is discussed:

1. Fusicoccin (FC) strongly promotes the electrogenic H^+/K^+ exchange in a number of higher plant tissues, and efficiently mimics the natural hormones in various important physiologic responses. The rapidly increasing knowledge of the mechanism of action of FC aids in the investigation of the more complex mode of action of the hormones.

2. The action of auxin and, in some cases, of cytokinins on cell enlargement is conveniently interpreted as due to the activation of the FC-sensitive H^+/K^+ exchange system. This also seems to be true for the inhibiting effect of abscisic acid (ABA) on stomata opening and seed germination, and probably, for some aspects of the seed dormancy-breaking effect of gibberellins.

3. For auxin, as well as for the other hormones, only a limited number of the characteristic physiologic responses seem to be mediated by changes of ion transport. Other important responses, notably those involving protein synthesis and differentiation, appear due to effects other than the activation of H^+/K^+ exchange. This suggests that natural hormones interact primarily with specific receptors different from that of FC and reach the ion transfer system through a relatively complex, branched series of intermediate steps.

B. Materials and Methods

The treatments of the plant materials mentioned in the following sections, and the physiologic and biochemical procedures employed have been described in detail in previous articles (2, 9, 16, 19, 26, 29, 32).

C. Results and Discussion

I. Physiologic Effects and Mechanism of Action of FC

The aim of this section is to: (1) summarize the characteristics of the effects of FC on cell enlargement, solute transport, stomata opening, seed germination, and metabolism; (2) present some very recent evidence on the capacity of FC to specifically bind to cell membrane preparations and to activate in vitro a plasmalemma-bound K^+-MG^{++}-dependent ATPase; and (3) propose a tentative model of the mechanism of action of FC.

1. Effects on Cell Enlargement, Monovalent Cation Uptake, and Transmembrane Electric Potential

FC has been shown to promote cell enlargement in stems and coleoptiles (5, 30, 32, 46) and also in IAA-insensitive tissues such as roots (16, 32, 36), isolated cotyledons (28), seed embryos (17, 18), and leaves (28): i.e., apparently, in practically all plant tissues in which the structure of the cell wall does not mechanically hinder the possibility of extension growth. In all cases investigated, this effect appears associated with an increase of the capacity to lower the pH of the incubation medium and an increase of the rate of uptake of K^+ (7, 9, 32, 41, 46). At high salt concentration in the medium, the uptake of other monovalent cations (Rb^+, Na^+, Cs^+) is also stimulated by FC. At low salt concentration in the medium, the effect of FC on K^+ uptake is much greater than that on Na^+ or other cations; thus FC characteristically increases the selectivity of the transport system for K^+ (9, 19, 26).

The presence of K^+ and, to a lesser extent, of other monovalent cations in the medium markedly enhances FC-induced proton extrusion: this effect is roughly proportional to the rate of cation uptake (9, 19) and decreases in the order $K^+ > Rb^+ > Na^+ = Li^+ = Cs^+$.

In all materials investigated (stems, coleoptiles, roots, isolated cotyledons, germinating seed embryos) FC-induced stimulation of H^+ extrusion and K^+ uptake is accompanied by an increase of the transmembrane PD (9, 29, 37; also S. Cocucci et al., unpublished data).

The effect of FC on cell enlargement and the associated increases of H^+ secretion, K^+ uptake, and PD are all markedly depressed by respiration and phosphorylation inhibitors (9, 29, 31) (CO, DNP, CCCP, FCCP, DCCD), by low temperature (9), and by anaerobiosis (Rasi et al., unpublished), thus showing that FC action on the foregoing processes depends on some metabolism-dependent mechanism. Cycloheximide and puromycin only partially depress the effects of FC on growth and ion transport even when used at concentrations high enough to block protein synthesis completely (Lado et al., unpublished). This suggests that protein synthesis, although necessary for the full functionality of the cell membrane, is not directly involved in the mechanism of action of FC on ion transport (see Sections 1, 10 and Fig.2).

2. Relationship Between the Effects of FC on Ion Transport and PD and Those on Cell Enlargement

FC does not significantly increases cell enlargement when cell extension is prevented either by structural characteristics of the wall (as in the fully grown first internode of pea seedlings) or by mannitol concentrations inducing high extracellular osmotic pressure close to the plasmolysis value (8, 9, 16, 30, 31). Even in these cases, however, the effects of FC on H^+/K^+ exchange and PD remain unchanged, thus showing that activation of ion transport is not a consequence of the effect on growth. On the other hand, those materials in which FC stimulates growth, also respond with an increase of cell enlargement to treatment with low pH-buffers. At low pH of the medium, the absolute effects of FC on growth are substantially decreased: namely, the effects of FC and of acid treatment are only partially additive. This suggests that FC-induced proton secretion is at least an important factor in determining wall loosening and thus cell enlargement (16, 32, 41).

3. Effects of FC on Transport of Other Solutes

Recent work in our laboratory shows that FC consistently stimulates in stem, coleoptile, and root segments the uptake of a number of solutes such as Cl^-, SO_4^{--}, succinate, leucine, aminoisobutyric acid, glucose, and 3 methyl-O-glucose, while no significant effect is detected for urea, thiourea, and tritiated H_2O (Radice et al., unpublished data).

In general, the present evidence suggests that FC is inactive on passive, noncarrier-mediated transport, while it influences to some degree a large number of carried-mediated, metabolic energy-dependent transport processes.

4. Effects of FC on Stomata Opening

FC induces stomata opening by increasing cell turgor of the guard cells, and it reverses the inhibiting action of ABA on this process. This effect is associated with, and presumably depends on, a marked increase of the capacity of K^+ uptake (43, 45) and of H^+ extrusion (Raschke, personal communication). Recent evidence indicates that physiologic opening of stomata is due to an increase of K^+ uptake, coupled to H^+ extrusion (38). It appears therefore reasonable to assume that also the FC-induced opening of stomata depends on the stimulation of H^+/K^+ exchange at the cell membrane level.

5. Seed Germination

FC is strikingly active in inducing germination of seeds either normally dormant, or made dormant by treatment with far-red light or with ABA. Breaking of seed dormancy by FC is accompanied by a very early increase of proton secretion (18) and K^+ uptake and by the hyperpolarization of the PD (S. Cocucci et al., unpublished data). A reasonable interpretation of these results is that FC activates primarily the H^+/K^+ exchange system, and thus water uptake and cell enlargement, thereby breaking the static equilibrium of the dormant seed and triggering the series of biosynthetic processes characteristic of seed germination (4, 18, 34).

6. Metabolic Effects of FC

In the pea internode segments (the material most thoroughly investigated in this regard) FC markedly increases respiration and depresses the rate of dissimilation to CO_2 of C_1 as compared to C_6 of ^{14}C-labeled glucose (C_1/C_6 ratio) (27, 32). Other important effects of FC are the increases in the rate of dark CO_2 fixation into malate, and the large increase in the pyruvate level (27, 32, and Lucchini et al., unpublished data). As discussed in the following, all of these changes can be at least tentatively interpreted as consequences of the activation of the K^+/H^+ exchange mechanism.

7. Kinetics of the FC Effect

The effects of FC on H^+ and K^+ transport, PD, and growth are extremely rapid (7, 9, 16, 32). In the case of H^+ extrusion and of PD hyperpolarization no lag period is observed, and the responses reach maximum intensity in less than 7 min (9, 32).

This suggests a small number of intermediate steps between the interaction of FC on its primary receptor and its final effect on the H^+/K^+ exchange system.

8. Binding of FC to Plasmalemma-enriched Membrane Preparations

Recent results obtained in our laboratory by Pesci et al. in collaboration with Dohrmann and Hertel in Freiburg indicate that FC specifically binds to some component present in plasmalemma-enriched membrane preparations from maize coleoptiles. The specific binding profile of FC corresponds to that of naphtylphtalamic acid, considered as a satisfactory marker for plasmalemma (15, 35). Saturation of FC specific binding occurs in the range 10^{-6} to 10^{-4} M, thus corresponding to the concentration saturating the effect on H^+/K^+ exchange in vivo.

9. Stimulating Effect of FC on K^+-Mg^{2+}-Activated ATPase of Plasmalemma Preparations

Recent work in our laboratory shows that, in vitro, FC can increase the

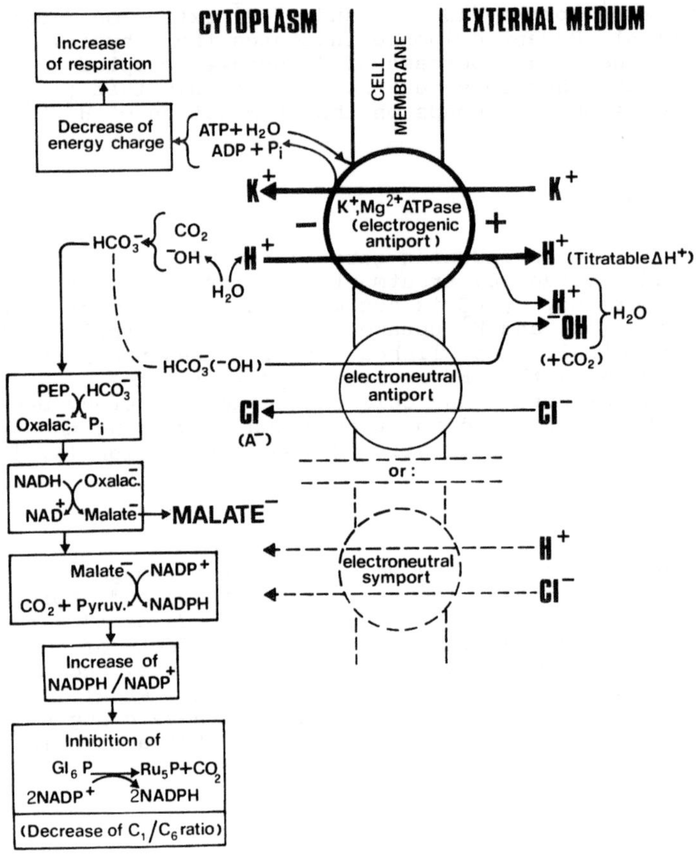

Fig. 1. Proposed model of mechanism of action of FC on ion transport. According to this model, stoichiometry of AT Pase-mediated H^+/K^+ exchange would be close to 1 but part of extruded protons would either re-enter cell together with Cl^- or associate with ^-OH exchanged with Cl^- (or other anion)

K^+, Mg^{2+}-dependent ATPase activity of plasmalemma-enriched preparations from maize coleoptiles and spinach leaves by 20 to 40%. Physiologically inactive derivatives of fusicoccin, such as DAK, do not induce the in vitro response. The concentration-activity curve of FC-induced activation of ATPase in membrane preparations closely agrees with that of the effect on K^+ uptake and H^+ extrusion in vivo (2).

The interest of this result is emphasized by the evidence suggesting that a K^+, Mg^{2+}-activated, plasmalemma-bound ATPase mediates the uptake of K^+ in plants (22, 23).

10. Mechanism of Action of FC

A model of FC mechanism of action has been proposed in a previous paper (26, see also 9). This model, presented in Figure 1, seems to cover most of the known biochemical and physiologic effects of FC. According to this model, the toxin would directly activate a plasma-

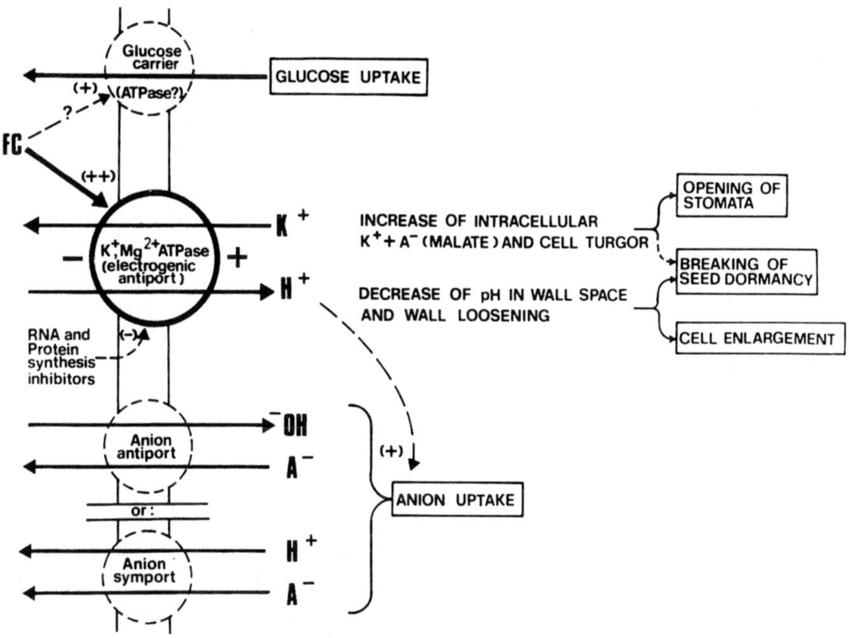

Fig. 2. Relationship between effect of FC on plasmalemma-bound AT Pase and effects on cell enlargement, opening of stomata, seed germination, and anion uptake. Effect on glucose uptake appears independent of proton gradient and of changes of PD and cation and anion uptake; it is thus tentatively interpreted as due to activation of a specific transport system

lemma-bound ATPase mediating the electrogenic K^+/H^+ antiport. The consequent alkalinization of the cytoplasm would be neutralized by the simultaneous operation of two mechanisms: (1) the exchange of intracellular ^-OH (or HCO_3^-) with an anion (Cl^-), or the symport in the cell of extracellular anion together with H^+; (2) the increased synthesis of organic acids, such as malate, in agreement with the pH-stat theory proposed by Davies (10) and by Raven and Smith (39). The increase of pyruvate and the inhibition of the glucose-6-P oxidative pathway would depend on the increase of malate, probably coordinated with allosteric effects. The stimulating action of FC on various carrier-mediated processes (see Section B.I.1), would depend on effects secondary to the activation of H^+/K^+ exchange, such as the formation of a H^+ gradient, and other changes of the intracellular as well as of the extracellular metabolite and ion distributions.

As shown in Figure 2, practically all of the known physiologic effects of FC would be explained by this model.

III. Effects of Natural Hormones on ion Transport. Analogies to and Differences from the Effects of FC

The main point discussed in this section is whether the effects of auxin, cytokinins, abscisic acid (ABA), and gibberellins on ion transport and associated physiological responses depend on their capacity to influence in some may the FC-sensitive H^+/K^+ exchange mechanism.

Table 1. Similarities between the effects of IAA and FC in pea internode segments

	Controls	IAA	FC
Cell enlargement	5	15	28
Q_{O_2}	180	225	280
C_1/C_6 ratio	0.99	0.93	0.73
Glucose-6-P	0.53	0.59	0.68
Pyruvate	0.060	0.088	0.350
Dark CO_2 fixation	27.5	33.0	43.6
K^+ uptake (1 mM)	0.48	0.96	2.20
Na^+ uptake (1 mM)	0.13	0.16	0.26
K^+/Na^+ uptake ratio	3.7	6.0	8.5
Cl^- uptake (10 MM)	1.7	2.6	3.9
SO_4^{-} uptake (5 mM)	0.59	0.67	1.31
Titratable H^+ extrusion (1 mM KCl)	0.16	0.40	0.89
H^+ extrusion activation by 10 mM KCl	+187%	+273%	+317%
H^+ extrusion activation by 10 mM NaCl	+37%	+67%	+60%
3-O-methyl glucose uptake (5 mM)	0.37	0.44	0.51
α-aminoisobutyric acid (5 mM) uptake	0.16	0.19	0.24
Transmembrane electric potential	-60	-77	-83

Cell enlargement as % increase of fresh wt in 2 h. Q_{O_2} as μl O_2 taken up /g fresh wt/h. Gl_6P and pyruvate as μmol/g fresh wt at 2 h of treatment. CO_2 fixation as nmol CO_2 incorporated in acid-soluble fraction /g fresh wt/h in 30 min. H^+ extrusion and uptake of ions, 3-O-glucose and AIBA as μEq/g/h. Transmembrane electric potential in mV

1. Auxins

When the effects of FC are compared with those of IAA or other auxins striking analogies as well as clear-cut differences become manifest.

As shown in Table 1, the analogies include: (1) the stimulation of cell enlargement; (2) all aspects of the stimulation of H^+, K^+, and, in general, ion transport, and the closely associated hyperpolarization of PD; (3) the effects on energy-dependent, carrier-mediated glucose and α-aminoisobutyric acid uptake; (4) a number of metabolic responses probably dependent on the activation of H^+/K^+ exchange, such as Q_{O_2}, dark CO_2 fixation, increase of Gl_6P, and pyruvate levels and decrease of the C_1/C_6 ratio; (5) the sensitivity to inhibitors of energy transfer processes (such as DCCD and CCCP). These analogies suggest that IAA action is mediated by the same mechanism influenced by FC.

On the other hand, the finding that all the effects of IAA shown in Table 1 are quantitatively much smaller than those of FC suggests a difference between the mechanisms of action of the two compounds. The analysis of other characteristics of the action of IAA and of FC confirms and extends this conclusion. In fact, Table 2 shows that: (1) FC is active in a number of materials that do not respond to IAA; (2) the kinetics of the effects of FC is extremely rapid, while that of IAA usually includes a lag period of minutes; (3) IAA at high concentrations induces ethylene synthesis in the pea stem, while FC does not; (4) in liquid tissue cultures FC stimulates K^+ uptake and cell enlargement, but does not significantly effect cell multiplication, a process characteristically promoted by IAA and other auxins; (5) in cell-free membrane preparations FC preferentially binds to the plasmalemma-enriched fraction, while IAA preferentially binds to the endoplasmic re-

Table 2. Differences between the effects of IAA and FC

	IAA	FC
Tissue specificity[a]	Stems and coleoptiles	Stems and coleoptiles, roots, leaves, cotyledons, seed embryos, stomata guard cells
Kinetics[b]	Lag period of minutes	No lag period
Ethylene synthesis[c]	Stimulation	No effect
Cell multiplication in tissue cultures[d]	Stimulation	No effect
Rooting[e]	Stimulation	No effect
Sensitivity to protein Synthesis inhibitors[f]	High	Low
Site of binding in vitro[g]	Endoplasmic reticulum-enriched fraction	Plasmalemma-enriched fraction

[a] For effects on cell enlargement, H^+ extrusion, and K^+ uptake

[b] For effects on electric potential difference and H^+ extrusion

[c] For concentrations of IAA and FC above 10^{-4} M (pea internode segments)

[d] Carrot storage tissue (Morpurgo, unpublished)

[e] Vine stems and bean hypocotyls

[f] Effects of actinomycin, cordycepin, cycloheximide, puromycin, and amino acid analogues on cell enlargement, H^+ extrusion, and K^+ uptake (pea internode and maize coleoptiles)

[g] Membrane preparations from etiolated maize coleoptiles

ticulum fraction; (6) RNA and protein synthesis inhibitors completely suppress the effects of IAA on growth and ion transport in the pea internode and maize coleoptiles segments, while a consistent fraction of the effect of FC is still observed even under conditions of practically complete block of protein synthesis.

All of these results suggest that the action of IAA starts from a primary receptor different from that of FC, and reaches the H^+/K^+ exchange system through a relatively complex series of intermediate steps, possibly involving the synthesis of some short-lived protein(s), as hypothesized in the alternative models of Figure 3.

2. Cytokinins

Promotion of cell enlargement in isolated cotyledons is considered a specific effect of cytokinins, because IAA and gibberellins have practically no effect on this material (24). On the other hand, FC promotes cell enlargement in isolated cotyledons, its action being more rapid and much more intense than that of cytokinins. The growth effect of FC in this material is associated, as usual, with the stimulation of proton secretion and the hyperpolarization of PD (28, 29). The finding that also cytokin-induced cell enlargement in cotyledons is accompanied by increased proton secretion and PD (28, 29) strongly suggests that the growth effect of cytokinins depends on the activation of the FC-sensitive mechanism of ion transport.

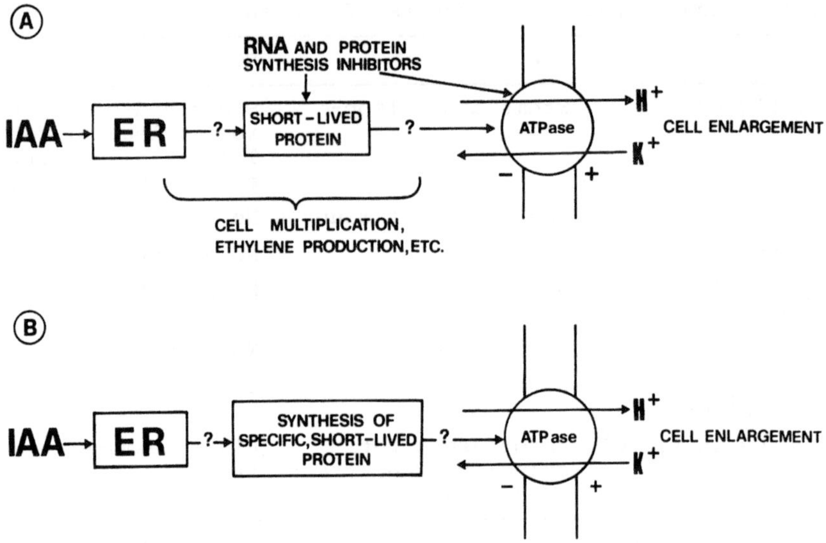

Fig. 3 A,B. Models of mechanism of IAA action on H^+/K^+ exchange. Primary receptor of IAA would be on endoplasmic reticulum (ER). Chain of events mediating effect of IAA on H^+/K^+ exchange system would comprise steps responsible for specific action of IAA on cell multiplication, ethylene synthesis, etc. (A) and (B) represent 2 alternative explanations of the inhibiting effect of protein synthesis inhibitors on IAA action. (A) presence of a short-lived protein, codified by a short-lived mRNA, would be included in chain of events mediating the effects of IAA; the synthesis of this protein would be independent of presence of IAA. (B) IAA would specifically induce synthesis of short-lived protein necessary for its physiologic action

On the other hand, FC appears unable to induce other important effects of cytokinins, such as the inhibition of senescence in leaf discs, the increase of anthocyanin in cotyledons (42), the rate of cell multiplication in tissue cultures. This suggests a scheme for cytokinin action on ion transport similar to that proposed for auxins: namely, cytokinin-induced activation of ion transport at the cell membrane level would be a consequence of a relatively long series of events in some of the complex metabolic changes primarily induced by the hormone.

3. Abscisic Acid

Abscisic acid (ABA) inhibits stomata opening and K^+ uptake in stomatal guard cells (43, 45), where K^+ uptake seems tightly coupled with H^+ extrusion (38). ABA also inhibits germination and the associated increase of proton extrusion (17, 18) and K^+ uptake (S. Cocucci et al., unpublished results). FC stimulates K^+ uptake (43, 45) and H^+ extrusion (Raschke, personal communication) associated with the opening of stomata and increases H^+ extrusion, K^+ uptake, and the hyperpolarization of PD associated with seed germination (18, M. Cocucci et al., unpublished data). In ABA-treated stomatal cells or seeds, the inhibiting effect of the natural hormone is efficiently reversed by FC. This suggests that these effects of ABA are mediated by its capacity to inhibit in some way the FC-sensitive mechanism of ion transport.

4. Gibberellins

The evidence concerning an effect of gibberellins on ion transport is incomplete and mainly indirect. However, recent studies on the effects of GA_3, ABA, and FC on seed germination suggest that part of GA_3 action can be mediated by the activation of the FC-sensitive H^+/K^+ exchange mechanism. In fact, recent results obtained by S. Cocucci et al. in this laboratory show that the promoting effect of GA_3 on germination is associated with an increase of K^+ uptake. Moreover, under appropriate experimental conditions, GA_3 can reverse the inhibiting effect of ABA on germination and the associated increases of proton secretion and K^+ uptake. FC is even more active than GA_3 in counteracting ABA-induced inhibition of these processes (17, 18, S. Cocucci et al., unpublished). These findings contribute to the developing feeling that ion transport properties of the cell membrane are of importance in controlling germination (34) and support the hypothesis that the opposite effects of GA_3 and FC, on one hand, and of ABA, on the other, on germination involve in some way the regulation of the same ion transport mechanism at the cell membrane level.

As in the case of auxin and cytokinins, the action of gibberellins on ion transport appears far more complex and indirect than that of FC and seems to include, as intermediary steps, the activation of specific biosynthetic pathways. In fact, Galli et al. in this laboratory have shown that both FC and GA_3 stimulate germination and cell enlargement in dormant seeds of *Haplopappus gracilis*: however, FC appears to act earlier on cell enlargement, while GA_3 acts earlier on DNA synthesis (12). In barley seeds both the synthesis of hydrolitic enzymes (34) and the activation of enzymes involved in the biosynthesis of lecithins (3) are among the earlier responses to GA_3. It thus seems reasonable to hypothesize that the effect of GA_3 in stimulating ion transport in seeds is just a terminal effect of a complex pattern of metabolic events, possibly involving de novo synthesis of lipid and protein components of cell membrane while the primary action of FC seems to consist of the direct activation of an already formed ion transport system.

III. Possible Role of the FC-sensitive H^+/K^+ Exchange Mechanism in Some Phytochrome-Regulated Physiologic Responses

Various important aspects of the physiology of higher plants are characteristically regulated by physical factors, such as light, temperature, and osmotic pressure of the medium. Growing evidence suggests that the control of ion transport is important in at least some of these physically regulated processes. In particular, the involvement of the FC-sensitive H^+/K^+ exchange mechanism in the action of light and of temperature on seed germination is suggested by the finding that FC efficiently substitutes red light in reversing far-red-induced dormancy of lettuce seeds (17). An even clearer indication that phytochrome-mediated control of cell enlargement involves the FC- and IAA-sensitive H^+/K^+ exchange mechanism comes from recent studies on the effects of red and far-red light. Growth by cell enlargement of the subapical zone of barley and wheat coleoptiles and of dark-grown pea seedlings is inhibited by red light (1, 21), and Lürssen has recently reported that in *Avena* coleoptiles growth inhibition by red light is associated with (and probably depends on) the capacity of red light to counteract IAA-induced proton extrusion (25). It thus appears likely that some of the phytochrome-mediated responses depend on the regulation of the proton/K^+ exchange system. On the other hand, other typical aspects of red and far-red light seem independent of the activation of proton extrusion. In radish cotyledons, for instance, the development of phenylalanine-ammonia lyase activity is strongly promoted by light treatment while it is not at all influenced by FC (44), although the toxin is very active in inducing proton secretion and cell enlargement in this material (28, 44).

This suggests that proton/K^+ exchange is just one of the mechanisms included in the wide physiologic spectrum of phytochrome action.

D. Conclusions

In my opinion, most of the available data on the effects of FC and of natural hormones on ion transport and associated phenomena can be interpreted as supporting the following working hypotheses:

1. A mechanism mediating the utilization of ∿P energy for the electrogenic H^+/K^+ exchange at the cell membrane level is present in almost all higher plant organs, where, under physiologic conditions, its rate of operation is regulated at a suboptimal level by internal controlling factors. The peculiar capacity of FC to activate this mechanism can be used as a tool to reveal its presence, its potential efficiency, and its characteristics.

2. Practically all biochemical and physiologic effects of FC, including monovalent ion transport, CO_2 fixation and malate synthesis, changes of metabolite levels and of respiratory pattern, promotion of cell enlargement, opening of stomata, and breaking of seed dormancy, can be interpreted as depending on the primary activation of this H^+/K^+ exchange mechanism.

3. Some important physiologic effects of plant hormones and of physical regulating factors, including those of IAA and cytokinins on cell enlargement, of ABA on stomata opening, of gibberellins and ABA on some aspects of seed germination, and also the phytochrome-mediated effects of light on cell enlargement and on seed germination, can be viewed in relation to the capacity of these hormones and physical factors to control the FC-sensitive H^+/K^+ exchange system. It seems clear, however, that the mechanism of action of natural hormones starts from primary receptors different from that of FC, and is mediated by a more complex chain of biochemical events, including branching points leading to various physiologic effects characteristics of each hormone and apparently independent of the regulation of the FC-sensitive H^+/K^+ exchange mechanism.

Acknowledgements

The contents of this paper represents the combined efforts of a group of individuals, including notably A. Ballarin-Denti, N. Beffagna, R. Cerana, R. Colombo, M. Cocucci, M.C. Cocucci, S. Cocucci, M.I.De Michelis, P. Lado, G. Lucchini, P. Pesci, M.C. Pugliarello, F. Rasi-Caldogno, and A. Scacchi. Thanks are due to Prof. A. Ballio of the Department of Biochemistry, University of Rome, for the generous gift of fusicoccin.

References

1. Anstis, P.J.P., Friend, J., Gardner, D.C.J.: Phytochemistry 14, 31 (1975)
2. Beffagna, N., Cocucci, S., Marrè, E.: Plant Sci. Lett., in press
3. Bien-Tal, Y., Varner, J.E.: Plant Physiol. 54, 813 (1974)

4. Chen, S.S., Thimann, K.V.: Science 153, 1537 (1966)
5. Cleland, R.: Proc. Natl. Acad. Sci. U.S. 70, 3092 (1973)
6. Cleland, R.: Planta 127, 233 (1975)
7. Cleland, R.: Biochem. Biophys. Res. Commun. 69, 333 (1976)
8. Cleland, R.: Planta 128, 201 (1976)
9. Cocucci, M., Marrè, E., Ballarin-Denti, A., Scacchi, A.: Plant Sci. Lett. 6, 143 (1976)
10. Davies, D.D.: Symp. Soc. Exptl. Biol. 27, 513 (1973)
11. Etherton, B.: Plant Physiol. 45, 527 (1970)
12. Galli, M.G., Sparvoli, E., Caroi, M.: Plant Sci. Lett. 5, 351 (1975)
13. Haschke, H.P., Lüttge, U.: Z. Naturforsch. 28c, 555 (1973)
14. Haschke, H.P., Lüttge, U.: Plant Physiol. 56, 696 (1975)
15. Hertel, A.: In: Membrane Transport in Plants, Zimmermann, U., Dainty, J. (ed.), Berlin: Springer-Verlag, 1974, p. 457
16. Lado, P., De Michelis, M.I., Cerana, R., Marrè, E.: Plant Sci. Lett. 6, 5 (1976)
17. Lado, P., Rasi-Caldogno, F., Colombo, R.: Physiol. Plantarum 31, 149 (1974)
18. Lado, P., Rasi-Caldogno, F., Colombo, R.: Physiol. Plantarum 34, 359 (1975)
19. Lado, P., Rasi-Caldogno, F., Colombo, R., De Michelis, M.I., Marrè, E.: Plant Sci. Lett., in press
20. Lado, P., Rasi-Caldogno, F., Colombo, R., Marrè, E.: Rend. Accad. Naz. Lincei 53, 583 (1972)
21. Lawson, V.R., Weintraub, R.L.: Plant Physiol. 56, 44 (1975)
22. Leigh, R.A., Wyn Jones, R.G., Williamson, F.A.: In: Membrane Transport in Plants, Zimmermann, U., Dainty, J. (ed.), Berlin-Heidelberg-New York: Springer , 1974, p. 457
23. Leonard, R.T., Hodges, T.K.: Plant Physiol. 52, 6 (1973)
24. Letham, D.S.: Physiol. Plantarum 25, 391 (1971)
25. Lürssen, K.: Plant Sci. Lett. 6, 389 (1976)
26. Marrè, E.: Proc. C.N.R.S. Int. Workshop, Transmembrane Ionic Exchanges in Plants, Rouen, 1976
27. Marrè, E., Colombo, R., Lado, P., Rasi-Caldogno, F.: Inform. Bot. Ital. 3, 97 (1972)
28. Marrè, E., Colombo, R., Lado, P., Rasi-Caldogno, F.: Plant Sci. Lett. 2, 139 (1974)
29. Marrè, E., Lado, P., Ferroni, A., Ballarin-Denti, A.: Plant Sci. Lett. 2, 257 (1974)
30. Marrè, E., Lado, P., Rasi-Caldogno, F., Colombo, R.: Plant Sci. Lett. 1, 179 (1973)
31. Marrè, E., Lado, P., Rasi-Caldogno, F., Colombo, R.: Plant Sci. Lett. 1, 185 (1973)
32. Marrè, E., Lado, P., Rasi-Caldogno, F., Colombo, R., Cocucci, M., De Michelis, M.I.: Physiol. Vég. 13, 797 (1975)
33. Marrè, E., Lado, P., Rasi-Caldogno, F., Colombo, R., De Michelis, M.I.: Plant Sci. Lett. 3, 365 (1974)
34. Mayer, A.M., Shain, Y.: Ann. Rev. Plant Physiol. 25, 167 (1974)
35. Normand, G., Hartmann, M., Schuber, F., Benveniste, P.: Physiol. Vég. 13, 743 (1975)
36. Pilet, P.E.: Plant Sci. Lett. 5, 137 (1975)
37. Pitmann, M.G., Schaefer, N., Wildes, R.A.: Planta 126, 61 (1975)
38. Raschke, K., Humble, G.D.: Planta 115, 47 (1973)
39. Raven, G.A., Smith, F.A.: Can. J. Botany 52, 1035 (1974)
40. Rayle, D.L.: Planta 114, 63 (1973)
41. Rayle, D.L., Cleland, R.: Current Top. Develop. Biol. 1976, in press
42. Servettaz, O., Castelli, D., Longo, C.P.: Plant Sci. Lett. 4, 361 (1975)

43. Squire, G.R., Mansfield, T.A.: Planta 105, 71 (1972)
44. Tomè, F., Bellini, E.: Plant Sci. Lett. 3, 413 (1974)
45. Turner, N.: Nature of London 235, 341 (1972)
46. Yamagata, Y., Masuda, Y.: Plant Cell Physiol. 16, 41 (1975)

Paper Demonstration References (see Section IX):

24 - 25 - 48 - 90 - 96 - 99 - 125 - 135

III. Hormone Analysis

8. The Application of High Performance Liquid Chromatography to the Analysis of Plant Hormones

A. CROZIER and D. R. REEVE

Abstract

A preparative high performance liquid chromatography system has been developed for the analysis of gibberellins, abscisic acid, and indoles. When used in conjunction with conventional analytical high performance liquid chromatography, it is an effective technique that enables small quantities of hormones in semi-purified plant extracts to be characterised by mass spectrometry. The procedures have been used to study the metabolism of (^{3}H)-GA_4 by *Phaseolus coccineus* seedlings.

A. Introduction

In recent years there has been much interest in the development of high performance liquid chromatography (HPLC) (6, 9). This technique offers excellent resolution, high peak capacity, and rapid speed of analysis, although because of the use of narrow bore columns the sample capacity is usually limited (14). Reference standards of abscisic acid (ABA), indole-3-acetic acid (IAA), and several cytokinins have been separated by HPLC and detected in nanogram quantities with an on-stream UV absorbance monitor (3, 4, 10). The application of HPLC to the analysis of endogenous plant hormones has, however, not been as successful. The restricted sample capacity of the technique means that without prior purification, only extracts representing relatively small quantities of plant tissue can be accommodated. The gross impurity and low level of hormones in such samples result in the UV monitor having to cope with a high background of extraneous peaks. To decide objectively which peaks, if any, represent hormones is a difficult if not impossible task. Faced with this problem some investigators have used bioassays to detect hormones in HPLC eluates (2). This is time-consuming and much of the practicality of HPLC is lost. The real value of HPLC lies in the analysis of extracts that have previously been subjected to extensive purification (5, 10). When used with such extracts, HPLC is a powerful tool with which to facilitate the isolation and subsequent characterization of plant hormones by mass spectrometry.

This report concerns a preparative HPLC technique suitable for the analysis and purification of acidic and neutral indoles, gibberellins (GA), and ABA in plant extracts prior to the application of analytical HPLC. The effectiveness of these two procedures is demonstrated in an investigation of the metabolism of (^{3}H)-GA_4 by *Phaseolus coccineus* seedlings. The acidic, ethyl acetate-soluble extract from (^{3}H)-GA_4-treated seedlings was purified by group separatory procedures prior to preparative HPLC. The metabolites isolated were subjected to benzyl esterification, then to analytical HPLC, and to direct probe mass spectrometry.

B. Materials and Methods

I. Preparative HPLC

The chromatograph, which has been described in detail elsewhere (14) utilized a 10 × 450 mm column containing either Partisil 10 or Partisil 20 coated with a 0.5 M formic acid stationary phase. This was eluted with a gradient of increasing amounts of ethyl acetate in hexane. Samples dissolved in mobile phase were introduced off-column via a six port sample valve with a 330 µl loop. The column effluent was first passed through an ISCO Model UA-5 UV absorbance monitor fitted with a 5-mm path length 25 µl flow cell and then an on-stream radioactivity monitor (14). Columns packed with Partisil 20 and run at 5 ml/min, corresponding to a solvent linear velocity of 1.5 mm/sec, generated 2800 theoretical plates for a solute with a capacity factor (k') of 1.5. This required an inlet pressure of 45 psi and gave a speed of analysis of 1.3 effective plates/sec. By comparison, Partisil 10 columns, operated at the same flow rate, had the following characteristics for a solute with a k' of 1.5: solvent linear velocity: 1.3 mm/sec; theoretical plates: 4200; inlet pressure: 130 psi; speed of analysis: 1.8 effective plates/sec. These performance figures compare favourably with those of narrow bore analytical HPLC systems.

II. Synthesis of GA Benzyl Esters (GABE)

Samples were dried and treated with a solution of dimethylformamide dibenzylacetal in dry dioxane for 2 h at 70°. The conversion of GAs to their respective benzyl esters was 80-90% effective for both standards and extracts. The GABEs thus formed were separated from the excess derivatizing reagents by preparative HPLC.

III. Analytical HPLC

The chromatograph consisted of an ARL constant pressure pump and an ISCO Model UA-5 UV absorbance monitor fitted with a 5-mm path length 10 µl flow cell. Two silica gel adsorption columns were used: a 4.6 × 500 mm prepacked Partisil 10 column eluted with 50% water-saturated tetrahydrofuran-dichloromethane (8:92 or 4:96) and a 4.6 × 250 mm prepacked Partisil 10 column run with a mobile phase of water-saturated diethyl ether. Samples were dissolved in 2-10 µl of mobile phase and injected on-column through a PTFE-faced silicone rubber septum. A flow rate of 1.6 ml/min was used for both columns. This required a pressure of 600 psi for the 500-mm column and 300 psi for the shorter system. The recovery of GABEs from both columns was 60-65%.

IV. Mass Spectrometry

Samples were introduced into an AEI MS 30 mass spectrometer by means of a direct probe with a glass capillary tip. The source temperature was 230°C but the probe itself was not heated. An electron beam energy of 70 eV was used, and spectra were recorded at a scan speed of 30 sec per decade.

V. Metabolism of (^{3}H)-GA_4 by *P. coccineus* Seedlings

Four-day-old light-grown seedlings of *P. coccineus* L. cv. Prizewinner were injected at the base of the stem with 5 µl of 80% aqueous ethanol

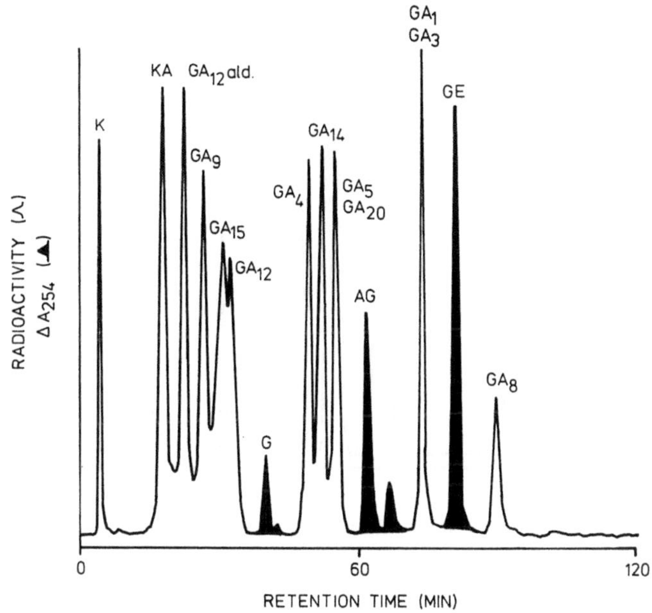

Fig. 1. Preparative HPLC of radioactive GAs and GA precursors. Column: 10 × 450 mm Partisil 20. Stationary phase: 40%, 0.5 M formic acid. Mobile phase: gradient of 0-100% ethyl acetate in hexane. Sample: approx. 24,000 dpm (^{14}C)-(-)-kaurene (*K*), 50,000 dpm (^{14}C)-*GA_3*, (^{3}H)-*GA_5*, (^{14}C)-*GA_{12}*, (^{14}C)-*GA_{15}*, and (^{3}H)-*GA_{20}*; 100,000 dpm (^{3}H)-(-)-kaurenoic acid (*KA*), (^{3}H)-*GA_1*, (^{3}H)-*GA_4*, (^{3}H)-*GA_8*, (^{3}H)-*GA_9*, (^{3}H)-*GA_{12}* aldehyde and (^{3}H)-*GA_{14}*; and uncalibrated amounts of gibberic acid (*G*), allogibberic acid (*AG*), and gibberellenic acid (*GE*). Detectors: radioactivity monitor 1800 cpm full scale delfection (FSD), UV monitor 0.5A FSD

containing 1,2(^{3}H)-GA_4 (specific activity 16.7 mCi/nmol; 1 mg, 50 μCi/200 seedlings). After 24 h the plants were extracted with methanol, and the acidic, ethyl acetate-soluble fraction was obtained. This was purified by polyvinylpyrrolidone (PVP) (8) and charcoal-celite column chromatography (1) and gel permeation chromatography (GPC) (13) before being analysed by preparative HPLC. The zones of radioactivity separated by preparative HPLC were derivatized, and the benzyl esters examined by analytical HPLC prior to direct probe mass spectrometry.

C. Results

I. Preparative HPLC

Chromatography of plant hormones on a silica gel partition column eluted with a hexane-ethyl gradient is a well-established procedure (7, 11, 12). As originally described, the method shows low efficiency, and retention volumes are not reproducible primarily because of the poor chromatographic properties of the support and the instability of the stationary phase (14). However, recent advances in liquid chromatography technology, especially the availability of efficient microparticulate silica gel supports (15), provided a means whereby the perform-

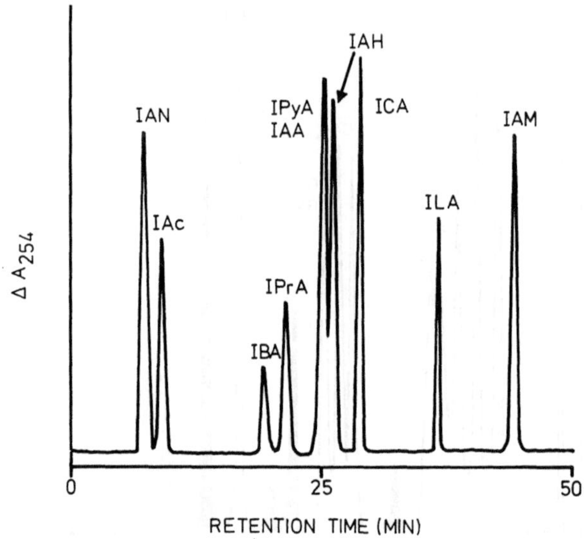

Fig. 2. Preparative HPLC of acidic and neutral indoles. Column: 10 × 450 mm Partisil 10. Stationary phase: 40%, 0.5 M formic acid. Mobile phase: gradient of 30-90% ethyl acetate in hexane. Sample: Indole acetamide (*IAM*), indole acetone (*IAc*), indole acetonitrile (*IAN*), indole-3-acetic acid (*IAA*), indole aldehyde (*IAH*), indole butyric acid (*IBA*), indole carboxylic acid (*ICA*), indole lactic acid (*ILA*), indole propionic acid (*IPrA*), and indole pyruvic acid (*IPyA*). Approx. 200 µg of each. Detector: UV monitor 0.5A FSD

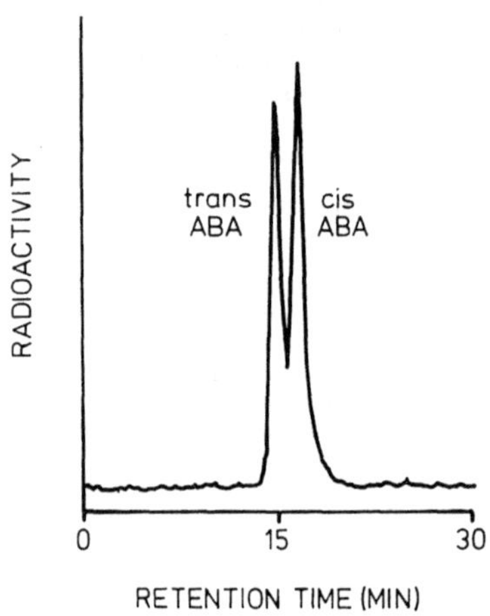

Fig. 3. Preparative HPLC of ABA. Column: 10 × 450 mm Partisil 10. Stationary phase: 40%, 0.5 M formic acid. Mobile phase: 55% ethyl acetate in hexane. Sample: (^{14}C)-cis, trans ABA and (^{14}C)-trans, trans ABA, 15,000 dpm. Detector: radioactivity monitor 400 cpm FSD

ance of this technique could be considerably improved (14). The separation of a number of radioactive GAs and GA precursors on this preparative HPLC system is illustrated in Figure 1. The solvent programme was designed to provide a rapid scan of a group of GAs differing widely in their polarity. The separation is determined primarily by the degree of hydroxylation. The UV absorption trace shows the elution profile of gibberic, allogibberic, and gibberellenic acids, which when run as internal standards allow very accurate estimations of the relative re-

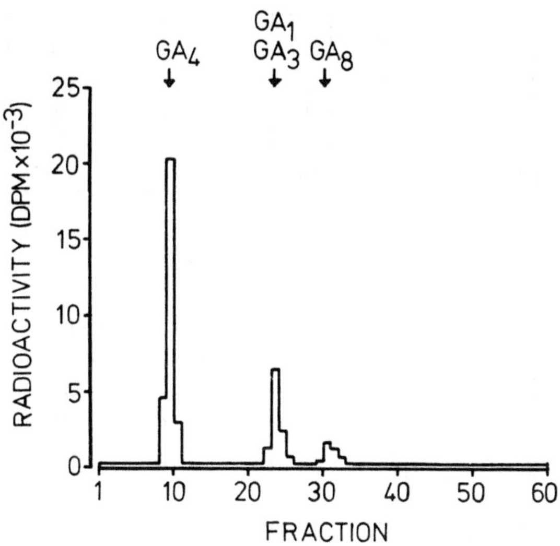

Fig. 4. Preparative HPLC of semipurified acidic, ethyl acetate-soluble extract obtained from *Ph. coccineus* seedlings 24 h after application of (^{3}H)-GA_4. Column: 10 × 450 mm Partisil 10. Stationary phase: 40%, 0.5 M formic acid. Mobile phase: gradient of 80-100% ethyl acetate in hexane. Sample: 64 mg, 10.8 × 10^6 dpm of extract. Detector: 60 fractions collected and $^1/200$ aliquots taken for estimation of radioactivity by scintillation counting. *Arrows* indicate elution points of GA_1, GA_3, GA_4, and GA_8

tention times of the radioactive GAs. However, even in the absence of these internal markers there is less than two percent run to run variation. With an adjustment of the solvent programme the chromatograph can be used to analyse acidic and neutral indoles (Fig.2) and ABA (Fig.3). In all instances the recovery of solutes from the column was greater than 95%. Individual compounds weighing up to 10 mg and multicomponent mixtures in excess of 100 mg can be chromatographed without loss of resolution (14). The chromatograph therefore has both the sample capacity and efficiency necessary for the analysis and purification of hormones in plant extracts. Its performance in this role is demonstrated in an investigation of the metabolism of (^{3}H)-GA_4 by *P. coccineus* seedlings.

II. Analysis of the Metabolism of (^{3}H)-GA_4 by *P. coccineus* Seedlings

1. Extraction, Purification, and Preparative HPLC

Seedlings were harvested 24 h after the application of 110 × 10^6 dpm of (^{3}H)-GA_4. The plants were extracted, and following solvent partitioning, an acidic, ethyl acetate-soluble fraction was obtained weighing 3.5 g and containing 36 × 10^6 dpm of radioactivity. After purification by PVP and charcoal-celite adsorption chromatography and GPC, the weight was reduced to 160 mg, and 27 × 10^6 dpm of radioactivity remained. Sixty-four mg of this semipurified extract was taken for preparative HPLC using a mobile phase of 80-100% ethyl acetate in hexane. Sixty fractions were collected and aliquots of each assayed for radioactivity. Three zones of activity were detected (Fig.4). The

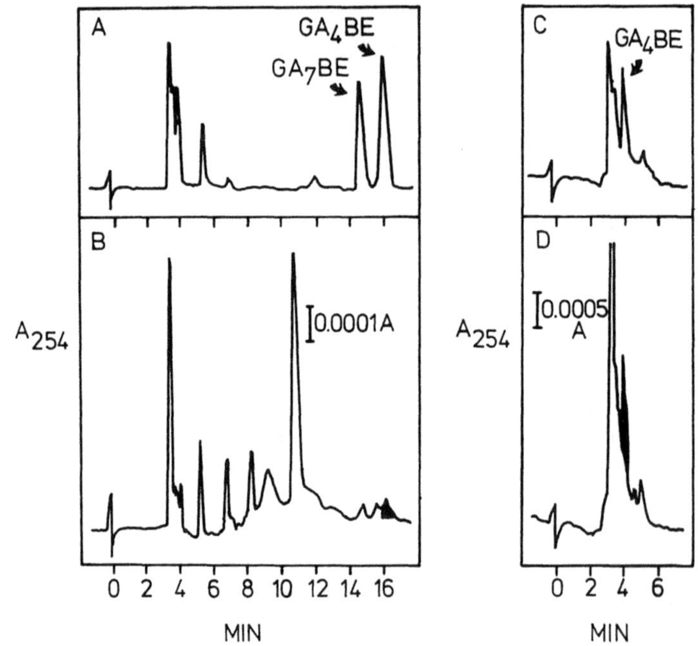

Fig. 5 A-D. Analytical HPLC of GABEs. (A) Column: 4.6 × 500 mm Partisil 10. Mobile phase: tetrahydrofuran-dichloromethane (4:96). Sample: GA_4BE and GA_7BE. Detector: UV monitor. (B) Column, mobile phase, and detector: as in (A). Sample: benzyl ester of preparative HPLC fractions 9-11 (0.62×10^6 dpm). Shaded UV-absorbing peak indicates location of radioactivity. (C) Column: 4.6 × 250 mm Partisil 10. Mobile phase: diethyl ether. Sample: GA_4BE. Detector: UV monitor. (D) Column, mobile phase, and detector: as in (C). Sample: UV-absorbing radioactive peak from B (0.36×10^6 dpm, retention time 16.2 min). Shaded UV-absorbing peak indicates location of radioactivity

compound in fractions 9-11 (10 mg, 6.0×10^6 dpm) co-chromatographed with GA_4, that in fractions 23-26 (1.0 mg, 1.9×10^6 dpm) with GA_1 and GA_3, while the material in fractions 30-33 (5 mg, 1.0×10^6 dpm) had an elution point identical to that of GA_8. The two main zones of radioactivity were benzyl esterified by treatment with dimethylformamide dibenzylacetal. This step is necessary because compounds with free carboxyl groups exhibit poor chromatographic behaviour in silica gel adsorption systems. For this reason and also because they cannot be detected by a UV monitor, GAs are not ideal solutes for analytical HPLC. Fortunately both these problems can be overcome by the conversion of GAs to their UV-absorbing benzyl esters.

2. Analytical HPLC and Mass Spectrometry of GABEs

A one-eight aliquot (7.5×10^5 dpm, 1.2 mg) of the GA_4 zone (fractions 9-11) was benzyl esterified and the reaction mixture chromatographed on the preparative HPLC using 35% ethyl acetate in hexane. This separated the presumptive (^3H)-GABE from the derivatizing reagent, residual free GA, and other impurities. The radioactivity (6.2×10^5 dpm) recovered from the preparative HPLC co-chromatographed with GA_4BE and was associated with less than 1 mg of dry weight. This semipurified

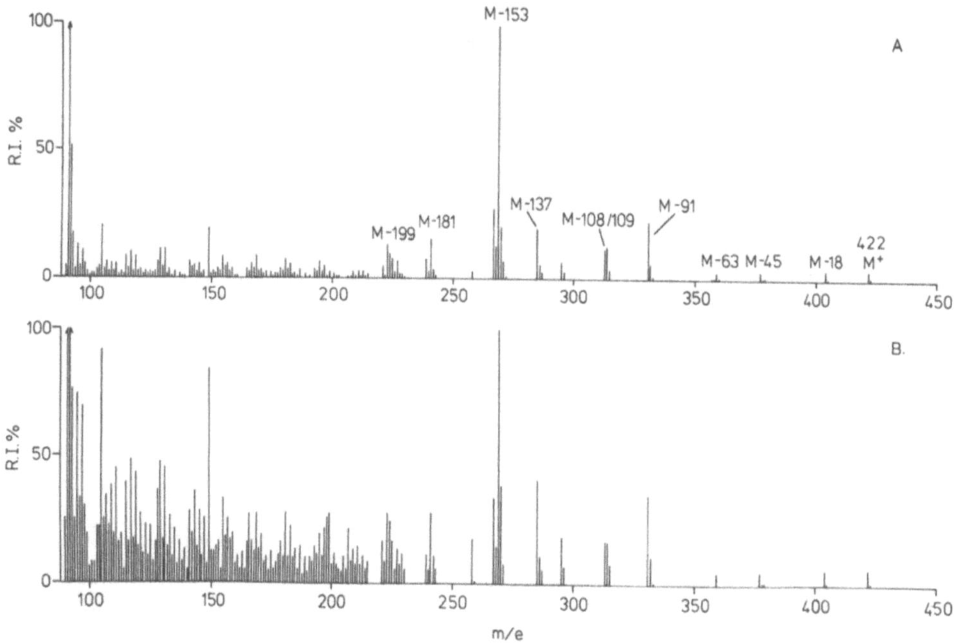

Fig. 6 A,B. Mass spectrum of (A), authentic GA_4BE and (B), purified benzyl ester of preparative HPLC fractions 9-11

fraction was subjected to analytical HPLC on a 4.6 × 500 mm Partisil 10 column eluted with 3% tetrahydrofuran in dichloromethane. A number of UV-absorbing peaks were observed, one of which had an identical retention time (16.2 min) to GA_4BE and contained 3.6 × 10^5 dpm of radioactivity. There was no indication of the presence of any GA_7BE (Fig. 5A,B). The GA_4BE-like peak was collected and, before being examined by mass spectrometry, was run on a well-conditioned 4.6 × 250 mm Partisil 10 column using a mobile phase of diethyl ether. This step was carried out to reduce background in the mass spectrum originating from solvent impurities. A UV-absorbing peak containing radioactivity and co-chromatographing with GA_4BE (Fig.5C,D) was collected from the HPLC and transferred to a glass capillary that was introduced into the mass spectrometer via a direct insertion probe. The spectrum obtained was identical to that of authentic GA_4BE (Fig.6).

The GA_1 zone (1.9 × 10^6 dpm, 1 mg), comprising fractions 23-26 from the preparative HPLC, was esterified and run on the preparative HPLC in 55% ethyl acetate in hexane to give a presumptive GABE (1.6 × 10^6 dpm, < 1 mg) with a retention time identical to that of GA_1BE. This fraction was subjected to analytical HPLC using a 4.6 × 500 mm Partisil 10 column and a mobile phase of 8% tetrahydrofuran in dichloromethane. Several UV-absorbing peaks were separated, and one of these had a retention time (14.9 min) identical to that of GA_1BE and contained 1.1 × 10^6 dpm of radioactivity (Figs. 7A,B). This peak was collected and run on the shorter Partisil 10 column in diethyl ether. Once again the radioactivity remained associated with a UV-absorbing component that co-chromatographed with GA_1BE (Figs.7C,D) and, when examined by mass spectrometry, a spectrum was obtained that was essentially identical to that of authentic GA_1BE (Fig.8)

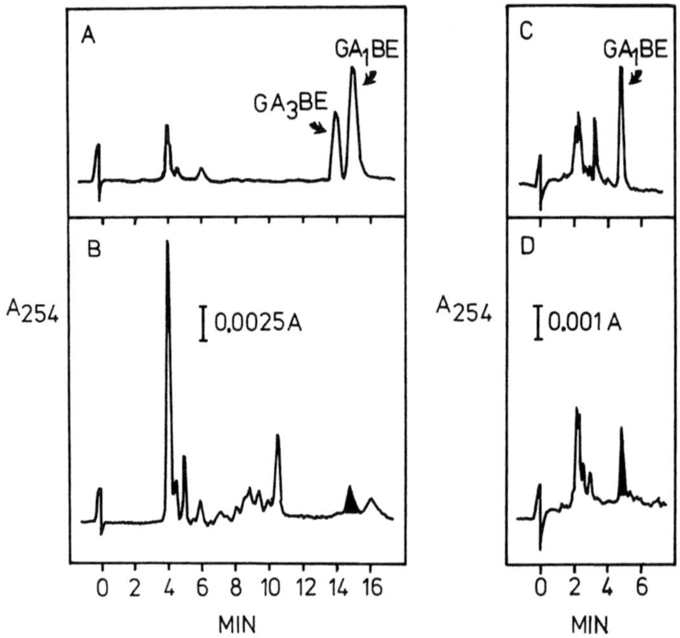

Fig. 7 A-D. Analytical HPLC of GABEs. (A) Column: 4.6 × 500 mm Partisil 10. Mobile phase: tetrahydrofuran-dichloromethane (8:92). Sample: GA_1BE and GA_3BE. Detector: UV monitor. (B) Column, mobile phase, and detector: as in (A). Sample: benzyl ester of preparative HPLC fractions 23-26 (1.6×10^6 dpm). Shaded UV-absorbing peak indicates location of radioactivity. (C) Column: 4.6 × 250 mm Partisil 10. Mobile phase: diethyl ether. Sample: GA_1BE. Detector: UV monitor. (D) Column, mobile phase, and detector: as in (C). Sample: UV-absorbing, radioactive peak from B (1.1×10^6 dpm, retention time 14.9 min). Shaded UV-absorbing peak indicates location of radioactivity

D. Discussion

Through the use of preparative and analytical HPLC and mass spectrometry, it has been possible to demonstrate conclusively the conversion of applied (^{3}H)-GA_4 to (^{3}H)-GA_1 by *P. coccineus* seedlings. The occurrence of GA_3 and GA_7 as major metabolites can be excluded. However, an additional metabolite was detected, and it had a retention time on preparative HPLC very similar to that of GA_8. Work is in progress to obtain the mass spectrum of this compound.

Although the 'background' was not subtracted from the mass spectra presented in Figures 6B and 8B, it would seem that the respective GABEs were more than 90% pure. The spurious ion at m/e 149 in Figure 6B is due to phthalate bleed from the analytical HPLC septum while that at m/e 207 in Figure 8B originates from the dimethyl silicone rubber septum at the gas chromatography-mass spectrometry (GC-MS) interface, which although not in use still contributed to the 'background'. The purity of the final extracts can also be gauged from the size of the UV-absorbing GABE peaks in Figures 5D and 7D. These correspond to 2 μg of GA_4BE and 6 μg of GA_1BE, respectively. This is in close agreement with the amounts calculated on the basis of the radioactivity present in the extracts.

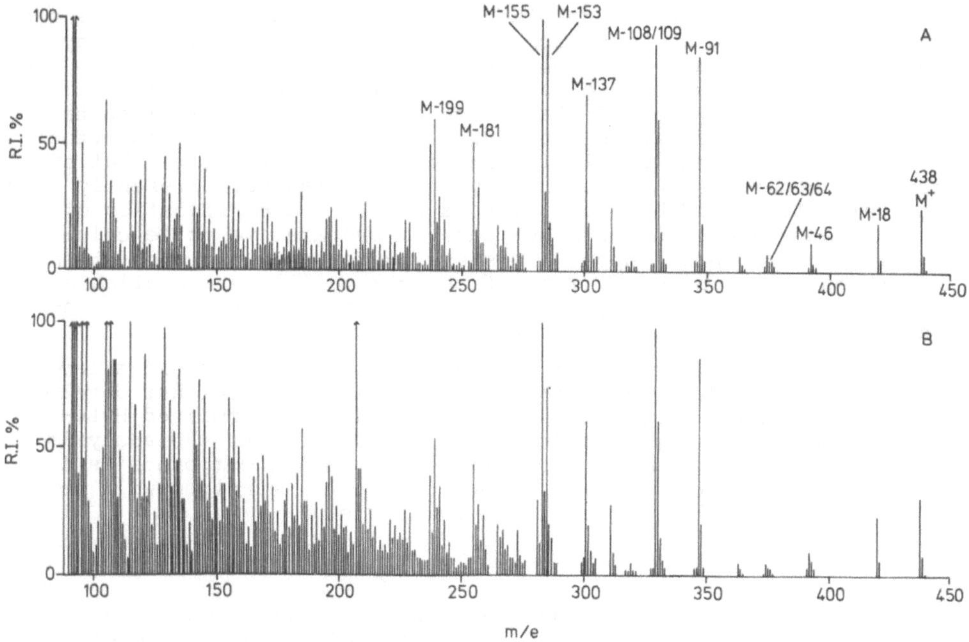

Fig. 8 A,B. Mass spectrum of (A), authentic GA_1BE and (B), purified benzyl ester of preparative HPLC fractions 23-26

The combined effects of the preparative and analytical HPLC systems resulted in a 4000-fold purification of GA_4 and a 10,000-fold improvement in the purity of GA_1 while there was a concomitant GA recovery of 25-30%. Because the levels of GA_1 and GA_4 in the semipurified ethyl acetate-soluble fraction were relatively high at 0.03% and 0.1%, respectively, the degree of purification was more than adequate to permit characterization by mass spectrometry. Clearly GAs in much lower concentrations could be characterised, if necessary, through the use of additional analytical HPLC procedures and GC-MS rather than direct probe-mass spectrometry.

It is only possible to use analytical HPLC effectively if it is complemented by a preparative HPLC system. The preparative chromatograph used in this study is less efficient than the average analytical HPLC but has 40 times the sample capacity. It is therefore well suited to reducing the dry weight of typical plant extracts to a level that is compatible with analytical HPLC. From Figure 2 and 3 it can be seen that the use of the preparative system need not be restricted to GAs. A number of indoles and ABA can be purified by this means, thereby allowing analytical HPLC and the more conventional techniques of gas liquid chromatography, and GC-MS to be employed with greater effectiveness.

I. Future Developments

Most of the commercial interest in HPLC has been focussed on analytical instrumentation and, given a sufficiently pure sample, it is a relatively simple matter to select a suitable system by reference to the copious literature that is available as a guide. In order to tackle the separation problems specific to plant hormone analysis, it will be necessary to devise HPLC systems with a high sample capacity. This may best be achieved through the application of current HPLC theory to the neglected field of preparative column technology. In particular, C18 and related chemically bonded stationary phases would be well suited to reverse phase purification procedures for GAs, indoles, and cytokinins. The data of Carnes et al. (2) indicate the possibility that such supports may supersede the slower alkylated dextran gels presently used in this role. It should be possible to approach the purification of GA-sugar conjugates and cytokinins by means of preparative scale columns packed with bonded phase supports possessing ion exchange properties. At the present time, however, many of these materials are available in columns suited only for analytical work, and some effort will be involved in adapting them to preparative systems.

While it can be anticipated that the preceding measures would alleviate some of the problems of plant hormone analysis, a further development that would assist this task is the technique of combined HPLC-mass spectrometry. The application of this procedure to the analysis of GA glucosyl conjugates and other thermally labile compounds is immediately apparent. However, the impact of such a technique will not be felt until a reliable interface is available.

Acknowledgements

We gratefully acknowledge the assistance of our colleagues Linda Nash and Takao Yokota, and the generous gifts of radioactive GAs received from Professor R.P. Pharis, Professor J.E. Graebe, and Dr. J. MacMillan. The investigation was supported by a Science Research Council (UK) grant to A.C. The Science Research Council also provided Professor M.B. Wilkins with funds for the purchase of the mass spectrometer. The on-stream radioactivity monitor was obtained with a grant to A.C. from the Royal Society.

References

1. Bowen, D.H., Crozier, A., MacMillan, J., Reid, D.M.: Phytochemistry 12, 2935 (1973)
2. Carnes, M.G., Brenner, M.L., Andersen, C.R.: J. Chromatog. 87, 271 (1973)
3. Carnes, M.G., Brenner, M.L., Andersen, C.R.: IN: Plant Growth Substances 1973, Tokyo: Hirokawa Publ. Co., 1974, p. 99
4. Challice, J.S.: Planta 122, 203 (1975)
5. Dekhuijzen, H.M., Gevers, C.T.: Physiol. Plantarum 35, 297 (1975)
6. Done, J.N., Kennedy, G.J., Knox, J.H.: Nature (London) 237, 77 (1972)
7. Durley, R.C., Crozier, A., Pharis, R.P., McLaughlin, G.E.: Phytochemistry 11, 3029 (1972)
8. Glenn, J.L., Kuo, C.C., Durley, R.C., Pharis, R.P.: Phytochemistry 11, 345 (1972)
9. Knox, J.H.: Laboratory Equipment Digest 12, 51 (1974)
10. Pool, R.M., Powell, L.E.: In: Plant Growth Substances 1973, Tokyo: Hirokawa Publ. Co., 1974, p. 93
11. Powell, L.E.: Plant Physiol. 39, 836 (1964)
12. Powell, L.E., Tautvydas, K.J.: Nature (London) 213, 292 (1967)
13. Reeve, D.R., Crozier, A.: Phytochemistry 15, 791 (1976)
14. Reeve, D.R., Yokota, T., Nash, L.J., Crozier, A.: J. Exp. Botany 27, 1243 (1976)
15. Southern, D.L.: Laboratory Equipment Digest 12, 67 (1974)

Paper Demonstration References (see Section IX):

9 - 20 - 35 - 91 - 114 - 120 - 130 - 133 - 143 - 185

9. Gas Chromatographic Identification and Quantitative Estimation of Natural Auxins in Developing Plant Organs

F. WIGHTMAN

Abstract

Evidence indicating the natural occurrence of the auxins, indole-3-acetic acid and phenylacetic acid, in a range of higher plants has been obtained from bioassay examination and gas chromatographic analysis of acidic ether fractions from vegetative shoots of these plants. Quantitative estimation of the relative amounts of each auxin in the different shoots was also achieved by gas chromatography. As part of a comprehensive study to examine the changes in amounts of different growth-regulating substances in plant organs during their growth and development, the identification and quantitative estimation of the auxins present in developing tobacco leaves have been determined by gas chromatography. The results demonstrate that in expanding leaves there is a close correlation between the endogenous levels of several auxin substances and the subsequent growth rate of this plant organ.

A. Introduction

Previous tracer and enzyme studies in this laboratory (1-5) have clearly demonstrated that the shoot systems of many higher plants contain enzyme systems capable of converting the aromatic amino-acids, L-tryptophan and L-phenylalanine, via similar reaction pathways to the auxins, indole-3-acetic acid (IAA) and phenylacetic acid (PAA). Chromatographic and bioassay evidence was also obtained in this earlier work to show that both PAA and IAA occur as native constituents of tomato and sunflower shoots and the endogenous level of each auxin was found to be sufficient to exhibit growth-promoting activity when the compounds were extracted from the shoot tissues, separated by paper chromatography, and then bioassayed with the wheat coleoptile segment test (5).

Further evidence indicating the natural occurrence of both IAA and PAA in a range of monocotyledonous and dicotyledonous plants has now been obtained by using similar chromatographic and bioassay procedures, and final identification and quantitative estimation of the relative amounts of each auxin in the shoots of these plants has been achieved by gas chromatography. This sensitive analytical technique has also been used by other recent workers to identify IAA in plant tissues (6) but the technique has not so far been employed to determine the changes in auxin levels in a plant organ during its growth and development. Since a close relationship of auxin level to growth rate is an important aspect of the auxin:growth-promotion concept, we decided to examine this relationship in a typical plant organ by employing gas chromatography to identify and quantify the auxin substances present in tobacco leaves at various stages of growth. The results of these recent studies to determine the identity and endogenous level of auxin substances in higher plant tissues are presented in this report.

B. Materials and Methods

I. Plant Material

The plant material used in these experiments included the shoots of 8-week-old tobacco (var. Mammoth), 6-week-old tomato (var. Michigan Ohio), 4-week-old sunflower (var. Russian Mammoth), and 2-week-old pea (var. Alaska), barley (var. Champlain), and corn (var. Golden Bantam) plants. All plants were grown in an air-conditioned greenhouse at 75°F under natural daylight supplemented to 18 h with incandescent lighting. Tobacco plants were 20-22 ins high and at the 12-14 leaf stage when the entire shoot, or individual leaves, were harvested for auxin extraction. Tomato and sunflower plants were 16-18 ins high and at the 7-8 leaf stage when the shoots were harvested, and the younger pea, barley, and corn plants were approximately 12 ins high when the shoots were excised for extraction.

II. Extraction Procedure, Chromatographic Identification, and Quantitative Estimation of Auxins in Vegetative Shoots

One-kg samples of shoots were homogenized in acetone in a blender, and the tissue was extracted for 24 h in a dark coldroom. After filtering, the acetone extract was concentrated in vacuo and the aqueous concentrate filtered through celite to remove precipitated chlorophyll and other pigments. The aqueous filtrate was then partitioned against three successive equal volumes of freshly distilled ether, first at pH 7.5 to give the neutral fraction, and then at pH 2.5 to give the acidic ether fraction. The auxins present in the acidic ether fraction from the different shoots were separated by paper chromatography using the solvent system, isopropanol:ammonia:water (8:1:1, vol/vol), and their growth-promoting activities were determined by bioassay of a half portion of each chromatogram using the wheat coleoptile segment test, according to the procedure described by Wightman (6). The presence of both IAA and PAA in the different acidic ether fractions was confirmed by gas chromatographic analysis of the compounds present in segments of the paper chromatograms that exhibited growth-promoting activity in the wheat coleoptile bioassay test. The compounds in these chromatogram segments were eluted with methanol, the eluate was evaporated to dryness, the residue was redissolved in 10% methanol in ether, and the compounds were then methylated with diazomethane. The methylated eluate from each chromatogram segment was then analysed by gas-liquid chromatography (GLC), first at 120°C to examine for the presence of Me-PAA, and then again at 220°C to check for the presence of Me-IAA. The gas chromatograph employed in these studies was a Pye Series 104 instrument fitted with dual-heated FID detectors and dual glass columns packed with either 3% OV-17, or 3% OV-225. The Me-IAA and Me-PAA peaks on the GLC scans were identified by co-injection of a sample of the methylated eluate with ng amounts of authentic methyl esters of the two auxin compounds. When the height of one peak in the scan was clearly enhanced in the precise region where the peak of the authentic compound was expected, then this was taken as good evidence for the presence of the compound in the sample.

Quantitative estimation of the levels of IAA and PAA in the different shoot ether fractions was achieved by using internal standards during GLC analysis of the paper chromatogram eluates. Known amounts of authentic Me-IAA or Me-PAA were co-injected with successive aliquots of one of the methylated chromatogram eluates, after which a standard curve plotting peak areas against ng amounts of the authentic ester injected was prepared for each auxin. During the final calculation of the amounts of IAA and PAA present in the different shoots, allowance was made for the difference in chromatographic losses known to occur with IAA (approx. 40% loss) and PAA (approx. 10% loss) during paper chromatography of acidic ether fractions in the isopropanol:ammonia:water solvent system used in this study.

III. Extraction and Gas Chromatographic Procedures for the Identification and Quantitative Estimation of Auxins in Developing Tobacco Leaves

For this investigation 100 8-week-old tobacco plants (var. Mammoth) were used, and the shoot apical bud and next 10 leaves down the stem were quickly harvested and collected into identical groups of tissue. The average length and fresh weight of the different sized organs were determined, and each batch of tissue was then homogenized in acetone. After filtration, each acetone extract was concentrated in vacuo, the aqueous concentrate was filtered through celite and then successively extracted with distilled ether, first at pH 8.0 to give a neutral ether fraction, and then at pH 2.5 to yield the acidic ether fraction. The latter fraction was expected to contain IAA and PAA and any other acidic auxins, plus abscisic acid and all free gibberellins that might be present in tobacco leaves. These acidic ether fractions were analysed by gas chromatography without any prior purification by paper or thin-layer chromatography. Each fraction was evaporated to dryness, the residue redissolved in a known volume of 10% MeOH in ether, and a sample was then taken for methylation with diazomethane. The methylated compounds in each sample were analysed by GLC, using linear temperature program procedures with dual glass columns packed with 3% OV-17. FID detectors were employed, and the most frequently used temperature program started at 95°C and went up to 310°C at 3° per minute. The different growth regulators were identified by using a range of internal standards during analysis of each fraction, and the amounts of IAA and PAA present were estimated quantitatively from standard curves calibrating peak areas with known amounts of the authentic methyl ester co-injected with the plant sample.

C. Results and Discussion

I. Occurrence and Quantitative Estimation of IAA and PAA in Vegetative Shoots

Since phenylacetic acid does not react with most of the known chromogenic reagents used to visualize aromatic compounds on thin-layer or paper chromatograms, the occurrence of PAA along with IAA in auxin fractions extracted from the vegetative shoots of several monocotyledonous and dicotyledonous plants was first demonstrated by paper chromatography followed by bioassay of the chromatograms. A solvent mixture containing isopropanol:ammonia:water was used for this analysis because preliminary chromatography of tomato ether fractions spiked with the authentic ^{14}C-labelled auxins showed that IAA and PAA are well separated on paper chromatograms developed in this solvent system (Fig.1). When the acidic ether fractions from tomato, tobacco, pea, sunflower, corn, and barley were chromatographed in this solvent and a portion of each chromatogram was then divided into 20 equal-sized transverse segments and bioassayed with the wheat coleoptile segment test, the results shown in Figures 2, 3, and 4 were obtained. Clearly, two regions with appreciable growth-promoting activity were present on

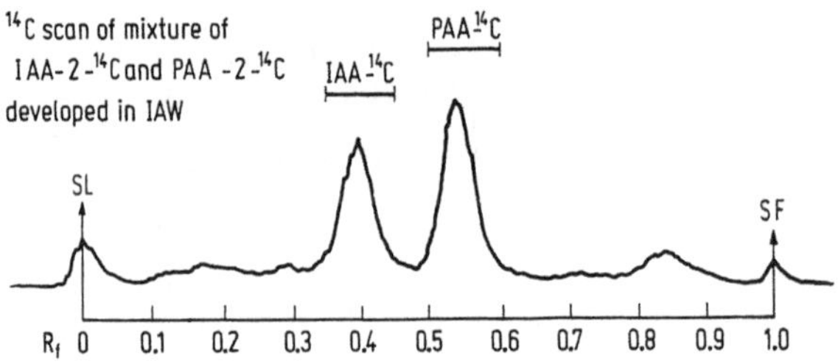

Fig. 1. Radioactivity scan of paper chromatogram of tomato acidic ether fraction containing μg amounts of authentic IAA-2-^{14}C and PAA-2-^{14}C. Chromatogram was developed in isopropanol: amonia: water solvent system

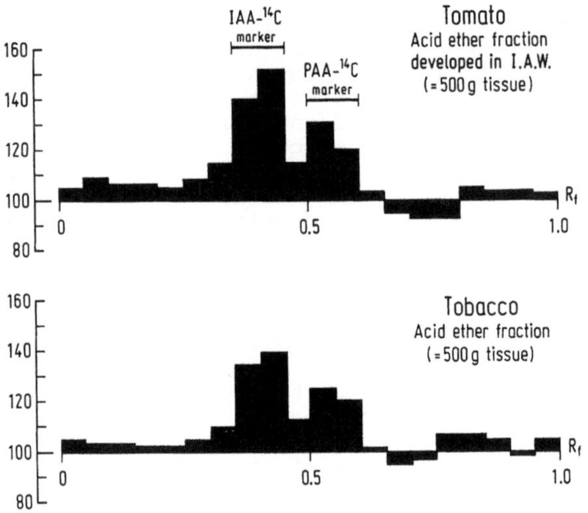

Fig. 2. Activity recorded in wheat coleoptile bioassay of paper chromatograms containing acidic ether fractions from tomato and tobacco shoots

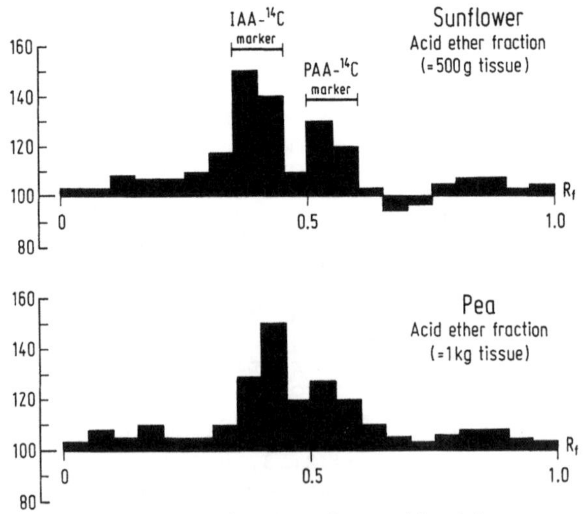

Fig. 3. Activity recorded in wheat coleoptile bioassay of paper chromatograms containing acidic ether fractions from sunflower and pea shoots

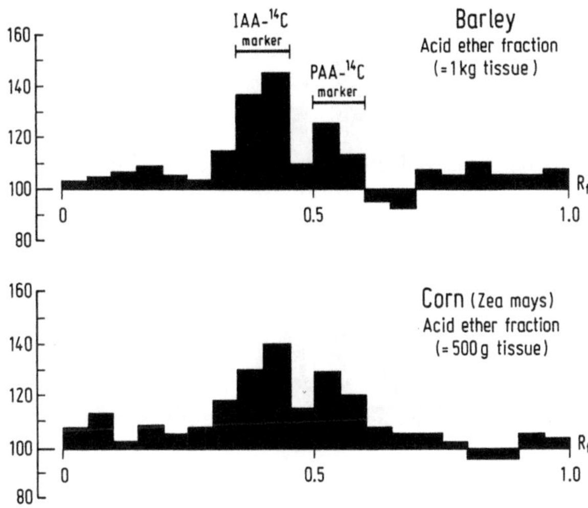

Fig. 4. Activity recorded in wheat coleoptile bioassay of paper chromatograms containing acidic ether fractions from barley and corn shoots

each chromatogram; the first between Rf 0.35 and 0.45 coincided with the position of marker IAA-2-^{14}C and the second, between Rf 0.50 and 0.60, was coincident with the position of marker PAA-2-^{14}C. That these two regions of growth-promoting activity were, in fact, due to the presence of significant amounts of IAA and PAA in the different acidic ether fractions was strongly supported by the results obtained when further portions of each fraction (equivalent to 500 g fresh weight of shoot tissue) were spiked with 50 µg IAA and 200 µg PAA and were then analysed by paper chromatography followed by wheat coleoptile bioassay. Chromatograms of all the spiked acidic fractions again showed only two major peaks of growth-promoting activity, and these were identical in Rf to the two peaks of activity exhibited by chromatograms from the unspiked fractions. A typical set of results obtained during this bioassay examination of chromatograms from spiked (bottom) and unspiked (top) acidic ether fractions is shown in Figure 5.

The presence of both IAA and PAA in all the acidic ether fractions examined in this study was confirmed by gas chromatographic analysis of the compounds present in the two regions of each paper chromatogram that showed appreciable growth-promoting activity. These two regions on the remaining half of each chromatogram were referred to as the "suspected IAA band" and the "suspected PAA band". They were excised from the chromatogram and eluted with hot methanol, and a sample of the eluate was taken for methylation with diazomethane. The series of scans shown in Figures 6, 7, 8, and 9 were obtained when these samples were analysed by GLC using columns containing either 3% OV-17 or 3% OV-225. In each figure, the two upper scans were produced by 2-5 µl injections of the appropriate chromatogram eluate, and the bottom scans were obtained when the same volume of the eluate was co-injected with ng samples of the authentic methyl ester of IAA or PAA. Thus, in Figures 6 and 8 a peak can be seen in each of the upper scans obtained from eluates of the suspected IAA band from tomato, tobacco, barley, and corn chromatograms, and this peak was clearly enhanced when 300 ng of authentic IAA methyl ester was co-injected with a further sample of each eluate. Similarly, in Figures 7 and 9, a peak is seen in each upper scan obtained from eluates of the suspected PAA band from the four

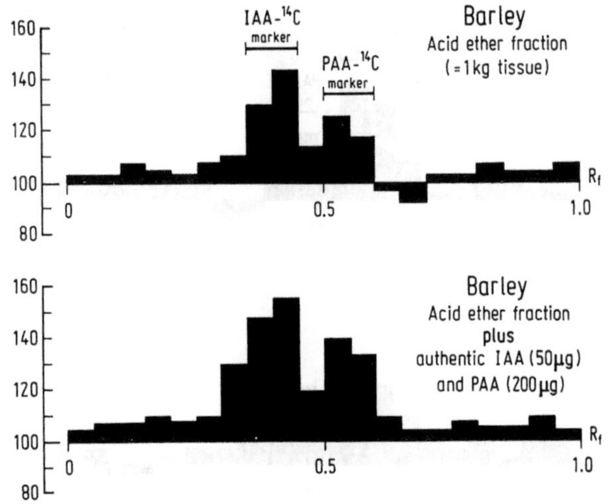

Fig. 5. Activity recorded in wheat coleoptile bioassay of paper chromatograms containing an acidic ether fraction from barley shoots (*top*) and an identical ether fraction spiked with µg amounts of authentic IAA and PAA (*bottom*)

chromatograms, which was strikingly enhanced when 200 ng of authentic PAA methyl ester was co-injected with further samples of each eluate. These GLC results thus provide the first conclusive evidence that PAA does indeed occur along with IAA as a natural auxin in vegetative shoots of tomato, tobacco, pea, sunflower, corn, and barley plants.

Since previous studies examining the auxin activity of indole-3- and phenyl-compounds in standard tests have shown that on a mol:mol basis, PAA is much less active than IAA in promoting extension growth (Table 1), it was important in the present study to estimate the quantitative levels of the two auxins in the different shoots to determine whether PAA occurs in amounts sufficiently high enough to exert growth-promoting activity. As described earlier, quantitative estimation of the levels of IAA and PAA in the different acidic fractions was achieved by co-injecting known amounts of the authentic methyl esters of each auxin during GLC analysis of the methylated eluates from the suspected IAA and PAA bands excised from the different paper chromatograms. Using the standard curves obtained in this procedure, it was possible to estimate the amount of IAA or PAA present in each chromatogram. From these values, the amount of each auxin per gram fresh weight of shoot tissue was calculated, and these results are shown in Table 2. Clearly, PAA was present in all the shoots examined in amounts several-fold higher than that of IAA, the amounts varying from as low as 70 ng/g fresh weight in barley shoots to as high as 220 ng/g fresh weight in tomato.

If we make the assumption that approximately 75% of the fresh weight of these different shoots represents the average water content of each system, then the values in Table 2 indicate that PAA is present in endogenous concentrations ranging from approximately 1×10^{-6} M in barley shoots up to about 3×10^{-6}M in tomato. Similar estimates of the internal concentration of IAA reveal lower levels for this auxin, ranging from approximately 1×10^{-7}M in barley up to about 4×10^{-7}M in tomato, tobacco and sunflower shoots. Such endogenous levels for PAA and IAA are somewhat lower than the concentrations of each com-

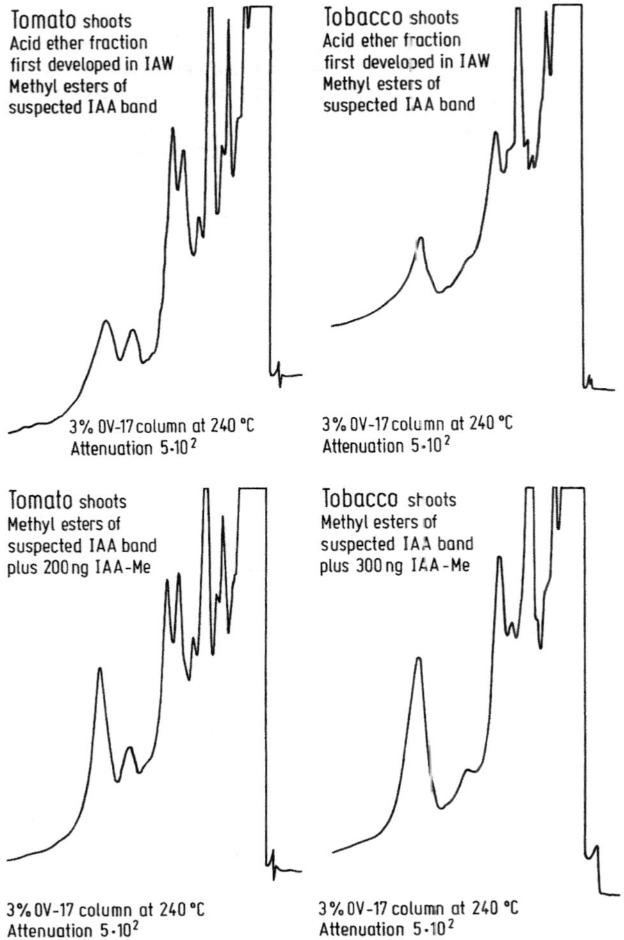

Fig. 6. Scans from GLC examination of methylated eluates from suspected IAA band on paper chromatograms of acidic ether fractions from tomato and tobacco shoots

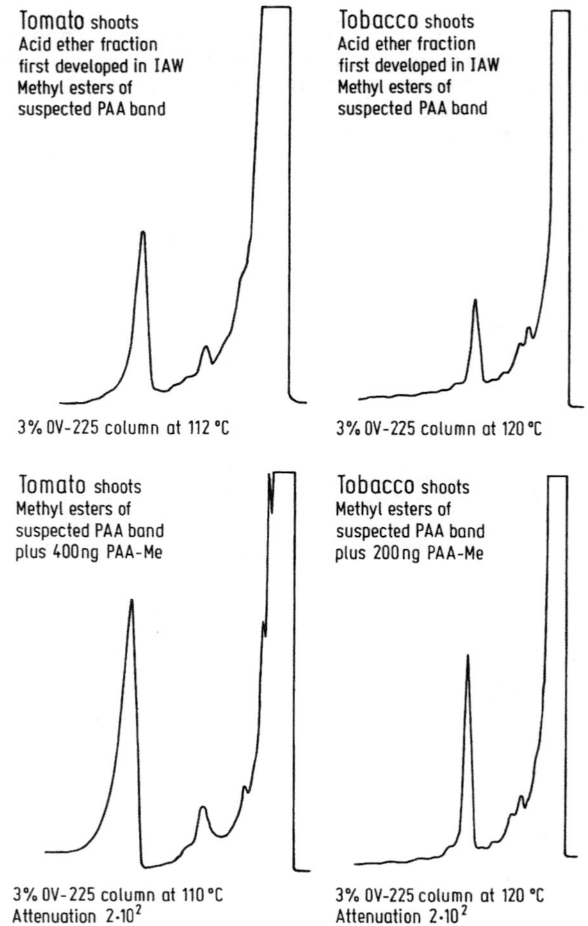

Fig. 7. Scans from GLC examination of methylated eluates from suspected PAA band on paper chromatograms of acidic ether fractions from tomato and tobacco shoots

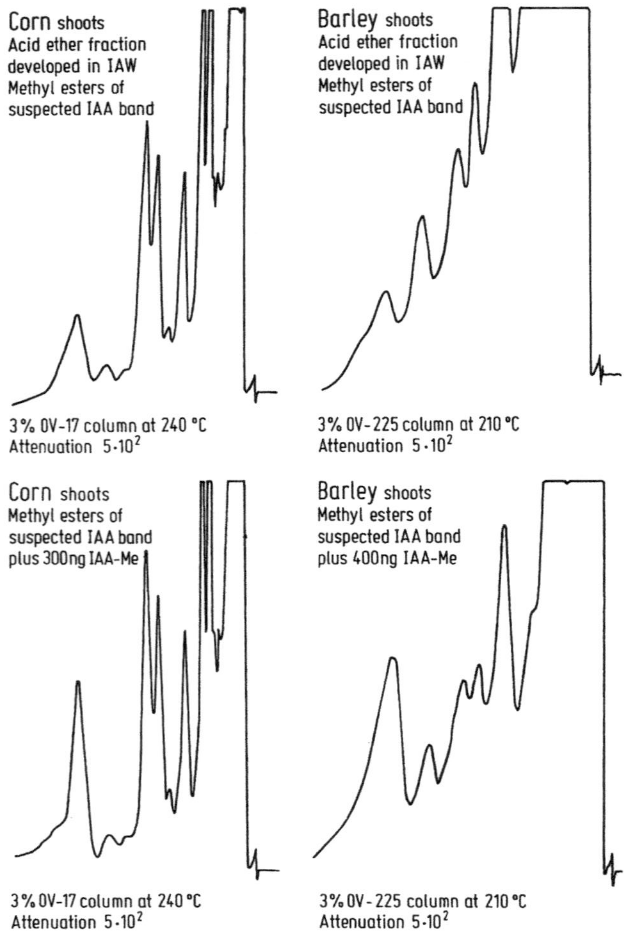

Fig. 8. Scans from GLC examination of methylated eluates from suspected IAA band on paper chromatograms of acidic ether fractions from barley and corn shoots

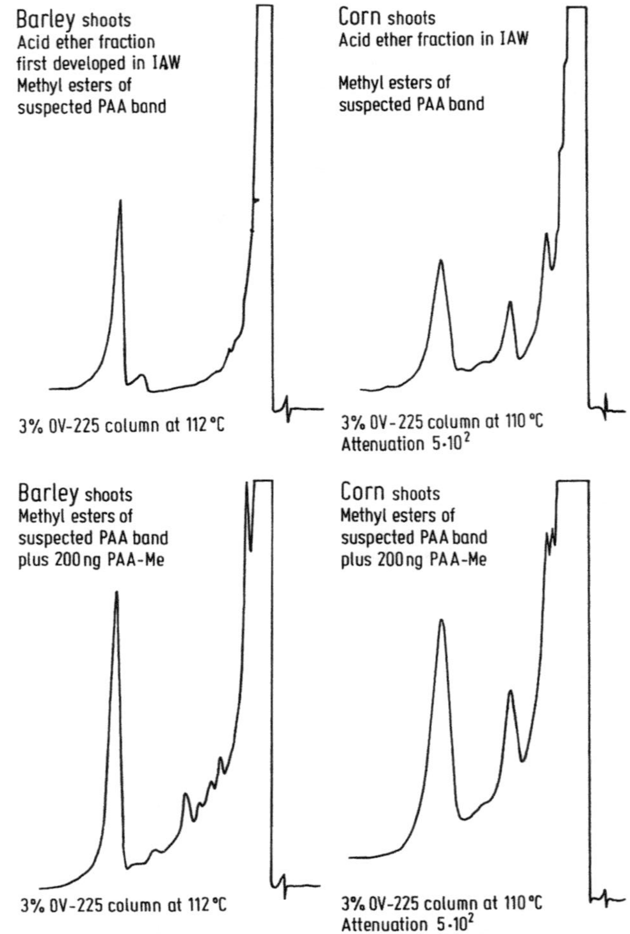

Fig. 9. Scans from GLC examination of methylated eluates from suspected PAA band on paper chromatograms of acidic ether fractions from barley and corn shoots

Table 1. Comparison of the activities of IAA and PAA in three standard auxin tests

		10^{-3} M	10^{-4} M	10^{-5} M	10^{-6} M
Wheat coleoptile	IAA	158[b]	160[b]	154[b]	130[b]
segment test[a]	PAA	145[b]	122[b]	101	102
Pea stem	IAA	130[b]	132[b]	126[b]	108
segment test[a]	PAA	130[b]	114[b]	100	102
Pea stem	IAA	6	4	2	1
curvature test	PAA	6	4	0	0

[a] Activity is expressed as percentage of final growth of water controls

[b] Activity statistically significant from water controls at 0.1% level

Table 2. Comparative amounts of IAA and PAA present in vegetative plant shoots, based on paper chromatography and GLC analysis of acid ether fractions

Plant material	Age of plant (weeks)	ng amount/g shoot tissue		IAA:PAA
		IAA	PAA	
Tomato (var. Michigan Ohio)	6	37	220	1:6
Tobacco (var. Mammoth)	8	40	168	1:4
Sunflower (var. Russian Mammoth)	4	43	202	1:5
Pea (var. Alaska)	2	22	86	1:4
Barley (var. Champlain)	2	11	70	1:6
Corn (var. Golden Bantam)	2	25	123	1:5

pound which exhibit optimum growth-promoting activity in standard auxin tests (Table 1); however, the values presented here are only estimates of the average concentration of each auxin in all parts of the vegetative shoot and it is most likely that the endogenous levels of IAA and PAA are much higher in growing organs of the system than in non-growing, mature regions of the shoot. To support this view, we now have evidence from a gas chromatographic analysis of the auxin substances present in different parts of both tobacco and tomato shoots which clearly demonstrates the wide range in endogenous levels of IAA and PAA in different regions of the developing system.

In summary, we can conclude from the present data (Table 2) that not only does PAA occur along with IAA as a natural component of many higher plants, but that its endogenous concentration is probably high enough in the growing regions of vegetative shoots to allow the substance to exert auxin activity.

II. Identification and Quantitative Estimation of the Auxins Present in Developing Tobacco Leaves

In an attempt to study the changes in concentrations of auxin substances in a typical plant organ at various stages during its growth and development, we have examined by gas chromatography the acidic ether fractions obtained from tobacco leaves at all stages of growth, from their earliest appearance below the apical bud to the fully expanded mature stage (see Fig.10). In these experiments, no attempt was made to partially purify the ether fractions by paper or thin-layer chromatography prior to GLC analysis since such preliminary purification usually leads to two problems, namely, (1) the partial degradation of certain substances, such as IAA, during the chromatography procedure, which makes their subsequent quantitative estimation unreliable, and (2), the possible failure to observe on the chromatogram the presence of new auxin substances not thought to be present in the ether fraction, and so certain regions of the chromatogram might not be chosen for further GLC examination. We have avoided these two major difficulties by analysing the unpurified acidic ether fraction from each leaf directly on the GLC, using linear temperature programming procedures. The entire ether fraction was first evaporated to dryness and redissolved in a fixed volume of 10% methanol in ether, and two samples were then taken for methylation with diazomethane. The methyl esters in each sample were then separated by GLC using glass columns packed with 3% OV-17 and temperature programs beginning at 95°C and rising to 270°C, or 310°C, at 3° per minute. With such linear programs, it is possible to separate and easily identify the methyl esters of a range of authentic phenyl and indole acids (Fig.11).

A typical scan obtained when a methylated sample from a tobacco leaf acidic ether fraction was examined by a similar temperature program procedure is shown in Figure 12. The scan obtained when a further sample of the same leaf fraction was co-injected with ng amounts of the authentic methyl esters of several auxin substances is shown in Figure 13. Clearly, based on the retention times of the four authentic methyl esters used in this study, there is very good evidence for the occurrence not only of IAA and PAA in these leaf ether fractions, but also of the auxin substances, indole-3-propionic acid (IPA) and indole-3-acrylic acid (IAcA). This conclusion was further confirmed in a second examination of the series of acidic ether fractions in which the samples taken for GLC examination were butylated rather than methylated. In the subsequent analysis, the peaks of the suspected butyl esters of IAA, PAA, IPA, and IAcA were all found to co-chromatograph with the butyl esters of the four authentic acids. Final confirmation of the presence of these four auxin substances in tobacco leaf ether fractions should come from a GLC-mass spectrometric examination of one of the leaf fractions, and this analysis is currently in progress.

Using the technique described in the previous section for preparing standard curves calibrating peak areas with known ng amounts of the authentic methyl esters, it has been possible to estimate the endogenous levels of the four auxin substances at each stage of leaf development. The quantitative results for IAA, PAA and IPA are shown in Figure 14, along with the curve showing the growth rate (elongation) of the different aged leaves during the 48 h period immediately following the time when samples of comparable leaves were taken for auxin analysis. The results show that IAA, PAA and IPA were present in tobacco leaves at all stages of development and that the changes in endogenous level of each auxin parallel each other during the entire period of development. Furthermore, at almost every stage of growth, the mole ratio of IAA:PAA was approximately 1:5, which is very similar to

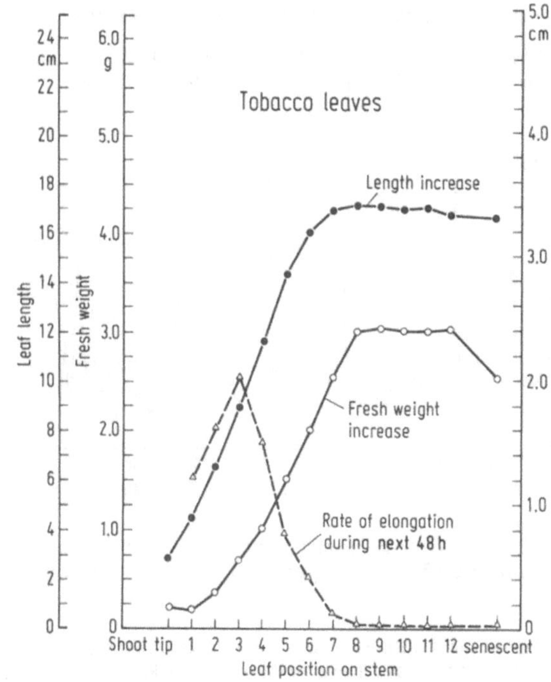

Fig. 10. Curves showing increase in length, fresh weight, and rate of elongation over next 48 h of tobacco leaves used in this investigation

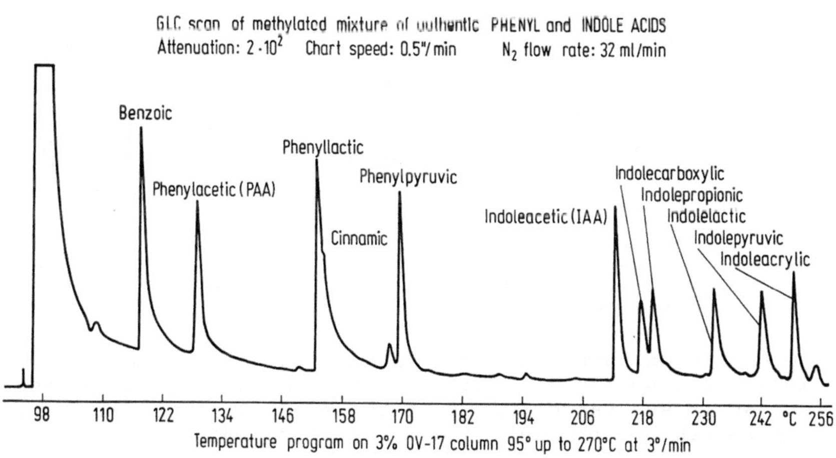

Fig. 11. GLC scan of sample of methylated mixture of authentic phenyl and indole acids analyzed by temperature programming on a 3% OV-17 column

the data reported earlier (Table 2) for the average levels of these two auxins in the entire vegetative shoot of young tobacco plants. Most importantly, the results in Figure 14 show that the highest levels of all three auxin substances were found in leaves Nos 2 and 3 on the stem, and these are the leaves which show the highest rate of ex-

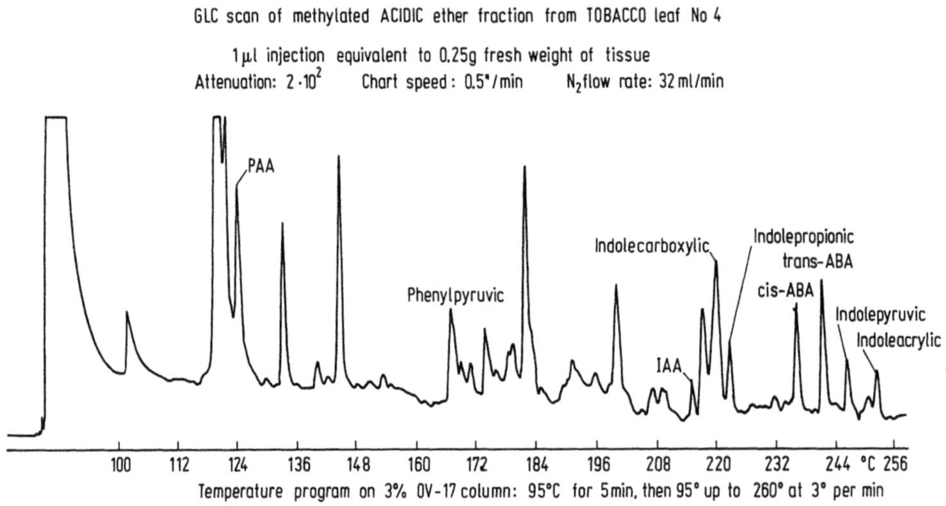

Fig. 12. GLC scan of sample of methylated acidic ether fraction from tobacco leaf No. 4 analysed by temperature programming on a 3% OV-17 column

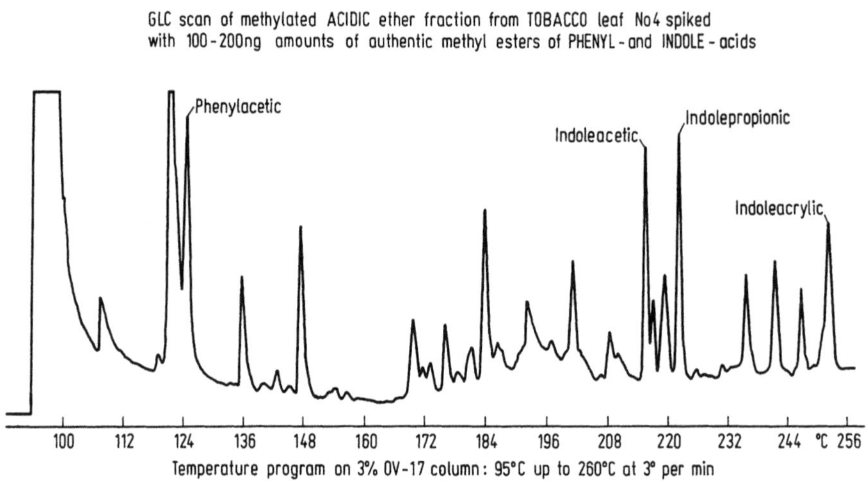

Fig. 13. GLC scan of sample of methylated acidic ether fraction from tobacco leaf No. 4 spiked with ng amounts of authentic methyl esters of the four auxin substances named on the scan. Analysis was carried out by temperature programming on a 3% OV-17 column

pansion growth during the next 48-h period. To our knowledge, this is the first clear demonstration of a close correlation between high auxin content and subsequent growth rate in a developing plant organ.

Procedures similar to the ones described in this paper are now being developed in my laboratory for the gas chromatographic analysis of plant fractions containing free gibberellins, cytokinins and inhibitory substances, in addition to auxin substances. It is hoped that

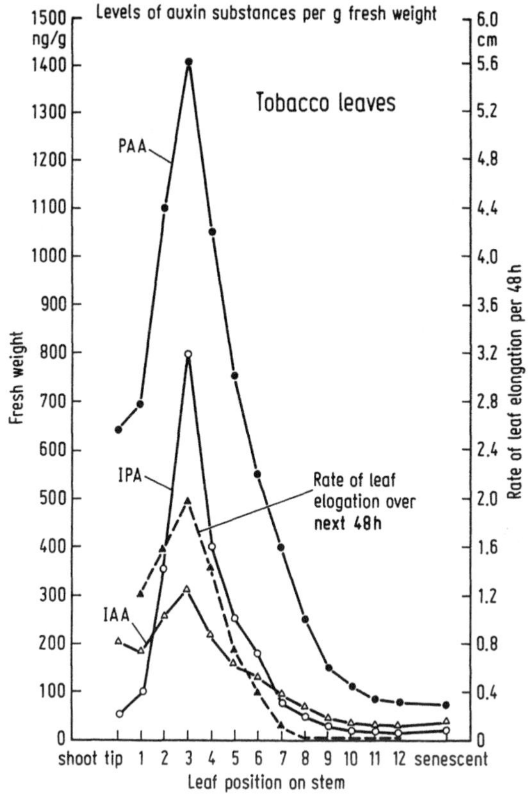

Fig. 14. Endogenous levels of indole-3-acetic acid (IAA), phenylacetic acid (PAA) and indole-3-propionic acid (IPA) in tobacco leaves at different stages of development

this comprehensive study will provide us with a clearer understanding of the control of growth and development in plant organs by the different kinds of growth-regulating substances synthesized in plants.

Acknowledgements

I would like to acknowledge the major contribution made to this work by my Visiting Research Associate, Dr. Jenny Phipps, and also by my research assistants B.S. Rauthan and B. Murray. We also wish to thank the National Research Council of Canada for supporting this work by providing Grant-in-aid of Research to the author.

References

1. Schneider, Elnora A., Gibson, R.A., Wightman, F.: In: Plant Growth Substances 1970. Carr, D.J. (ed.) Sydney: Springer, 1972, pp. 82-90
2. Gibson, R.A., Schneider, Elnora A., Wightman, F.: J. Exp. Botany 23, 381 (1972)
3. Gibson, R.A., Barrett, G., Wightman, F.: J. Exp. Botany 23, 775 (1972)

4. Wightman, F.: In: Nitrogen metabolism in plants. Biochem. Soc. Symp. 38, 247 (1973)
5. Wightman, F., Rauthan, B.S.: In: Plant Growth Substances 1973. Tokyo: Hirokawa Publ. Co. Inc., 1974, pp. 15-27
6. See: Powell, L.E.: Plant Physiol. 39, 836 (1964)
   Stowe, B.B., Vendrell, M., Epstein, E.: In: Biochemistry and Physiology of Plant Growth Substances, Wightman, F., Setterfield, G. (eds.) Ottawa: Runge Press Ltd., 1968, pp. 173-182
   Bayer, Margaret H.: Plant Physiol. 44, 267 (1969)
   Alden, T.: Physiol. Plantarum 25, 54 (1971)
   Bridges, I.G., Hillman, J.R., Wilkins, M.B.: Planta (Berl.) 115, 189 (1973)
   Elliott, M.C., Greenwood, M.S.: Phytochemistry 13, 239 (1974)
   Rivier, L., Pilet, P.E.: Planta (Berl.) 120, 107 (1974)
   Bandurski, R.S., Schulze, A.: Plant Physiol. 54, 257 (1974)
   Shindy, W.W., Smith, O.E.: Plant Physiol. 55, 550 (1975)
   Champault, A.: C. R. Acad. Sci. (Paris) 280, 591 (1975)
7. Wightman, F.: Can. J. Botany 40, 680 (1962)

Paper Demonstration References (see Section IX):

9 - 35 - 130 - 133

IV. Auxins and Root Growth Inhibitors

10. Interactions in Translocation of Growth Regulators in Shoots with Special Reference to Auxin Transport as Affected by Two Inhibitors, a Morphactin and Lycoricidinol

H. KALDEWEY, U. GINKEL, R. KARMANN, and I. PALAND

Abstract

Observations on auxin transport as affected by different factors are reviewed briefly. So far there seems to be no definite evidence that interactions in auxin transport result from a competition between transport-affecting compounds and auxin transport sites. There are, however, reports to show that inhibition of auxin transport is linked to an increase in immobilization or degradation of auxin and vice versa. Hence, the control of auxin transport by any factor may involve an alteration in the availability of auxin molecules by affecting systems other than the auxin transport system itself. This assumption is supported by presenting the results of indole-3-acetic acid (IAA) transport experiments in the presence of two transport inhibitors, the morphactin 2-monochloro-9-hydroxyfluorene-9-carboxylic acid methylester (CFMe) and lycoricidinol, a substance occurring naturally in *Lycoris* and *Narcissus* (cf.12). It is shown that similar effects on auxin transport are brought about by different mechanisms.

A. Introduction

As early as 1934, just after publication of his classic paper on auxin transport (72), van der Weij reported on interactions in auxin transport (73). He submitted sections of *Avena* coleoptiles, apically fed with an auxin donor, to an atmosphere enriched with ethyl ether at different concentrations. Generally, he observed a decrease in the amount of auxin, delivered to basal agar receivers within unit time. This means that ether caused a decrease in the so-called transport intensity. From the plot of auxin accumulation in the basal receivers versus transport duration, van der Weij deduced a linear increase of auxin activity with time. The straight lines drawn through the data obtained at the different ether concentrations and "thought to be the probably valid curve(s)" ("die m.E. wahrscheinlich geltende Kurve", 73, p. 834) of auxin transport, cut the time axis at the same point. Therefore, van der Weij concluded that ether did not influence the velocity of the auxin stream. Hence, the decrease observed in the transport intensity had to be traced to a decrease in the density of the auxin stream.

Although the calculations of the least-squares straight lines of best fit from the data presented by van der Weij do, in fact, point to a reduction of the transport velocity in the presence of ether, and despite the dubiousness of using van der Weij's "intercept method" in describing the properties of an auxin transport system (25, 50), the results of van der Weij did point to the necessity of actively metabolizing living cells in order to maintain basipolar auxin transport.

This has been confirmed by a number of experiments with various narcotics or inhibitors of cell metabolism, such as 2,4-dinitrophenol and cyanides (11, 51, 61), iodoacetate and sodium azide (43, 61, 77), or sodium fluoride (77); further, it was shown that a reduction of the O_2-concentration strongly inhibits auxin transport (13, 17, 76), whereas an enhancement was observed after addition of ATP (61); for further references see 14, 15, 21, 41, 42, 68, 80.

It might be questioned whether the preceding phenomena should be looked at as interactions sensu stricto of the auxin transport system with the substances applied. They may rather result from an overall effect on the metabolic activity of the tissue. A similar interpretation may hold for the promotion of auxin transport at raised temperatures (17, 19, 23, 49, 72, 75) and for its inhibition at decreased hydrature, brought about either by application of different ions (61) or by water stress (29).

Probably, more specific effects on auxin transport have been obtained with so-called transport inhibitors, such as 2,3,5-triiodobenzoic acid (TIBA) (2, 9, 18, 19, 32, 37, 42, 43, 45, 51, 65, 78, 81), morphactins (5, 28, 35, 36, 43, 46, 55, 58, 63, 65), 1-N-naphthylphthalamic acid (NPA) (38, 43, 44, 64, 65), and some other compounds (4, 6, 8, 27, 34, 47); there are also reports on interactions with auxins (39, 53, 60, 62) and other naturally occurring growth regulators: gibberellins (3, 22, 31, 54, 56), abscisic acid (ABA) (3, 8, 48, 59, 70), ethylene (1, 7, 40, 53), and cytokinins (52, 53).

Despite the numerous papers reporting an influence on auxin transport, little is known on the nature of such interactions. Still, it seems that "little definite evidence supports the view that auxins and structurally related molecules compete with IAA for specific sites" (14, p. 351), as was propounded by Goldsmith in 1968. It appears rather, especially from experiments with membrane fractions (20, 38), that transport inhibition "may result from some effect on the transport mechanism other than a competition for the auxin attachment site" (66, p. 269). There is, however, evidence that the amount of auxin transported is inversely related to immobilization or degradation of auxin (8, 16, 24, 26, 30, 31, 57, 69, 71). Hence a competition for the auxin molecules between immobilization or degradation systems and the transport system might well be a crucial factor in the regulation of auxin transport. This point of view is supported by the results of auxin transport experiments obtained with two effective transport inhibitors, namely, the morphactin CFMe and lycoricidinol (Lyc), the latter isolated from shoots of *Narcissus pseudonarcissus* (see 12).

B. Material and Methods

The methods used were essentially the same as described elsewhere (27-31). In short, 6-mm hypocotyl sections cut from the apical region of light-grown seedlings of *Citrullus vulgaris* Schrad (seed was kindly supplied by Prof. Yusuf Vardar, Ege University, Izmir, Turkey) were used. IAA-2-^{14}C (NH_4-salt, sp. act. 58 mCi/mmol, Radiochemical Centre, Amersham, U.K.) was supplied in agar donor blocks at a concentration of 10^{-5} M to the apical cut surface of the sections either alone or together with Lyc (3×10^{-5} M)[1] or CFMe (10^{-6} M). The time course of

---

[1] We are grateful to Dr. Yasuo Komoda, Medical and Dental University, Tokyo, Japan for the isolation of Lyc from *Narcissus* shoots while he stayed at our laboratories as a fellow of the HUMBOLDT-Stiftung. He also ran TLC checks mentioned in Conclusions

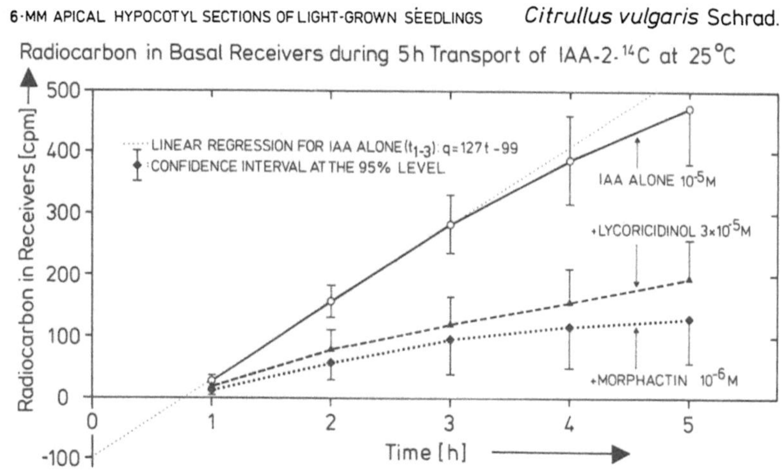

Fig. 1. Effect of Lyc and CFMe on auxin transport in hypocotyl sections of *Citrullus*, placed upright onto basal agar receivers. Compounds were applied to apical cut surface in donor blocks together with IAA-2-^{14}C. Receivers were replaced every hour and their radiocounts summed up to give time course of ^{14}C-accumulation. Average of eight replications of controls and four replications of inhibitor treatments

auxin accumulation in basal receivers was followed over a 5-h period. The sections were then divided transversely into four subsections of equal length and placed twice for 1 h onto basal agar receivers of a volume similar to that of the subsections. By this means, the radiocarbon fractions were determined that had remained diffusible, yet not necessarily transportable. The subsections were then prepared for counting of the immobilized radiocarbon fractions.

C. Results and Discussion

I. Time Course of Transport

The inhibition of basipetal auxin transport by both Lyc and the morphactin was first detected after 1 h of transport and became significant after 2 h (Fig.1). The inhibitors decreased the release of radiocarbon into basal receivers to less than half that of the controls, which had been apically supplied solely with IAA-2-^{14}C. The inhibitory effects of the 10^{-6} M CFMe and of the 30-fold concentrated Lyc did not differ significantly.

The decline of the slopes of the time course curves in the presence of the inhibitors as compared with the time course of radiocarbon accumulation in the basal receivers of the control hypocotyl sections clearly indicated that the so-called transport intensity (25, 50, 72) was reduced.

A steady release of radiocarbon by the control sections appeared during the first 3 h. From the least-squares straight line of best fit for that period, the time required by the auxin front to reach the base of the section was estimated at 0.78 h, corresponding to a velocity of the front of 7.7 mm/h (see however critics in 50). A calculation of these parameters in the presence of the inhibitors seemed of little use since

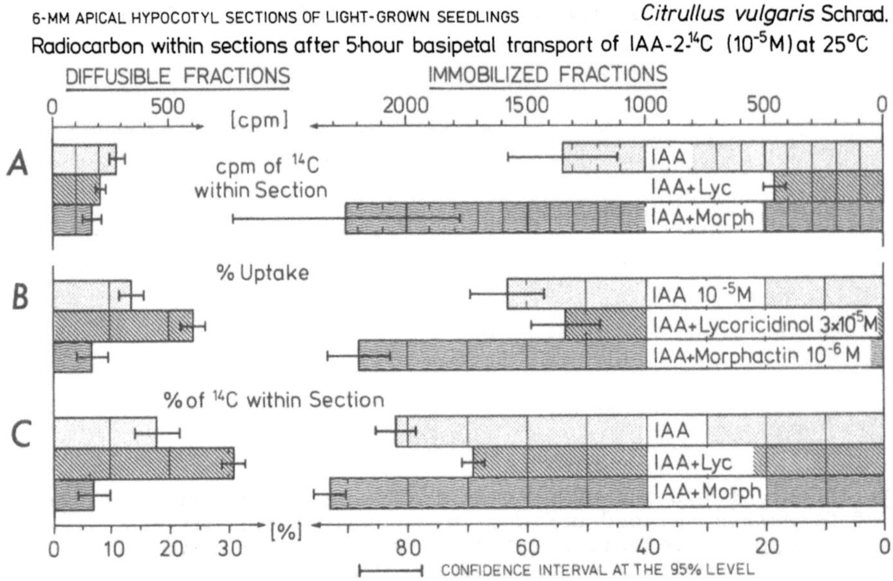

Fig. 2 A-C. Effect of Lyc and CFMe on content and distribution of radiocarbon that was immobilized or remained diffusible in hypocotyl sections of *Citrullus* after basipetal transport of IAA-2-^{14}C. Further details in legend of Figure 1

the slopes of the arrival curves declined up to 3 h and 5 h, in presence of Lyc and CFMe, respectively. This points to an increase in inhibition with time. However, while the release of radiocarbon in unit time appeared to become constant after 3 h in presence of Lyc, the slope of the arrival curve continued to decline when CFMe had been added to the IAA donor blocks.

II. Uptake, Immobilization, and Conversion of IAA

From the latter observation, one may assume differences in the mode of action of the two compounds. This assumption was confirmed by measurements of the uptake of radiocarbon from the donor blocks. The uptake values, which did not differ significantly whether they were calculated as loss from donors or as the recovery of radiocarbon within the transport sections and the agar receivers, were 2097 ± 187, 855 ± 64, 2542 ± 398 (cpm ± confidence interval at the 95% level), in the controls, in presence of Lyc and of CFMe, respectively. Thus, Lyc strongly reduced auxin uptake whereas the morphactin increased it.

In order to get an idea of the fate of auxin within the tissue, the sections, after the transport period, were handled as described (see Material and Methods). This method allowed the separation of the radiocarbon fractions that had remained diffusible (yet not necessarily transportable in the sense of active auxin transport) from those that had been immobilized during the transport period. The data given in Figure 2A clearly indicate that ^{14}C-immobilization was strongly and significantly increased by CFMe and decreased by Lyc. Both inhibitors significantly diminished the diffusible fractions. In any case, the immobilized fractions were larger than the diffusible ones, but the ratio between the two fractions was differentially susceptible to inhibitor treatment. While CFMe significantly increased the immobilized fraction expressed as the percentage of total ^{14}C in the sections, Lyc had the opposite effect (Fig.2C); consequently, the diffusible fractions were reversely affected. To obtain an impression of the activity of the immobilization system, the immobilized fractions were further calculated as the percentage of total uptake (Fig.2B). The calculations showed that most of the ^{14}C that had been taken up during the 5-h period always stayed with the immobilized fractions. Moreover, they confirmed that CFMe significantly enhanced the immobilization activity; the reduction caused by Lyc was not significant.

From the results one may deduce that the morphactin weakened the activity of the auxin transport system whereas Lyc did not, inhibition of auxin transport by Lyc being just a consequence of the diminished uptake. This assumption seems to be supported by the calculations of the transport values as the percentage of uptake: in the controls and Lyc-treated sections, 23% of the radiocarbon taken up during the 5-h period reached the basal receivers, whereas this portion was as little as 5% in presence of CFMe.

However, these interpretations provide no explanation for the reduction of auxin uptake caused by Lyc; further, if they were true, one would expect the morphactin to diminish auxin uptake. Inhibition of auxin uptake has in fact been reported as an effect of morphactins in transport experiments with different species (5, 55, 58). Though not statistically significant, the contrary was true in our system. The following observations appear suitable for providing more information.

III. Distribution of Radiocarbon Along the Sections

Figures 3 and 4 show the effects of Lyc and CFMe on the distribution along the hypocotyl sections of the radiocarbon fractions that were immobilized and still diffusible. In Figure 3 the experimental data were calculated as the percentage of the total amount of radiocarbon found within the sections after 5 h of transport. The calculations show that most of the radiocarbon, with either treatment including the controls, was immobilized near the donor blocks. However, the morphactin strongly and significantly increased immobilization in this region whereas Lyc did not. Yet, both inhibitors caused a significant decrease in the portion immobilized within the lowermost halves of the hypocotyl sections. In the presence of CFMe, this was due to the increase of immobilization in the apical subsection, whereas Lyc caused a significant increase of the proportion that remained diffusible. Thus, both inhibitors caused an increase in the percentage of total radiocarbon that remained near the donor blocks. Consequently, less radiocarbon could move toward the base of the sections. A calculation of the radiocarbon fractions found in the different subsections, separately, either as the percentage of total diffusible or of total immobilized, made this even more illustrative (Fig.4). The curves clearly show that, in the presence of the inhibitors, the portions that remained near the donors were significantly higher than in the controls. The contrary was true in the lower halves of the hypocotyl sections.

The observations appear appropriate to illustrate the effects of both inhibitors.

The morphactin obviously activates the immobilization system. This leads to a significant increase in the immobilized radiocarbon fraction, either taken absolutely or calculated as the percentage of uptake or of total radioactivity within the sections (Fig.2). Since im-

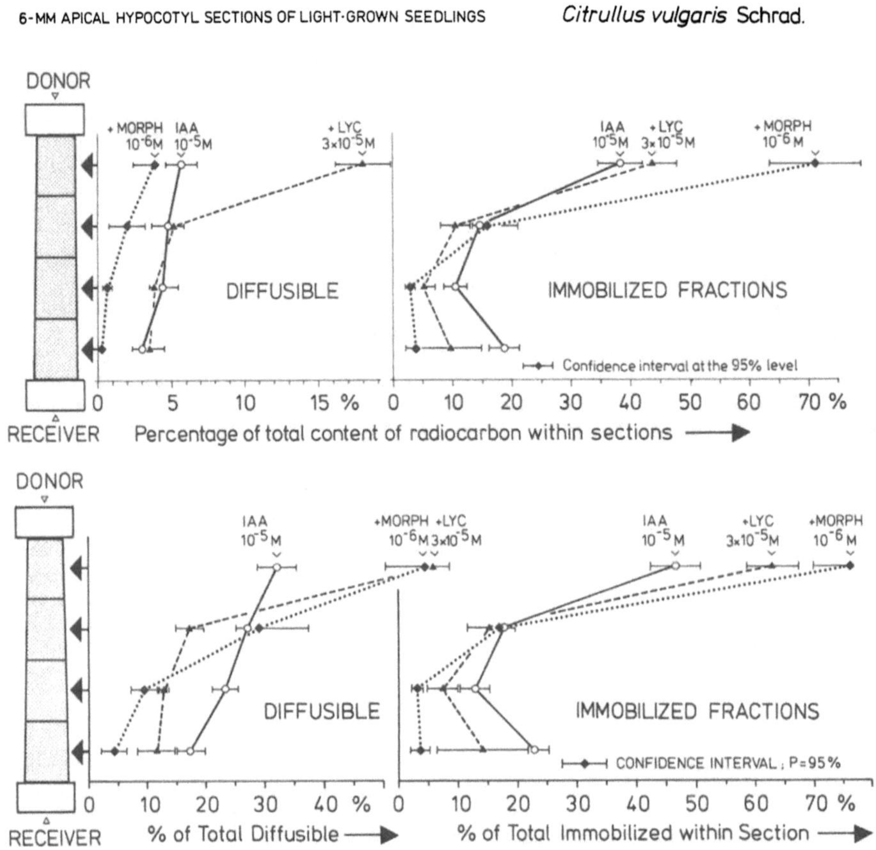

Figs. 3 and 4. Effects of Lyc and CFMe on distribution of radiocarbon along hypocotyl sections of *Citrullus* after basipetal transport of IAA-2-^{14}C. Further details in legend of Figure 1

mobilization is significantly increased, especially near the donor, the amount being transported toward the base of the sections and then accumulating in the receivers is reduced. However, increased immobilization and probably metabolization of the auxin maintain the concentration gradient for IAA between the donor block and the adjacent tissue. Thus, the first step of penetration of auxin into the section, namely, physical diffusion, and hence auxin uptake, is increased. The observed inhibition of auxin transport into the basal receivers is thus rather a consequence of the high level of immobilization activity, whereby less auxin is available for the transport system, the activity of which might not be affected at all.

The effects of Lyc on auxin transport appear to be quite different. The auxin, having diffused into the sections, seems to be maintained in a diffusible stage, yet much of it is not basipetally transported. Hence, auxin stays within the tissue near the donor (see Fig.3) and thus diminishes the concentration gradient between the donor block and the nearby tissue. Consequently, less auxin diffuses into the hypocotyl section and less moves downward, although the percentage of auxin taken up and transported into the receivers is not changed. One might deduce then, that Lyc somehow hinders the attachment of the auxin molecules to the basipolar auxin transport system. The activity of the immobilization system, however, is probably not affected, since the portion of auxin uptake immobilized in the presence of Lyc does not significantly differ from that of the controls. Thus, the decrease in the absolute amount of immobilized radiocarbon seems to result from the decrease in auxin uptake. The significant acropetal shift of immobilization may be traced to the fact that the auxin, having left the donor blocks, possibly due to a disturbance of the polarity of the transport system, remains within the apical parts of the sections for a longer period and thus becomes available for the immobilization system.

D. Conclusions

The results of the experiments reported do not as yet permit a decision as to whether the inhibitors primarily affect the system of auxin transport or that of immobilization. Increased immobilization in the presence of the morphactin might be interpreted as the consequence of a weakening of basipolar transport, leading to more auxin molecules becoming available for immobilization; however, the probable increase of auxin uptake does not fit with an interpretation of this kind. On the other hand, the primary effect of Lyc might be seen as a weakening of the auxin immobilization system. In this case, auxin uptake is expected to be reduced, as was observed. Auxin transport, due to diminished competition of the weakened immobilization system for the auxin molecules, should then be promoted rather than inhibited, whereas the latter was true. More information especially on the kinetics of auxin immobilization and metabolization after diffusion into the transport sections, is necessary to discriminate the different processes involved. In fact, the results of preliminary experiments do indicate that auxin metabolism is differentially affected by the two inhibitors. As compared with the controls, the fractions immobilized within the transport sections in the presence of CFMe, different from the findings in *Zea* coleoptiles (5), contained a higher percentage of a substance chromatographically identical with indole-3-acetylaspartic acid whereas the contrary was true when Lyc had been added to the donor blocks. In contrast, a radiocarbon compound that moved to the Rf of indole-3-acetyl glucose appeared to be decreased in the presence of CFMe, whereas an increase was observed when Lyc had been applied. Since the conjugation of IAA with glucose probably precedes its conjugation with the aspartate (10, 79), the preliminary results point to an involvement of auxin immobilization or degradation in the effects on auxin transport of both the morphactin and the lycoricidinol.

Acknowledgements

Thanks are due to the Deutsche Forschungsgemeinschaft for a grant to the senior author. The calculations were done in the Rechenzentrum der Universität des Saarlandes and we wish to thank Alfred Bingert for helpful assistance.

References

1. Abeles, F.B.: Plant Physiol. 41, 946 (1966)
2. Audus, L.J., Thresh, R.: Ann. Botany (London) 20, 439 (1956)
3. Basler, E.: Plant Cell Physiol. (Tokyo) 15, 351 (1974)
4. Beyer, E.M.: Plant Physiol. 50, 322 (1972)

5. Bridges, I.G., Wilkins, M.B.: J. Exp. Botany 24, 711 (1973)
6. Brown, B.T., Johansen, O., Sasse, W.H.F.: Experientia 28, 1290 (1972)
7. Burg, S.P., Burg, E.A.: Plant Physiol. 42, 1224 (1967)
8. Chang, Y.P., Jacobs, W.P.: Am. J. Botany 60, 10 (1973)
9. Christie, A.E., Leopold, A.C.: Plant Cell Physiol. (Tokyo) 6, 337 (1965)
10. Davies, P.J.: Physiol. Plantarum 28, 95 (1973)
11. DuBuy, H.G., Olson, R.A.: Am. J. Botany 27, 401 (1940)
12. Edelbluth, E., Kaldewey, H.: Planta 131, 285-291 (1976)
13. Goldsmith, M.H.M.: Plant Physiol. 41, 15 (1966)
14. Goldsmith, M.H.M.: Ann. Rev. Plant Physiol. 19, 347 (1968)
15. Goldsmith, M.H.M.: In: The Physiology of Plant Growth and Development. Wilkins, M.B. (ed.). London: McGraw-Hill, 1969, p. 127
16. Goldsmith, M.H.M., Thimann, K.V.: Plant Physiol. 37, 492 (1962)
17. Gregory, F.G., Hancock, C.R.: Ann. Botany (London) 19, 451 (1955)
18. Hay, J.R.: Plant Physiol. 31, 118 (1956)
19. Hertel, R., Leopold, A.C.: Planta 59, 535 (1963)
20. Hertel, R., Thomson, K.-S., Russo, V.E.A.: Planta 107, 325 (1972)
21. Jacobs, W.P.: In: Plant Growth Regulation. Klein, R.M. (ed.). Ames: Iowa State Univ. Press, 1961, p. 397
22. Jacobs, W.P., Case, D.B.: Science 148, 1729 (1965)
23. Kaldewey, H.: Ber. Deut. Botan. Ges. 78, 128 (1965)
24. Kaldewey, H.: Wiss. Z. Univ. Rostock, Math.-Nat. Reihe 16, 487 (1967)
25. Kaldewey, H.: In: (68), pp. 1-23 (1968)
26. Kaldewey, H.: In: Plant Stimulation. Popoff, K.L. (ed.). Sofia: Bulg. Acad. Sci. Press, 1969, p. 211
27. Kaldewey, H., Ginkel, U.: In: Plant Growth Substances 1973. Tokyo: Hirokawa Publ. Co., 1974, p. 1079
28. Kaldewey, H., Ginkel, U., Lehmann, I., Seiwert, R.: Proc. Res. Inst. Pomology, Skierniewice, Poland, Ser. E 3, 215 (1973)
29. Kaldewey, H., Ginkel, U., Wawczyniak, G.: Ber. Deut. Botan. Ges. 87, 563 (1974)
30. Kaldewey, H., Jacobs, W.P.: Am. J. Botany 61, 699 (1974)
31. Kaldewey, H., Jacobs, W.P.: Biochem. Physiol. Pflanzen 168, 401 (1975)
32. Kaldewey, H., Kraus, Lj.: In: (33), p. 137 (1972)
33. Kaldewey, H., Vardar, Y. (eds.): Hormonal Regulation in Plant Growth and Development. Weinheim: Verlag Chemie 1972
34. Katekar, G.F., Geissler, A.E.: Plant Physiol. 56, 645 (1975)
35. Krelle, E.: Biochem. Physiol. Pflanzen 161, 299 (1970)
36. Krelle, E., Libbert, E.: Planta 80, 317 (1968)
37. Kuse, G.: Mem. Coll. Sci, Univ. Kyoto, Ser. B 20, 207 (1953)
38. Lembi, C.A., Morré, D.J., Thomson, K.-S., Hertel, R.: Planta 99, 37 (1971)
39. Leopold, A.C., Lam, S.L.: Physiol. Plantarum 15, 631 (1962)
40. Lyon, Ch.J.: Plant Physiol. 45, 644 (1970)
41. McCready, C.C.: Ann. Rev. Plant Physiol. 17, 283 (1966)
42. McCready, C.C.: In: (74), p. 1005 (1968a)
43. McCready, C.C.: In: (67), p. 108 (1968b)
44. Morgan, D.G., Söding, H.: Planta 52, 225 (1958)
45. Morris, D.A., Kadir, G.O., Barry, A.J.: Planta 110, 173 (1973)
46. Naqvi, S.M.: Experientia 27, 232 (1971)
47. Naqvi, S.M.: Z. Pflanzenphysiol. 71, 1 (1974)
48. Naqvi, S.M., Engvild, K.C.: Physiol. Plantarum 30, 283 (1974)
49. Newman, I.A.: Australian J. Biol. Sci. 16, 629 (1963)
50. Newman, I.A.: In: Mechanisms of Regulation of Plant Growth. Bieleski, R.L., Ferguson, A.R., Cresswell, M.M. (eds.). Wellington: The Royal Soc. of New Zealand, 1974, p. 665
51. Niedergang-Kamien, E., Leopold, A.C.: Physiol. Plantarum 10, 29 (1957)

52. Osborne, D.J., Horton, R.F., Black, M.K.: In: (67), p. 79 (1968)
53. Osborne, D., Mullins, M.G.: New Phytologist 68, 977 (1969)
54. Palmer, J.H., Halsall, D.M.: Physiol. Plantarum 22, 59 (1969)
55. Parups, E.V.: Physiol. Plantarum 23, 1176 (1970)
56. Pilet, P.E.: Nature (London) 208, 1344 (1965)
57. Pilet, P.E.: Wiss. Z. Univ. Rostock, Math.-Nat. Reihe 16, 479 (1967)
58. Pilet, P.E.: Experientia 26, 608 (1970)
59. Pilet, P.E.: Physiol. Plantarum 25, 28 (1971)
60. Rayle, D.L., Ouitrakul, R., Hertel, R.: Planta 87, 49 (1969)
61. Reiff, B., v. Guttenberg, H.: Flora (Jena) 151, 44 (1961)
62. Sachs, T.: Planta 127, 201 (1975)
63. Schneider, G.: In: (33), p. 317 (1972)
64. Thomson, K.-S.: In: (33), p. 83 (1972)
65. Thomson, K.-S., Hertel, R., Müller, S., Tavares, J.E.: Planta 109, 337 (1973)
66. Thomson, K.-S., Leopold, A.C.: Planta 115, 259 (1974)
67. Vardar, Y. (ed.): The Transport of Plant Hormones. Amsterdam: North-Holland Publ. Co. 1968a
68. Vardar, Y.: In: (67), p. 156 (1968b)
69. Veen, H.: Planta 73, 281 (1967)
70. Veen, H., Frissel, M.J.: Physiol. Plantarum 34, 208 (1975)
71. Veen, H., Jacobs, W.P.: Plant Physiol. 44, 1157 (1969)
72. Weij, H.G. van der: Rec. Trav. bot. néerl. 29, 379 (1932)
73. Weij, H.G. van der: Rec. Trav. bot. néerl. 31, 810 (1934)
74. Wightman, F., Setterfield, G. (eds.): Biochemistry and Physiology of Plant Growth Substances. Ottawa: Runge Press Ltd. 1968
75. Wilkins, M.B., Cane, A.R.: J. Exp. Botany 21, 195 (1970)
76. Wilkins, M.B., Martin, M.: Plant Physiol. 42, 831 (1967)
77. Wilkins, M.B., Whyte, P.: In: (74), p. 1051 (1968)
78. Winter, A.: In: (74), p. 1063 (1968)
79. Zenk, M.H.: In: Régulateurs naturels de la croissance végétale. Nitsch, J.P. (ed.). Paris: CNRS, 1964, p. 241
80. Ziegler, H.: Fortschr. Botan. 35, 67 (1973)
81. Zwar, J.A., Rijven, A.H.G.C.: Australian J. Biol. Sci. 9, 528 (1956)

Paper Demonstraion References (see Section IX):

23 - 57 - 83 - 106 - 124

11. Auxins and the Regulation of Root Growth

M. C. Elliott

Abstract

Data presented are compatible with the proposal that elongation and georesponses of the primary root are regulated by the interaction of acropetally moving, growth-promoting indole-3-acetic acid (IAA) derived primarily from the fruit and shoot system, with abscisic acid (ABA) (and possibly other inhibitors) moving in a basipetal direction from the root cap. The primary action of IAA in promotion of cell elongation may involve the stimulation of a proton pump in the cell membranes.

A. Introduction

The classical Cholodny-Went theory of tropisms (11, 59), with its emphasis on the basipetal transport of auxin from the site of synthesis at the root tip to the zone of cell elongation, where it achieves a supra-optimal concentration, provided a picture that was so beautifully integrated with the one obtained for the auxin regulation of coleoptile growth that there has been some resistance to rejecting it. However, some of the results of Pilet (42) and of Audus and Brownbridge (2, 3) challenged the theory's applicability to roots. Once Pilet's demonstration (43) that there was an acropetal polarity of IAA transport through *Lens culinaris* root segments had been confirmed and extended to a wide range of species (53), the way was clear for the surge of experimentation that has shown clearly that the geotropic response of the root is a consequence of the release of a growth inhibitor(s) (not IAA) from the root cap and its transport to the elongation zone (1, 46, 61). This work left considerable doubt about the role (if any) of indole-3-acetic acid (IAA) in the regulation of root growth. The difficulty of unequivocally characterising IAA in roots (53, 56), the complexities introduced by use of non-sterile tissue (32), and the tendency of some workers to ignore the differences in hormone economy of roots of different species, which are implied by excised root culture experiments (56), have slowed progress. However, the concentration on the hormone relations of the maize root, which has followed the demonstration by Juniper et al. (25) of its suitability for micro-manipulation experiments, has led to much progress being achieved. Our use of mass spectrometry for the identification of root indoles (17) led to the unequivocal identification of IAA from the roots of sterile maize seedlings (19), which was quickly confirmed in the laboratories of Wilkins (8) and Pilet (52). This paper will consider the role of IAA in the regulation of maize root growth.

B. Materials and Methods

I. Plant Material

Seeds of *Zea mays* L. cv Giant White Horsetooth were surface-sterilised (19) and then germinated in sterilised petri dishes for 3 days at 25°C in darkness. Samples were tested for sterility (19), and seedlings were

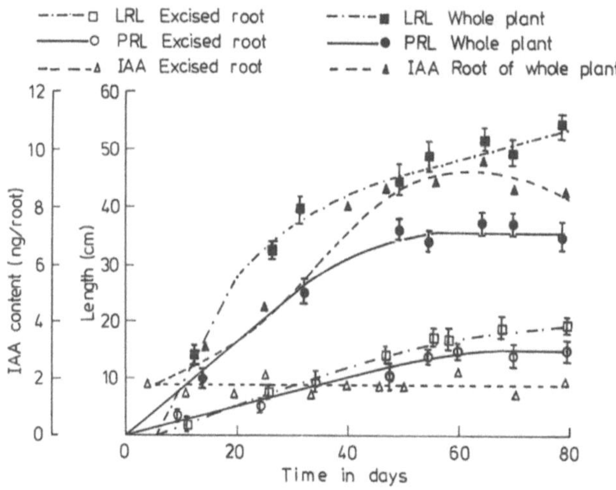

Fig. 1. Comparison of primary root length (*PRL*), lateral root length (*LRL*), and IAA content of roots of whole maize plants cultured in White's inorganic nutrients with excised roots cultured in White's medium (2% glucose)

selected for uniformity of length of the primary roots (17 ± 1 mm) and separated into two series. The apical 15 mm were cut from the roots of one series of seedlings and placed (one root per flask) into 250 ml Ehrlenmeyer flasks containing 100 ml of modified White's culture medium containing ferric sodium ethylenediaminetetraacetate (FeEDTA) (55) and 2% D(+) glucose as carbon source. The roots were incubated in darkness at 25°C. The second series of seedlings was kept intact, and their roots were immersed in a solution containing the inorganic components of White's medium, which was contained in blackened 250-ml flasks. This series of plants was maintained at 25°C under constant illumination. In each case the culture medium was changed weekly.

At intervals, at least 15 specimens were taken from each series, and determinations were made of the length of the primary root, the total length of the laterals on the primary root, the capacity for geotropic response of the primary root, and the IAA content of the primary root plus its laterals (27). Sterility checks were carried out at the same time.

For determination of the capacity for geotropic response, the apical 15 mm were removed from the primary root, and the cut surface was pushed into an agar solidified (1.5% agar) White's medium containing 6% glucose supported in a glass tube. The tube was placed in a flask of White's medium so that the root was in a horizontal position just below the surface of the liquid. The flasks were incubated at 25°C in the light for 24 h; then the angle of curvature was recorded by making shadow photographs.

C. Results and Discussion

Figure 1 shows a comparison of the growth of the primary roots of the intact plants with that of the excised roots. Growth of the primary root and its laterals in an intact plant is clearly far superior to the growth of the excised primary root and its laterals. It is noteworthy

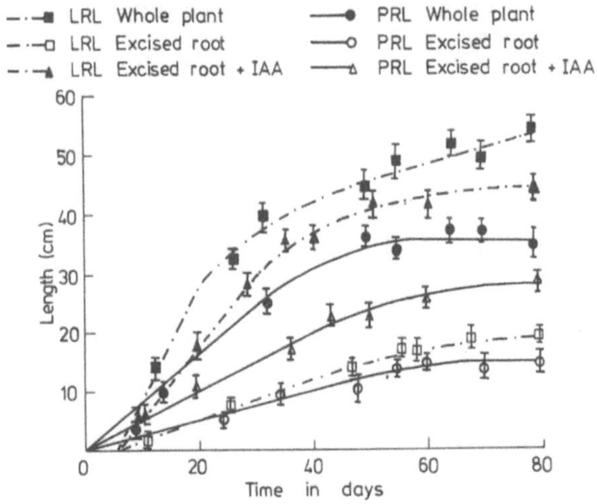

Fig. 2. Comparison of primary root length (*PRL*) and lateral root length (*LRL*) of roots of whole maize plants cultured in White's inorganic nutrients with excised roots cultured in White's medium (2% glucose) and excised roots cultured in White's medium (2% glucose) containing 10^{-11} M IAA.

that the form of the root (i.e., the relative proportions of lateral length:main axis length and the disposition of the laterals along the main axis) is the same after growth in excised culture as on the intact plant. The difference is that the rate of growth of the excised root and the final size achieved are much lower. From Figure 1 it can be seen that the IAA content per primary root system of the intact plant increases dramatically, whereas that of the excised root remains more or less constant. The fact that the higher growth rate of the root of the intact plant is coupled with a high IAA content posed the problem of whether the higher growth rate is a consequence of the higher IAA content or the higher IAA content is a consequence of the higher growth rate. If the former is true, it should clearly be possible to enhance the growth of the excised root by adding IAA to the culture medium. Figure 2 shows a comparison of the growth of the roots of the intact plants and the excised roots in standard medium with that of excised roots in culture medium made 10^{-11} M with respect to IAA by the addition of filter-sterilised IAA solution to the standard culture medium. The culture medium was changed at weekly intervals. The IAA significantly enhanced the growth of the excised roots, strongly suggesting that excised maize root growth in culture is limited by lack of IAA. In other words the auxin content of the roots is markedly sub-optimal. Our failure to boost the growth of the excised root up to the level of the root of the intact plant may arise from a number of causes. Firstly, the stability of IAA under the culture conditions is very limited, and secondly, the IAA is being made available to the root all over its surface instead of by way of the normal transport system to the elongation zone. It has been clearly shown that the polar metabolism-dependent transport of IAA in maize roots, like that in roots of other species, is predominantly acropetal (62). It would be expected, therefore, that application of IAA through the cut surface of the root by the technique of Raggio and Raggio (49) would lead to a more marked stimulation of growth by added IAA. In some experiments we have indeed found this to be the case, but the variability between experiments and between replicates in the same experiment has been very high. Neverthe-

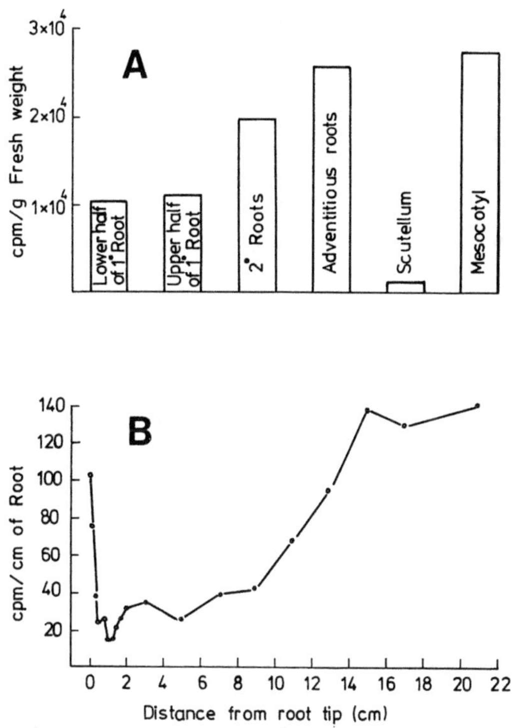

Fig. 3 A,B. (A) Distribution of radioactivity 24 h after application of 180,000 cpm of ^{3}H-IAA (23.5 Ci/mmol) in 2 µl of solution (containing 0.1% "ethylan") to shoot apex of 6-day-old maize plants

(B) Profile of radioactivity along primary root 24 h after application of 180,000 cpm of ^{3}H-IAA (23.5 Ci/mmol) in 2 µl of solution (containing 0.1% "ethylan") to shoot apex of 6-day-old maize plants

less the evidence that excised maize root growth is limited in comparison with the root of the intact plant, by the low levels of IAA present, is good and poses the problem of the reasons for the differences in IAA content of the two series of roots. In view of the numerous demonstrations of the transport of IAA applied to a shoot into the root (53) and demonstrations of the occurrence of root IAA predominantly in the stele (8, 22), it was natural to speculate that in the intact maize plant the IAA content of the root is maintained at high levels by transport from a site of synthesis in the shoot. Accordingly we have determined the fate of ^{3}H-IAA applied to the coleoptile, fruit endosperm, and shoot apex of maize seedlings. There is little movement of label from the coleoptile into the root (we would not expect a transient protective structure to have this role), but as shown in Figure 3A (36), labelled IAA applied to the shoot apex of six-day-old seedlings moves into the mesocotyl, adventitious roots, and into the primary root and its laterals during a 24-h transport period. The accumulation of label in the mesocotyl - if we assume it reflects a tendency for endogenous IAA to accumulate there - may explain the proliferation and rapid growth of adventitious roots in maize. Figure 3B (36) shows the distribution of radioactivity along the primary root. The accumulation of label at the apex is of great interest in view of the demonstration (52) that endogenous IAA can be found in the root cap. Application of IAA to the fruit endosperm at this stage gives a similar distribution.

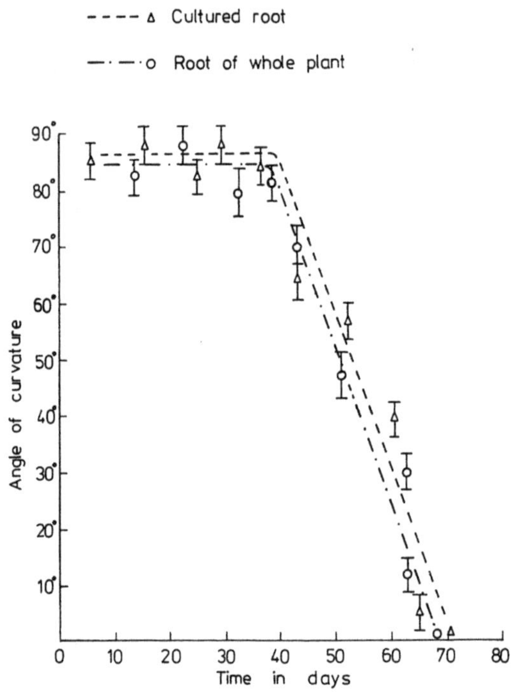

Fig. 4. Georesponse of apical 1.5 cm of cultured excised maize roots (White's medium, 2% glucose) and roots of whole maize plants (White's inorganic nutrients) after varying periods in culture. *Vertical bars*: standard error of the mean

Batra et al. (5) have used rather different techniques but have results that are essentially the same. In our experiments (36), application of 2% 2, 3, 5-triiodobenzoic acid (TIBA) in lanolin paste as a ring around the primary root base reduces by 70% the movement of labelled IAA from the shoot system into the primary root. We are presently testing the hypothesis that during the early stages of development IAA required for root growth is provided by the fruit and that during ontogeny there is a gradual shift until the major source of root IAA is the shoot. Hence correlation of root/shoot growth may be partly achieved by the rate of IAA supply from the shoot.

If the root IAA came exclusively from the shoot the fact that the amount of IAA per root system in the excised root remained more or less constant during the culture period (Fig.1) would be very surprising, since it has been established (22) that IAA is rapidly metabolised in root tissues. Our suspicion that there must be some endogenous IAA synthesis in the root was confirmed by incubating excised roots for 24 h in standard culture medium containing ^{14}C-labelled D,L trytophan. The metabolites detected included a compound that was chromatographically indistinguishable from IAA. In view of the similarity in form of the cultured roots and roots of higher plants, it seems that root-synthesised IAA fulfills the same regulatory role as that supplied by the shoot, i.e., the root-produced IAA merely reinforces the shoot-produced IAA. We still have to demonstrate whether IAA biosynthesis occurs in the root of the intact plant, however, and it may be that removal of the normal source of root IAA (the shoot system) derepresses the

IAA biosynthesising enzymes in the root, which then produce IAA, but at a sub-optimal level.

Since IAA levels in the cultured maize roots were sub-optimal, it was very interesting to compare the geotropic response capacities of the excised roots with those of the roots of the intact plants (Fig.4). It can be seen that the response capacities are virtually identical and that, essentially, as long as the root is growing, it gives a normal geotropic response. Other workers have clearly shown that geotropic responses may occur in roots that have a sub-optimal auxin concentration (2, 3, 42). A more detailed examination of the response in our experiments revealed that the roots of the intact plants reached their final angle of curvature more quickly than the excised roots. Hence the data that we have is compatible with the view that the regulation of root elongation growth and geotropism is a consequence of the interaction of an acropetal flow of growth-promoting IAA, synthesised predominantly in the shoot, with a basipetal flow of growth-inhibiting material (probably including ABA, 31, 44, 45, 60) released from the root cap (1, 18, 47, 48, 61). The significance of the weak basipetal IAA transport in roots (28, 38, 41, 43) remains unclear. Pernet and Pilet (40) have shown clearly in maize that movement of IAA from the root cap to the extending zone is too feeble to be of significance in the georeaction, but there is still a possibility that the major acropetal flow that occurs in morphogenetically insignificant tissues (7, 23, 54) is followed by the presentation of IAA to the reactive tissues by way of a basipetal transport stream. In any case the accumulated IAA in the root cap (52) presumably represents the excess of IAA (derived from the acropetal stream) over that required by the root for growth processes. Even if these complications do exist, they would not, however, change the basic thesis of the regulation of elongation by the interaction of growth-promoting IAA with growth-inhibiting ABA but would simply mean that the physiologically effective molecules were coming to the elongating cells from the same direction rather than from opposite directions. Demonstrations in several laboratories (5, 23, 29) of the capacity of acropetally transported auxin to stimulate growth of root segments provide strong support for the latter view (18).

It seems likely that an understanding of the mode of action of IAA in regulation of root elongation will come from experiments on the rapid growth reactions that have received much attention recently (21).

Burstrom et al. (9) have demonstrated an auxin-induced increase in elastic extensibility of pea root sections within 15 min of treatment. More precisely stated, their resonance frequency method of measurement of extensibility shows an auxin-induced reduction in the normal time-dependent decrease in elastic extensibility. More rapid growth responses have been reported by Jackson (24) and List (33), but it seems likely that all these effects may be consequences of a more fundamental effect of auxin. Currently, work of much promise in Scott's laboratory (5, 14, 15) is concerned with comparing the "acid-growth" reactions of the root with those of the coleoptile. Work on coleoptile and other seedling stem tissue (21) has led to the hypothesis that auxin initially stimulates elongation by inducing hydrogen ion secretion into the cell wall. Edwards and Scott (14, 15) have shown that the rapid growth responses of the maize primary root to hydrogen ions differ in several respects from those of shoot tissue. The magnitude of the root response may be ten times greater than that of shoot tissue. Scott and Edwards have shown that at pH 7.0, IAA initially reduced the elongation rate of 2-mm maize root segments in citrate-phosphate buffered and unbuffered solution. IAA at 0.1 μM in buffer significantly (though briefly) promoted the elongation rate, but this promoted rate represented only about 20% of the growth rate achieved with only buffer at pH 4.0. IAA in pH 4.0 buffered and unbuffered solutions partially inhibited the elongation of the root segments. In some comparative experiments with 2-mm maize coleoptile segments, it was found that 10 μM IAA at pH 6.8 promoted the elongation rate to a level equal to or greater than the maximal hydrogen ion-induced rate.

The development of auxin-induced growth promotion in coleoptile segments was consistent with that found by other workers (4, 12, 50); the elongation rate increased rapidly after a 10- to 15-min lag period and reached a steady state by 30 min. In root segments, a lower IAA concentration caused an increase in elongation rate 20-40 min after application and was transient, arriving and decaying in a 5-min interval.

Marre and his colleagues showed that the promotion of cell enlargement in etiolated pea stem segments by fusicoccin (FC) was associated with the extrusion of hydrogen ions (34) and an increase in the negative transmembrane potential (35). Pilet (47) found that FC strongly stimulated the elongation of decapped maize root segments and that it interacted with the growth inhibitors (which include ABA) produced by the root cap. Further studies of the type described by Scott and his colleagues and Pilet may lead us to an understanding of the modes of action and interaction of IAA and ABA in regulation of root growth.

D. Conclusions

The data presented in this paper are compatible with the proposal that in the maize plant the elongation of the primary root and the georesponses are regulated by the interaction of growth-promoting, acropetally moving IAA derived primarily from the fruit and shoot system, with ABA (and possibly other inhibitors) moving in a basipetal direction from the root cap. The primary action of IAA in promotion of cell elongation may involve the stimulation of a proton pump in the cell membranes.

The temptation to generalise about roots other than those of maize has been resisted because the variations in sterile culture characteristics of roots of different species suggest that there may be differences in the auxin regulatory systems involved (10). The major difference between the roots of monocotyledonous plants, whose roots have proved extremely resistant to continuous culture and whose growth during a limited number of passages is routinely enhanced by added auxin, and dicotyledons (e.g., tomato, *Lycopersicon esculentum*, Mill.), which may maintain a high growth rate through hundreds of subcultures in auxin-free medium, cannot be ignored. Despite this, however, it is possible to cite experiments that suggest similarities between the maize root biology described here and the biology of root systems of dicotyledonous plants. Acropetal polar transport of IAA was first described in segments of dicotyledonous plant roots (43), widely confirmed (6, 26) and demonstrated in roots of intact seedlings (23, 29, 30). The transport of auxin from a site of application in the shoot system has been clearly shown (16, 20, 39), and IAA of shoot origin has been implicated in regulation of root elongation (37), lateral root formation (51, 57), and regulation of cambial activity (13, 58).

It is possible then that the basic regulatory interactions are indeed similar in all roots, and the differences in culture characteristics of roots of various species (where these reflect differences in auxin requirements) are simply an indication of a difference in balance in the intact plants between the proportions of IAA synthesised in the root and shoot systems, or a difference in the capacity of the excised root to adapt to the lack of shoot-supplied auxin. It is noteworthy that secondary thickening in isolated dicotyledonous roots is induced with difficulty (10), and this may well be a consequence of the removal of the normal source of auxin.

Limitations of space have prevented discussion of the interactions of auxin with gibberellins and cytokinins, although these must clearly have been of importance in the regulation of certain aspects of root growth (53).

Clearly these conclusions imply non-acceptance of the Cholodny-Went hypothesis of the regulation of root elongation and geotropism. Although Davies and Mitchell (38) have interpreted their observations as evidence for the involvement of supra-optimal concentrations of auxin in the regulation of root georeactions, further support for this idea seems unlikely. Nevertheless, as Wilkins (61) has observed, the geotropic response involves in principle the mechanism suggested by Cholodny (11) and Went (59) in that an apically synthesised growth regulator undergoes lateral displacement towards the lower side of the organ where it reduces growth. The essential difference is that the inhibitor is not IAA.

References

1. Audus, L.J.: In: The Development and Function of Roots. Torrey, J.G., Clarkson, D.T. (eds.). New York-London: Academic Press, 1975, p. 327
2. Audus, L.J., Brownbridge, M.E.: J. Exp. Botany 8, 105 (1975)
3. Audus, L.J., Brownbridge, M.E.: J. Exp. Botany 8, 235 (1975)
4. Barkley, G.M., Leopold, A.C.: Plant Physiol. 52, 76 (1973)
5. Batra, M.W., Edwards, K.L., Scott, T.K.: In: The Development and Function of Roots. Torrey, J.G., Clarkson, D.T. (eds.). New York-London: Academic Press, 1975, p. 299
6. Bonnett, H.T., Torrey, J.G.: Plant Physiol. 40, 813 (1965)
7. Bowen, M.R., Wilkins, M.B., Cane, A.R., McCorquodale, I.: Planta 105, 273 (1972)
8. Bridges, I.G., Hillman, J.R., Wilkins, M.B.: Planta 115, 189 (1973)
9. Burstrom, H.G., Uhrstrom, I., Olausson, B.: Physiol. Plantarum 23, 1223 (1970)
10. Butcher, D.N., Street, H.E.: Botan. Rev. 30, 513 (1964)
11. Cholodny, N.: Jahrb. Wiss. Botan. 65, 447 (1926)
12. Dela Fuente, R.K., Leopold, A.C.: Plant Physiol. 46, 186 (1970)
13. Digby, J., Wangermann, E.: New Phytologist 64, 168 (1965)
14. Edwards, K.L., Scott, T.K.: Planta 119, 27 (1974)
15. Edwards, K.L., Scott, T.K.: Planta (1976)
16. Eliasson, L.: Physiol. Plantarum 27, 412 (1972)
17. Elliott, M.C.: New Phytologist 70, 1005 (1971)
18. Elliott, M.C.: Plant Physiol. 56, p. 211 (1975)
19. Elliott, M.C., Greenwood, M.S.: Phytochemistry 13, 239 (1974)
20. Eschrich, W.: Planta 78, 144 (1968)
21. Evans, M.L.: Ann. Rev. Plant Physiol. 25, 195 (1974)
22. Greenwood, M.S., Hillman, J.R., Shaw, S., Wilkins, M.B.: Planta 109, 369 (1973)
23. Hejnowicz, Z.: Acta Soc. Botan. Polon. 37, 451 (1968)
24. Jackson, W.T.: Physiol. Plantarum 13, 36 (1960)
25. Juniper, B.E., Groves, S., Landau-Schacher, B., Audus, L.J.: Nature (London) 209, 93 (1966)

26. Kirk, S.C., Jacobs, W.P.: Plant Physiol. 43, 675 (1968)
27. Knegt, E., Bruinsma, J.: Phytochemistry 12, 753 (1973)
28. Konings, H.: Acta Botan. Neerl. 16, 161 (1967)
29. Konings, H.: Acta Botan. Neerl. 18, 528 (1969)
30. Konings, H., Gayadin, A.P.: Acta Botan. Neerl. 20, 646 (1971)
31. Kundu, K.K., Audus, L.J.: J. Exp. Botan. 25, 479 (1974)
32. Libbert, E., Wichner, S., Schiewer, U., Risch, H., Kaiser, W.: Planta 68, 327 (1966)
33. List, A.L., Jr.: Planta 87, 1 (1969)
34. Marré, E., Colombo, R., Lado, P., Rasi-Caldogno, F.: Plant Sci. Lett. 2, 139 (1974)
35. Marré, E., Lado, P., Ferroni, A., Ballarin-Denti, A.: Plant Sci. Lett. 2, 257 (1974)
36. Martin, H.V., Wangermann, E., Elliott, M.C.: (Unpublished data)
37. McDavid, C.R., Sagar, G.R., Marshall, C.: New Phytologist 71, 1027 (1970)
38. Mitchell, E.K., Davies, P.J.: Physiol. Plantarum 33, 290 (1975)
39. Morris, D.A., Briant, R.E., Thompson, P.G.: Planta 89, 178 (1969)
40. Pernet, J.J., Pilet, P.E.: Planta 128, 183 (1976)
41. Pilet, P.E.: Bull. Soc. Bota. Suisse 61, 410 (1951)
42. Pilet, P.E.: Bull. Soc. Vaud. Sci. Nat. 65, 409 (1953)
43. Pilet, P.E.: Nature (London) 204, 561 (1964)
44. Pilet, P.E.: Planta 106, 169 (1972)
45. Pilet, P.E.: Planta 111, 275 (1973)
46. Pilet, P.E.: What's New in Plant Physiology, 6, 10 (1974)
47. Pilet, P.E.: Planta 122, 299 (1975)
48. Pilet, P.E.: Plant Sci. Lett. 5, 137 (1975)
49. Raggio, M., Raggio, N.: Physiol. Plantarum 9, 466 (1956)
50. Rayle, D.L., Cleland, R.: Plant Physiol. 46, 250 (1970)
51. Richardson, S.D.: Physiol. Plantarum 11, 698 (1958)
52. Rivier, L., Pilet, P.E.: Planta 120, 107 (1974)
53. Scott, T.K.: Ann. Rev. Plant Physiol. 23, 235 (1972)
54. Shaw, S., Wilkins, M.B.: J. Exp. Botany 25, 199 (1974)
55. Sheat, D.E.G., Fletcher, B.H., Street, H.E.: New Phytologist 58, 128 (1959)
56. Street, H.E., Bullen, P.M., Elliott, M.C.: In: Wachstum Regulatoren bei Pflanzen. Libbert, E. (ed.). Rostock: Gustav Fischer, 1967, p. 407
57. Torrey, J.G.: Plant Physiol. 12, 873 (1958)
58. Torrey, J.G.: Symp. Soc. Exp. Biol. 17, 285 (1963)
59. Went, F.W.: Proc. Kon. Akad. Wetensch. Amsterdam 30, 10 (1926)
60. Wilkins, H., Wain, R.L.: Planta 121, 1 (1974)
61. Wilkins, M.B.: Current Adv. Plant Sci. 13, 317 (1975)
62. Wilkins, M.B., Scott, T.K.: Nature (London) 219, 1388 (1968)

Paper Demonstration References (see Section IX):

11 - 39 - 41 - 66 - 86 - 133 - 142 - 143 - 155 - 174

12. Root Growth Inhibitors

R. L. WAIN

Our interest in this subject arose from the discovery (43) of the selective herbicides, 3,5-diiodo-4-hydroxybenzonitrile (ioxynil) and the dibromo derivative (bromoxynil), both of which are widely used in agriculture. This discovery stimulated research on their mode of action and metabolism, and investigations on chemical structure in relation to herbicidal activity were also initiated (14, 15, 42, 44, 58). One of the compounds examined in this work was 3,5-diiodo-4-hydroxybenzoic acid (DIHB). In translocation experiments the roots of cress seedlings exposed to light in culture solution containing 10^{-5} M DIHB were found to grow to about three times the length of control roots and were the same length as untreated roots grown in darkness (46). Thus the well-known inhibitory effect of light on root growth (5, 19, 20, 29, 41) could, in the case of cress roots, be removed by DIHB treatment. This growth-promoting effect of DIHB in the light was shown to arise by promotion of cell elongation and not by stimulating cell division (45).

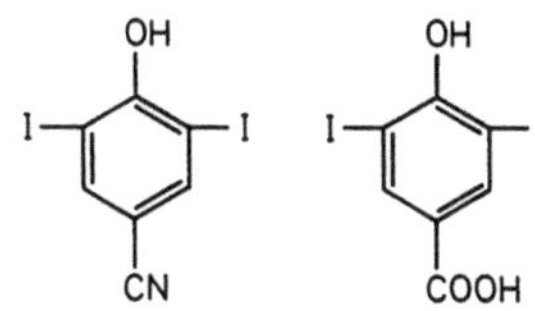

Ioxynil DIHB

At this stage *Zea mays* seedlings were chosen for detailed studies on the effect of light on root growth. The inhibitory effect was shown by Wilkins to be an almost instantaneous response, and a short flash of light (1-2 sec) was found to cause inhibition that persisted for a period of at least 6 h (50, 51).

The roots of light-exposed *Zea* seedlings were observed to be pigmented near the apex. Violaxanthin and neoxanthin were among the compounds identified when the pigments were extracted (51). These two pigments have been shown in previous work at Wye to be converted to the hormone inhibitor xanthoxin when exposed to light (37). Xanthoxin, however, was not detected in the light-exposed *Zea* roots, though Audus found traces in a different variety of maize (2). In a further study, Wilkins identified abscisic acid (ABA) and 6-methoxy-2-benzoxazolinone as growth inhibitory compounds in the primary roots of *Zea mays* (49). Kundu and Audus (16, 17) have also reported the presence of an ABA-like compound in maize roots, and we have found ABA present in greater quantities in light-exposed root caps than in those from dark-grown seedlings (53). Our next investigations involved interaction experiments between DIHB and xanthoxin using wheat seedlings. It was shown that these compounds were antagonistic towards each other, and the inhibitory effect of xanthoxin on root growth was partially removed in the presence of DIHB (52).

Removal of the root cap from maize seedlings (decapping) completely eliminated the inhibiting effect of light on root growth, showing that this region of the root was essential for the response (51). Experiments were carried out in which the roots of dark-grown *Zea* seedlings were decapped in weak green light. It was shown that resiting a detached cap on to the decapped root apex had little effect on subsequent growth in darkness. When a decapped root was exposed to light and then supplied in weak green light with a dark-exposed root cap, no inhibition occurred when incubated in darkness. However, when a root cap which had been briefly exposed to light was affixed to a decapped root maintained in the dark, inhibition of growth resulted (53). It seemed reasonable to assume that such inhibition was produced by chemicals which had been produced in the root cap in response to light exposure. It was then decided to investigate the time required for such compounds to be produced and to move out of the illuminated cap into the whole root in quantities sufficient to cause inhibition. This was done by exposing the whole root to white light for 2 min and returning the seedlings to darkness for increasing lengths of time before decapping. Root elongation was then measured after a standard incubation period of 4 h. It was found that a period of 2.5-3 h was required to establish an effect (53). In further experiments it was shown that a period of 5 h was required for a decapped root to regain its capacity to perceive light (53).

At this stage in the investigations it became evident that there was a similarity between the role of the root cap in detecting light and its function in the geotropic response. For example (a) the cap is solely responsible for perception of gravity (12, 23) and light (53), (b) several hours after decapping the root can regain its capacity to perceive gravity (12, 23) and light (51), and (c) the root cap produces growth-inhibiting factor(s) in response to gravity (7, 25, 33) and light (53). The subject of gravity perception is well discussed in the reviews by Audus (2) and Wilkins (47, 48).

In view of the foregoing, experiments were undertaken to establish whether light exposure of the root cap was important in the mechanism of the geotropic response. For this work, a variety of *Zea mays* (LG 11) was chosen which was not geotropically responsive in the dark. When a dark-exposed decapped root was supplied with a light-exposed root cap, both from seedlings grown vertically in darkness, a geotropic response was readily detected during a period of incubation with the root held horizontally in darkness. When, however, a dark-exposed cap was used, the geotropic response was negligible whether it was applied to a light- or dark-exposed decapped root (54).

It would appear likely from these studies that the root cap is capable of producing chemicals which are effective in inhibiting root elongation in the light and that these same chemicals can also operate in the downward bending of roots which occurs in response to gravity (see also Pilet (27)). Since ABA is known to be present in light-exposed root caps it was decided to examine the effect of applying ABA to the root tips of LG 11 maize seedlings in darkness. When these seedlings, together with controls, were held with their roots horizontally in the dark it was found that only the ABA-treated roots gave a geotropic response (55). The role of the root cap in the perception of light and gravity and its chemical implications have been the subject of elegant investigations by Pilet (21, 23-28, 30).

As already noted, two xanthophyll pigments, violaxanthin and neoxanthin, both of which are precursors of xanthoxin, are present in the root apex of light-exposed *Zea mays* seedlings (51). From previous investigations carried out in this laboratory, it would seem logical that these pigments would be converted to xanthoxin when the root was illuminated (37). Traces of xanthoxin were in fact found by Audus (2), but it was

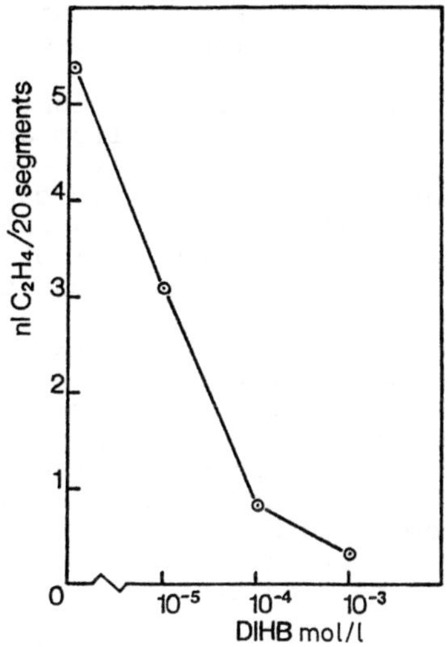

Fig.1. Effect of DIHB on ethylene produced by 20 apical cress root segments (10 mm) after incubation at 25°C for 16 h

not detected in our experiments using a different maize variety. The fact that no xanthoxin was found in our light-exposed roots, however, does not rule out its involvement, for its rate of utilization in inhibitory processes might be so rapid that the quantities remaining in the tissues would be too small to be detected. It is known that xanthoxin can be converted to ABA in plant tissues (38), so another possibility is that any xanthoxin formed by photooxidation in the root becomes rapidly converted to ABA, which, as already noted, has been detected in the light-exposed root caps (53).

Since DIHB treatment removes the inhibitory effect of light on root growth, the possibility of interactions between DIHB and xanthoxin or DIHB and ABA must be considered. Such an interaction between DIHB and xanthoxin has been established in our laboratory, and there is evidence from current investigations that DIHB can alleviate some of the inhibitory effects of ABA on root growth.

More recent work has established that, in the light-induced growth inhibition of roots immersed in culture solution, factors other than ABA and xanthoxin can also operate. Thus we have shown that when the culture solution is aerated over the period of the experiment, the light-induced inhibition of root growth is greatly reduced, and under these conditions the effect of DIHB in promoting root growth is also much less pronounced (18). The observation that roots growing under non-aerated conditions were thicker and shorter than aerated roots (18) led us to suspect that ethylene might be involved, since roots exposed to ethylene are known to exhibit similar morphologic characteristics (11, 18, 35). In agreement with this, we found that when the ethylene precursor methionine was present in the culture solution it restricted root growth, and this inhibitory effect was shown to be largely removed by DIHB (18). In view of these results, it was of interest to find that the production of ethylene by intact roots of cress (*Lepidium sativum*) seedlings was greater in the light than in the dark (31). Furthermore, light-exposed cress roots released much less ethylene into the culture solution when DIHB was present. This capacity of DIHB to reduce endogenous ethylene production was also detected in similar experiments using cress root segments (Fig.1). The possibility of IAA involvement in ethylene production by root tissues and means by which

DIHB might interact with this system formed the basis of separate investigations, which have been reported by Robert (32).

The capacity of DIHB to counteract ethylene effects on root growth does not seem to arise from any chemical interaction between DIHB and ethylene (4). Thus it may be concluded that DIHB operates in some way to reduce the biosynthetic capacity of the root tissues to produce ethylene. However, one finding that still requires explanation is that the inhibitory effect on root growth produced by large amounts of ethylene in culture solution can to some extent be reduced by DIHB (18).

Plants growing under natural conditions have their roots in the dark, so the somewhat unique properties of DIHB would not appear to be of much agricultural significance. However other factors are known to operate in restricting root growth in the soil. One of these is the mechanical resistance of the soil. It is well known, for example, that in highly compacted soils, root growth is much less prolific than in non-compacted soils. Indeed, high mechanical resistance of soils, related to naturally occurring hard pans, or to compaction by machinery and untimely cultivation (10, 36), often reduces crop yields through poor development of root systems (1, 22, 36, 39). In the laboratory, progressive increases in mechanical resistance of the growth medium cause reduced root elongation (3, 9, 57), stunting and thickening of root systems (6, 10), and reduced shoot growth (40). Under certain conditions, abnormally high levels of ethylene can occur in compacted soils (34) and, as already noted, ethylene can restrict root growth. Furthermore, the production of ethylene by plant tissues is known to be enhanced if the tissues encounter a mechanical restriction to their growth (8, 13). It was logical therefore to examine DIHB for its capacity to improve root growth of seedlings in compacted soil. These studies have been rewarding. Thus, for example, with pea seedlings, addition of 10^{-4} M or 10^{-6} M solutions of DIHB to soil, followed by compaction to give a precise mechanical resistance (42.4 bar) reduced the level of root growth inhibition. The overall lengths of the roots were 19 and 31% greater, respectively, than those of corresponding roots in compacted soil treated only with water. A reduction in the amount of root bending and thickening induced by compaction was also observed in these experiments. No such beneficial effects were obtained from DIHB treatment when the seedlings were grown in loose soils. More information on this aspect of our work is provided by the demonstration "Growth promotion of mechanically-stressed seedlings", which is being staged at this conference by Dr. Henry Wilkins.

The significance of these findings has been discussed by Wilkins (56). Field investigations are now being undertaken to assess the agricultural significance of DIHB treatment in compacted soils. Encouraging effects have already been observed with cereals, where some response is apparent in shoot growth. No yield data, however, are yet available. In laboratory experiments with culture solutions containing $^{32}PO_4$ an increased uptake of the nutrient was demonstrated when DIHB was present in the culture solution.

At the time of writing, it would seem that enhanced seedling performance can be achieved under adverse soil conditions, and this finding could open exciting possibilities in practical agriculture by helping to combat the effects of soil compaction on plant growth, using chemical rather than mechanical means.

<u>References</u>

1. Adams, E.P., Blake, G.R., Martin, W.P., Boelter, D.H.: 7th Intern. Conf. Soil Sci. 1960, p. 607

2. Audus, L.J.: In: The Development and Function of Roots. Torrey, J.G., Clarkson, D.T. (eds.). New York-London: Academic Press, 1975, p. 327
3. Barley, K.P., Farrell, D.A., Greacen, E.L.: Australian J. Soil Res. 3, 69 (1965)
4. Burden, R.S.: Private communication
5. Bürström, H.: Physiol. Plantarum 13, 597 (1960)
6. Cannell, R.Q.: MAFF Tech. Bull. 29, 439 (1975)
7. Gibbons, G.S.B., Wilkins, M.B.: Nature (London) 226, 558 (1970)
8. Goeschl, J.D., Rappaport, L., Pratt, H.K.: Plant Physiol. 41, 877 (1966)
9. Gooderham, P.T., Fisher, N.M.: MAFF Tech. Bull. 29, 469 (1975)
10. Goss, M.J., Drew, M.C.: ARC Letcombe Lab. Ann. Rept. 1971, 1972, p. 35
11. Harvey, E.M., Rose, R.C.: Botan. Gaz. 60, 27 (1915)
12. Juniper, B.E., Groves, S., Landau-Schachar, B., Audus, L.J.: Nature (London) 209, 93 (1966)
13. Kays, S.J., Nicklow, C.W., Simons, D.H.: Plant Soil 40, 565 (1974)
14. Kerr, M.W., Wain, R.L.: Ann. Appl. Biol. 54, 441 (1964)
15. Kerr, M.W., Wain, R.L.: Ann. Appl. Biol. 54, 447 (1964)
16. Kundu, K.K., Audus, L.J.: Planta 117, 183 (1974)
17. Kundu, K.K., Audus, L.J.: J. Exp. Botany 25, 479 (1974)
18. Larqué-Saavedra, A., Wilkins, H., Wain, R.L.: Planta 126, 269 (1975)
19. Masuda, Y.: Physiol. Plantarum 15, 780 (1962)
20. Ohno, Y., Fujiwara, A.: Plant Cell Physiol. (Tokyo) 8, 141 (1967)
21. Pernet, J.J., Pilet, P.E.: Planta 128, 183 (1976)
22. Phillips, R.E., Kirkham, D.: Agron. J. 54, 29 (1962)
23. Pilet, P.E.: Nature (London) 233, 115 (1971)
24. Pilet, P.E.: Planta 106, 169 (1972)
25. Pilet, P.E.: Planta 111, 275 (1973)
26. Pilet, P.E.: Planta 122, 299 (1975)
27. Pilet, P.E.: Physiol. Plantarum 33, 94 (1975)
28. Pilet, P.E.: Planta 130, 245 (1976)
29. Pilet, P.E., Went, F.W.: Am. J. Botany 43, 190 (1956)
30. Rivier, L., Pilet, P.E.: Planta 120, 107 (1974)
31. Robert, M.L., Taylor, H.F., Wain, R.L.: Planta 126, 273 (1975)
32. Robert, M.L., Taylor, H.F., Wain, R.L.: Planta 129, 53 (1976)
33. Shaw, S., Wilkins, M.B.: Planta 109, 11 (1973)
34. Smith, K.A., Dowdell, R.J., Hall, K.C., Crees, R.: ARC Letcombe Lab. Ann. Rept. 1973, 1974, p. 35
35. Smith, K.A., Russell, R.S.: Nature (London) 222, 769 (1969)
36. Swain, R.W.: MAFF Tech. Bull. 29, 189 (1975)
37. Taylor, H.F., Burden, R.S.: Proc. Roy. Soc. (London) Ser. B 180, 317 (1972)
38. Taylor, H.F., Burden, R.S.: J. Exp. Botany 24, 873 (1973)
39. Taylor, H.M.: In: Compaction of Agricultural Soils. Monog. Am. Soc. Agric. Engineers, 1971, p. 292
40. Taylor, H.M., Ratliff, L.F.: Soil Sci. 108, 113 (1969)
41. Torrey, J.G.: Plant Physiol. 27, 591 (1952)
42. Ugochukwu, E.N., Wain, R.L.: Chem. and Ind. p. 35 (1965)
43. Wain, R.L.: Nature (London) 200, 28 (1963)
44. Wain, R.L.: Proc. VIIth Brit. Weed Contr. Conf. 1964, p. 306
45. Wain, R.L., Intarakosit, P., Taylor, H.F.: Meded. Rijks. Landbouw. Gent 33, 1341 (1968)
46. Wain, R.L., Taylor, H.F., Intarakosit, P., Shannon, T.G.D.: Nature (London) 217, 870 (1968)
47. Wilkins, M.B.: In: Current Adv. Plant Sci., Pergamon Press No. 13, 1975, p. 317
48. Wilkins, M.B.: Sci. Prog., Oxf. 63, 187 (1976)
49. Wilkins, H., Burden, R.S., Wain, R.L.: Ann. Appl. Biol. 78, 337 (1974)

50. Wilkins, H., Larqué-Saavedra, A., Wain, R.L.: Proc. 8th Int. Conf. Plant Growth Subs. 1973, p. 1231
51. Wilkins, H., Larqué-Saavedra, A., Wain, R.L.: Ann. Appl. Biol. 78, 169 (1974)
52. Wilkins, H., Larqué-Saavedra, A., Wain, R.L.: Nature (London) 248, 449 (1974)
53. Wilkins, H., Wain, R.L.: Planta 121, 1 (1974)
54. Wilkins, H., Wain, R.L.: Planta 123, 217 (1975)
55. Wilkins, H., Wain, R.L.: Planta 126, 19 (1975)
56. Wilkins, S.M., Wilkins, H., Wain, R.L.: Nature (London) 259, 392 (1976)
57. Zimmerman, R.P., Kardos, L.T.: Soil Sci. 91, 280 (1961)
58. Zaki, M.A., Taylor, H.F., Wain, R.L.: Ann. Appl. Biol. 59, 481 (1967)

Paper Demonstration References (see Section IX):

44 - 92 - 100 - 102 - 133 - 150 - 160 - 166 - 182

13. Growth Inhibitors in Growing and Geostimulated Maize Roots

P. E. PILET

Abstract

The root cap of maize produces or releases growth-inhibiting substances (one of which could be abscisic acid). Removing the cap increased the root growth velocity, but only for the first hours after decapping. These growth inhibitors are not species specific; they move basipetally and also, in the apex only, laterally. Their concentration increases in the lower part of the geostimulated root segments. This is not only due to the downward transport of the inhibitors inside the apex, but also to the production of these growth regulators, which was found to be greater in the lower half of the cap from a horizontally placed root. The redistribution of the growth-inhibiting substances may simply explain the difference in elongation of the upper and lower sides of geostimulated roots. Experiments made by using exogenous indole-3-acetic acid and abscisic acid (both detected inside the root tip when using the gas chromatography-mass spectrometry technique) and fusicoccin, clearly indicated that the root growth and the georeaction of the root are controlled by at least two opposite flows of endogenous regulators. The first is a basipetal transport of ABA and other growth inhibitors, and the second corresponds to an acropetal movement of auxin.

A. Introduction

Very early, it appeared that growth was the consequence of a "balance between growth-promoting and growth-inhibiting agents" (57). In roots, biochromatography was used to show the presence, beside auxins, of at least one growth inhibitor; Bennet-Clark et al. (5) called it the β-inhibitor. It has also been observed in roots of *Triticum* (22), *Pisum* (3), and *Lens* (32). Masuda (23) and Pilet (32) showed that this inhibitor was more abundant in the root tips maintained in light. Using Sycamore cells, it was subsequently proved (10, 24, 52) that the most active compound of this growth-inhibiting factor was abscisic acid (ABA). On the other hand, it has been observed that the overall growth rate of geostimulated roots is depressed and that the downward bending arises from the growth rate of the lower part being reduced to a greater extent than that of the upper part (1, 6, 17, 21, 53). This was attributed to the increase of the level of the inhibitor. In recent years, there have been a number of investigations involving microsurgery of the root apex (decapped and half-decapped roots, decapitated and half-decapitated roots) (61). They clearly indicate (14, 55, 60) that the root cap produces at least one growth inhibitor that is transported basipetally to the extension zone of the growing roots. Such an inhibiting factor is responsible for the georeaction. The aim of this report is to discuss some of our recent results about the physiologic properties of these inhibiting substances produced by the maize caps, by comparing them to results published by several colleagues.

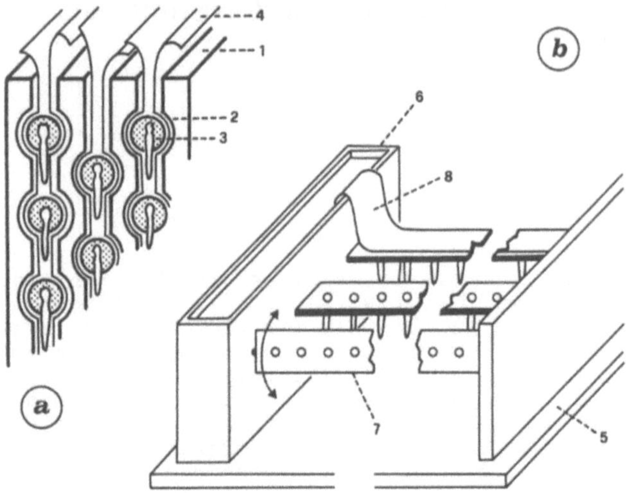

Fig. 1 a,b. Preparation of maize roots: (a) frames for germinating caryopses; (b) special support for apical root segments. *1*: plastic frame; *2*: hole for caryopse (*3*); *4*: moist filter paper; *5*: plastic support; *6*: cup for buffered solution; *7*: movable frame; *8*: filter paper

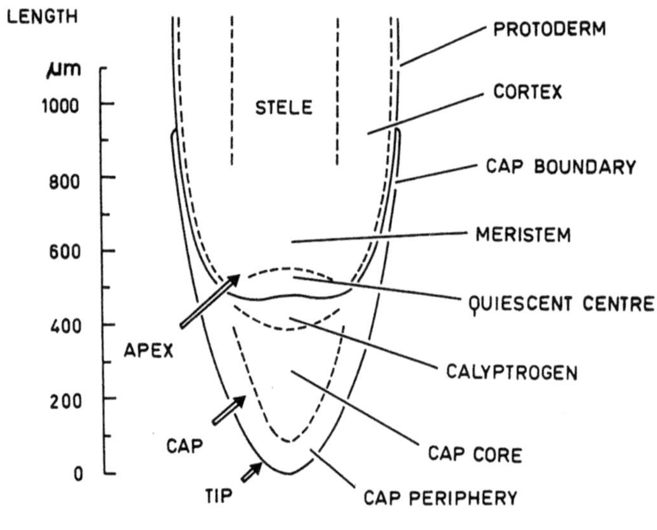

Fig. 2. Outline of tissues of first mm of primary maize root

B. Material and Methods

Two varieties of *Zea mays* L. were used: cv. Kelvedon 33 and cv. Anjou 210. Selected caryopses were grown in darkness (22°C), and, between plastic frames (Fig.1a), the roots elongated vertically on moist paper towels (37). Then they reached 15 ± 3 mm in length, 10 ± 0.2 mm apical segments were prepared and mounted in plastic frames with their basal cut ends in contact with moist filter paper (Fig.1b). The frames were then placed in closed boxes (22 ± 1°C) in which a humid atmosphere

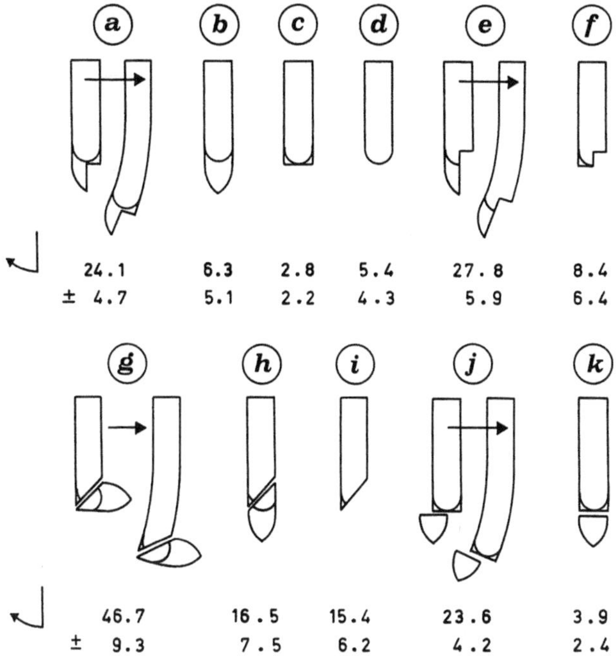

Fig. 3 a-k. Curvature (in degrees ± standard error) after 14 ± 0.5 h of apical maize (cv. Kelvedon) root segments vertically maintained

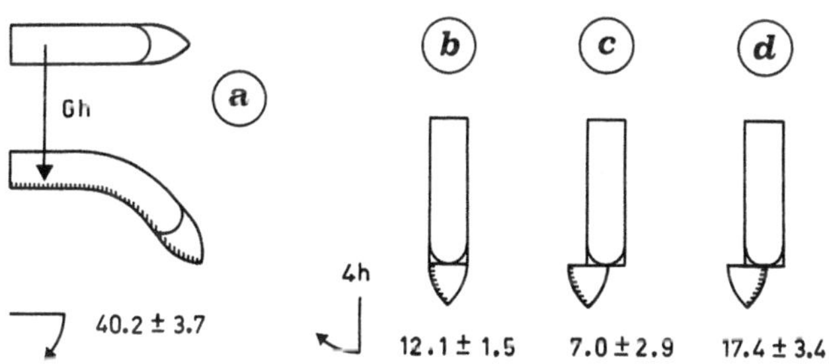

Fig. 4 a-d. Curvature (in degrees ± standard error) of horizontal and vertical apical maize (cv. Kelvedon) root segments. Caps were removed and replaced by caps from 6-h geostimulated roots

(90 ± 5%) was maintained (42). For the light assays, the segments were immediately brought into white fluorescent light ($0.9 \pm 6 \times 10^{-2}$ J/m^2/sec). Curvature was recorded by making shadow photographs, and standard errors of the mean were calculated and significant differences assessed by the t test.

The root tips used for these experiments were formed by several kinds of well-characterized (4, 16, 28) tissues (Fig.2): at first, the cap with its meristem (calyptrogen), the core, and periphery cells; and then, the quiescent center, which is the site of a founder cell population from which all cells of the root can be derived. With the meristem, it forms the root apex. For the maize roots, one has to notice the cap boundary that exists between the cap and root proper and makes possible the decapping assays first introduced by Juniper et al. (16).

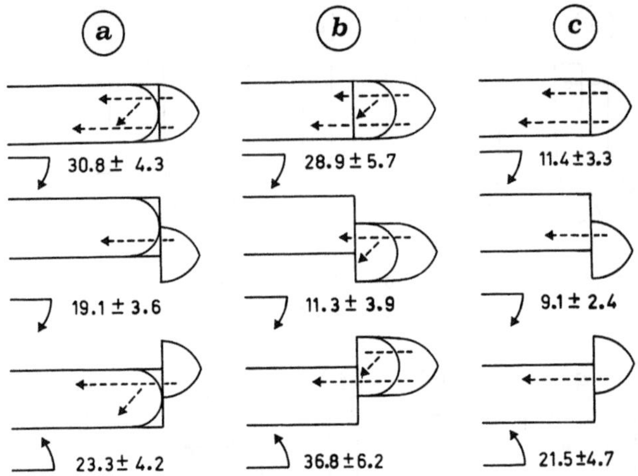

Fig. 5 a-c. Curvature (in degrees ± standard error) after 6 h of horizontally placed maize (c.v. Kelvedon) root segments. Caps (a) or tips (b) were removed and replaced on cut surfaces. Apex was removed and only cap replaced (c). *Broken arrows*: relative transport of growth inhibitors from cap

C. Results and Discussion

I. Occurrence and General Considerations

Several assays (35, 38-40), when using vertical maize segments with their tips cut in different ways (Fig.3), clearly indicate that the cap is the source of at least one growth inhibitor which moves basipetally through the apex into the extending zone. The principle of each assay (and of the control tests) is to establish an asymmetry in the tip, so that the presence of the inhibitor, going from the cap into the apex, might be shown.

II. Production

Root segments were placed horizontally (Fig.4) and, after 6 h, developed a strong curvature (Fig.4a). The caps were then removed and replaced on vertical decapped root segments, covering completely (Fig. 4b) or half (Fig.4c,d) the apical cut surface of these segments. After 4 h, a bending occurred that was stronger for (b) and (d) than for (c). This indicates that the lower half of the cap from geostimulated roots produced inhibitors in larger amounts (or activity) than the upper half (47).

III. Longitudinal and Lateral Transport

Root segments were then maintained in a horizontal position (Fig.5) with the cap (Fig.5a) or the tip (Fig.5b) removed and replaced exactly on the apical cut surfaces or re-fixed on the lower and on the upper half of these surfaces. The curvature (after 6 h) indicates that the production of growth inhibitors was greater in the lower part of the cap than in the upper part (47) and that these inhibitors could move

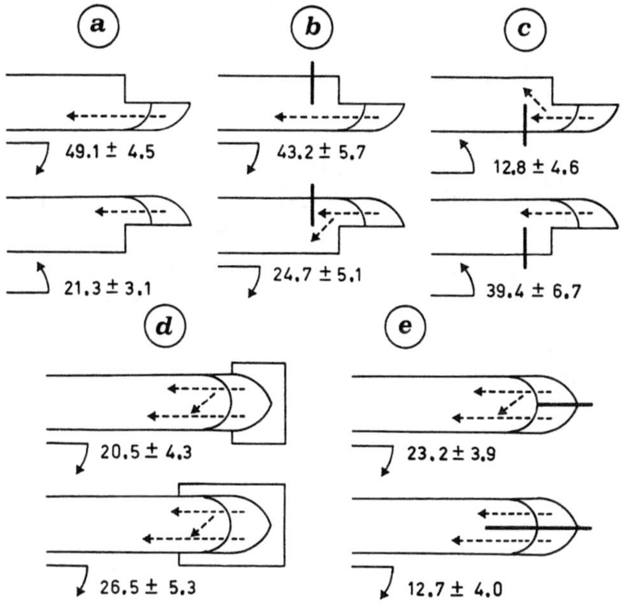

Fig. 6 a-e. Curvature (in degrees ± standard error) after 14 ± 0.5 h of half-decapitated maize (cv. Kelvedon) roots with (b,c) or without (a) a mica barrier inserted, and after 6 h of root segments into which mica was inserted vertically (d) or horizontally (e) inside the cap or tip

longitudinally and laterally, but only inside the apex (38, 39, 43). When this apex was removed and the cap replaced directly on the cut surface (c), similar results were obtained.

In another series of assays (Fig.6), half-decapitated root segments (Fig.6a) developed a bending towards the side on which the remaining half-tip was located. The insertion of a mica barrier into the side of the root with (or without) the half-tip present (Fig.6b,c) causes an inversion of the curvature (39, 43). Here again, it has to be concluded that inhibiting substances move basipetally from the cap to the extending zone of roots and also, but only inside the apex, laterally.

Then the mica was inserted either vertically (Fig.6d) or horizontally (Fig.6e) into the tips of horizontal segments. The barrier was introduced into the cap only or into the tip. Of course, the presence of the mica reduced the 6-h bending, but no differences were observed based on whether the vertical mica was inside the cap or the tip. In contrast, a significant decrease of curvature was obtained when the horizontal barrier was inserted into the cap and the apex. This confirms the lateral movement of the inhibitors inside the apex.

IV. Elongation and Role of the Cap

The production of inhibitors from the cap can be tested when analysing, for short periods of time, the root elongation (38). The growth velocities (Fig.7) of intact (Fig.7a) and decapped (Fig.7b) vertical root segments were analysed. As can be seen, removal of the cap significantly enhanced the growth velocity, but only during the first 3 h after

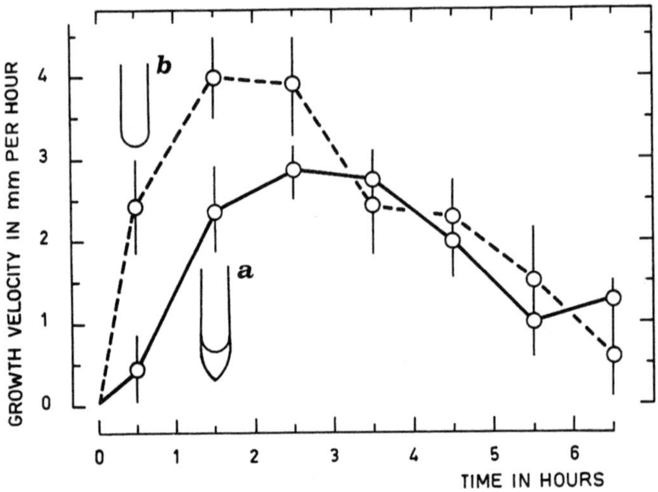

Fig. 7. Growth velocities of vertical maize (cv. Kelvedon) roots with time when cap is intact (*a*) and removed (*b*). *Points* and *vertical lines*: mean value of root sample ± standard deviation, respectively. Initial length: 15.3 ± 0.9 mm

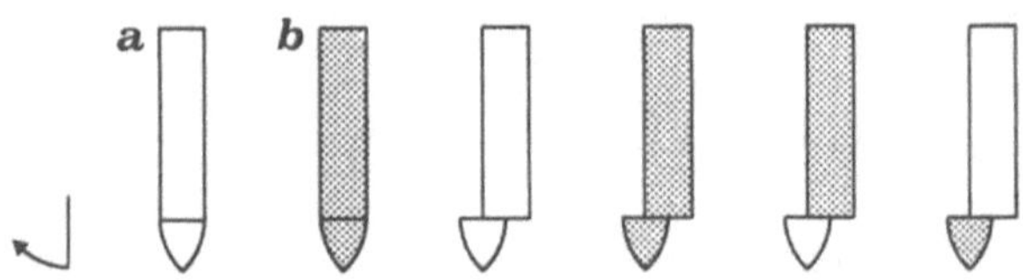

LIGHT	6.1	4.3	23.6	25.9	27.2	20.6
±	3.5	2.9	4.2	4.7	9.1	7.0
DARK	2.6	4.8	9.1	21.8	4.4	22.7
±	2.0	3.7	3.2	5.2	3.5	8.2

Fig. 8. Curvature (in degrees ± standard error) after 14 ± 0.5 h of apical root segments prepared from two maize varieties: *a*: cv. Kelvedon; *b*: cv. Anjou. All segments were decapped and cap was replaced on stump, symmetrically or asymmetrically

the decapping. This clearly confirms that the cap is the source of growth inhibiting substances.

V. Light Effects

It has been reported (42) that the apical root segments of maize from the Anjou variety were georeactive both in darkness and in light, while those of the Kelvedon variety were reactive - at least for the first few hours of horizontal presentation - only in light. The bending of decapitated roots whose caps were replaced symmetrically or asymmetrically was analysed after 14 h, in the vertical position in the light or in the dark (Fig.8), using the Kelvedon (Fig.8a) or Anjou (Fig.8b) varieties.

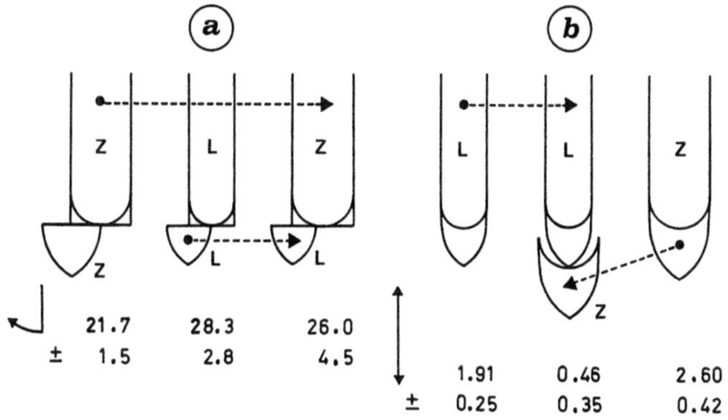

Fig. 9 a,b. Diagram showing experiments for testing non-species specificity of growth inhibitors. (a), curvature after 14 ± 0.5 h (in degrees ± standard error) of decapped root segments with cap replaced asymmetrically; (b), elongation after 3.5 ± 0.2 h (in mm/h ± standard error) of intact roots; *Z*: maize (cv. Kelvedon) roots; *L*: lentil roots

It can be seen that in light, significant bending was obtained for all the assays. In contrast, in darkness, the curvature was observed only when the tips were prepared from the Anjou variety, while the segments with an Anjou stump and a Kelvedon cap did not bend significantly (47). Consequently, the effect of light on the formation of the growth-inhibiting regulators can be related only to the root caps. Similar conclusions were obtained by Wilkins and Wain (59, 60), who found that the inhibitors produced in the cap on exposure to white light could be used by the root to produce bending in response to gravity in darkness, i.e., under conditions in which dark-exposed roots do not normally respond to a gravitational stimulation.

VI. Growth Specificity

It was of interest to ascertain whether the inhibiting substances produced by the maize root cap were specific. In other words, could the inhibitors formed in maize caps also depress the growth velocity of other roots (38)? In a first series of assays (Fig.9), maize (Z) and lentil (L) roots were decapped, and their caps were replaced asymmetrically (Fig.9a). It can be seen that the lentil cap fixed on the maize stump may also induce a bending. In a second set of experiments (Fig.9b), the elongation of the roots was analysed 3.5 h after the maize caps were placed on intact lentil roots. A sharp decrease of elongation was obtained. These findings indicate that the growth-inhibiting substances are not species specific.

VII. Chemical Nature

Several kinds of experiments and assays were performed to try to better understand the chemical identity of these inhibiting substances.

1. Water solubility

The geotropic curvatures developed by decapped and intact roots (35) were then compared. Mean bendings are plotted against time (Fig.10).

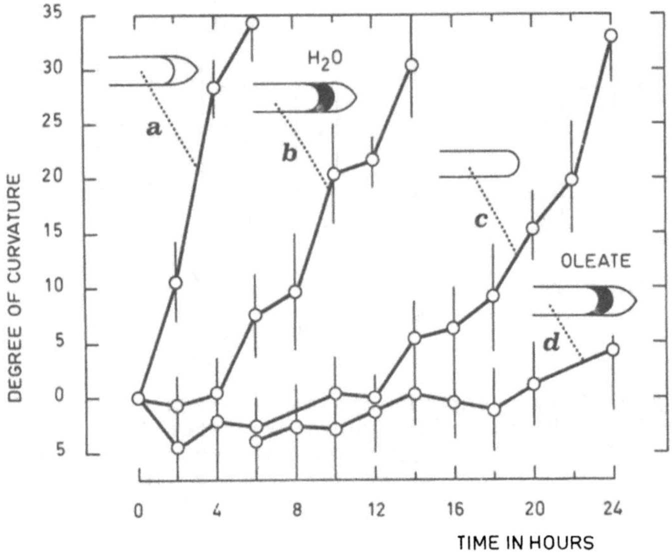

Fig. 10. Georeaction (in degrees ± standard error) of horizontally placed (0 h) maize (cv. Kelvedon) roots intact (*a*), decapped (*c*), and decapped with cap replaced and H_2O (*b*) or oleate (*d*) between cap boundary and cap cells

Table 1. IAA content (GC-MS determinations) of maize root tips

Sections (in mm from tip)	IAA in µg per kg of FW	IAA in µg per g of DW
0.0-0.5 (Cap)	356.6 ± 16.2	2.67 ± 0.12
0.5-1.0 (Apex)	179.9 ± 23.4	1.16 ± 0.15
1.0-4.0	76.5 ± 28.3	0.56 ± 0.21

FW: fresh weight DW: dry weight
Each result given with ± standard error

When intact roots (Fig.10a) showed a strong geocurvature, decapped roots (Fig.10c) continued growing horizontally for a maximum of 11 h, after which geotropism reappeared. When the root cap was replaced on the cap boundary, with a hydrophilous film between the two tissues (Fig.10b), the geocurvature resumed after a recovery time of about 5 h. In contrast, the presence of a lipophilous film (oleate) between the cap and the cap boundary (Fig.10d) cancelled the geotropic responses. Consequently, the inhibiting substances can be considered as water-soluble compounds.

2. Auxin content

The original Went-Cholodny theory (9, 26, 31, 36, 58) predicted that auxin may act, controlling root elongation, at a supraoptimal level for growth. In fact, indole-3-acetic acid (IAA) has been definitely found in the root (8, 13, 31, 32) and in the root cap (51). As indicated in Table 1, the concentration of IAA (tested by gas chromato-

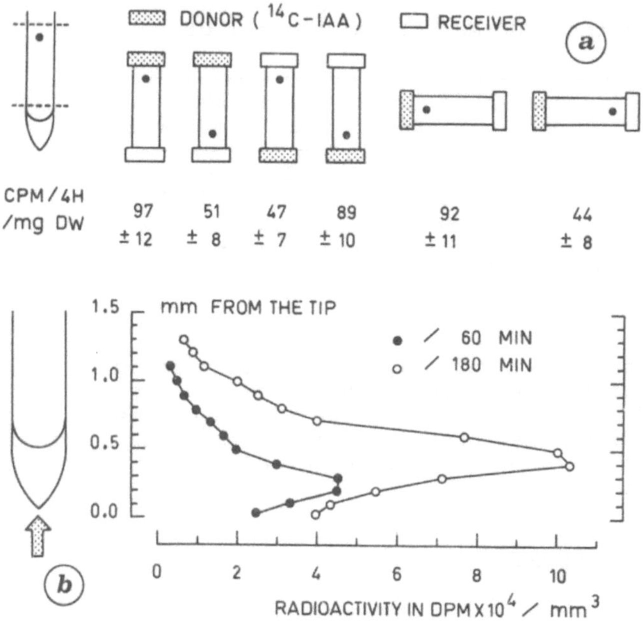

Fig. 11 a,b. Measurements of labelled IAA transport in root. (a) lentil root (5 ± 0.2 mm segments decapped at 3 mm) with ^{14}C-IAA (labelled in -COOH, sp. radioact.: 8 mCi/mM; 12 µM); (b), maize (cv. Kelvedon) root (10 mm apical segments) with (5-^{3}H) IAA (sp. radioact.: 27 Ci/mM; 10^{-7} M)

graphy-mass spectrometry (GC-MS)) in the root tip was quite large. But this may indicate, moreover, that (1) IAA moves basipetally and is produced by the root cap, and (2) IAA moves acropetally and is accumulated inside the root cap. Only the first conclusion may indicate that the growth inhibitor could be IAA.

3. Auxin transport

A bidirectional transport of IAA in roots has been observed previously (18, 31) and confirmed under special conditions (11, 12, 26, 27). The polar movement of IAA through maize root segments has, however, been extensively investigated (33, 54, 62), and it now seems clear that an acropetal transport of IAA occurred with greater efficiency in stellar tissues, whereas basipetal movement occurred only to a slight extent in both the stele and the cortex (7, 56).

In a first series of assays (Fig.11) using segments of lentil roots and ^{14}C-IAA (Fig.11a), it can be seen that IAA moves preferentially in the acropetal direction (33). This finding was confirmed by the fact that, when applied on the root cap (Fig.11b), ^{3}H-IAA entered the root tip and moved basipetally inside the root cap (29). However, from the cap into the apex, the basipetal transport of IAA seems to be very slow. On the other hand, the radioactivity found in the apex could also be attributed to the presence of a few cap cells in which IAA might still accumulate and to a possible IAA movement in the aqueous film covering the root protoderm.

Table 2. ABA content (GC-MS determinations) of maize root tips

Sections (in mm from tip)	ABA in µg per kg of FW	ABA in µg per kg of DW
0.0-0.5 (Cap)	36.1 ± 7.6	0.271 ± 0.057
0.5-1.0 (Apex)	66.5 ± 18.4	0.428 ± 0.118
1.0-4.0	33.3 ± 0.9	0.245 ± 0.006

Explanation as in Table 1

Consequently, the relatively high IAA content found in the root cap cannot be due to the IAA production inside the cap. It can only be the consequence of a predominantly acropetal IAA movement in roots (30, 31, 33). These findings also clearly indicate that IAA cannot be a component - at least not the essential one - of the inhibitor complex produced by the root cap.

4. Abscisic acid content

Abscisic acid (ABA), which strongly inhibits root elongation (34, 35), has also been identified in the maize cap (19, 20). As shown in Table 2, the level of ABA (tested by the GC-MS technique) in the maize root tip was also large (50). As already discussed for IAA (see "Auxin content" above), these results can be attributed either to an acropetal ABA movement or to a basipetal ABA transport. Meanwhile, when comparing the IAA (Table 1) and ABA (Table 2) levels in root cap, it can be noticed that the content of IAA is considerably larger than that of ABA. On the other hand, it seems clear that IAA is acropetally transported and accumulates inside the cap in which the endogenous level is at a maximum. In contrast, the ABA content in the cap is not optimal, which probably means that ABA can move only from the cap. This is in agreement with all previous observations related to the basipetal movement of the growth-inhibiting substances produced by the root cap.

5. ABA transport: interactions with IAA

The relative bending response of vertical decapped root segments (control agar blocks fixed on both apical and basal cut surfaces) was determined after 7 ± 0.5 h (Fig.12).

Agar blocks containing IAA or ABA were used, and a mica barrier was also inserted in some cases (43). The bending of such segments clearly indicates that ABA can move in the basipetal direction and also laterally; its transport properties are thus very similar to those of the growth-inhibiting substances produced by the cap. On the other hand, IAA, moving acropetally, may increase the bending due to the accumulation of ABA and some growth inhibitors in the lower zone of geostimulated roots. These findings were confirmed by a last set of assays (Fig. 13). When using vertical decapped roots whose caps were replaced asymmetrically on the cut surfaces (Fig.13a), the 14-h curvature indicated that exogenous IAA, when transported in the acropetal direction, may stimulate the bending caused by the growth-inhibiting substances. On the other hand, ABA, when applied asymmetrically to the apex (Fig.13b), may induce a bending similar to that produced by the root cap. Finally,

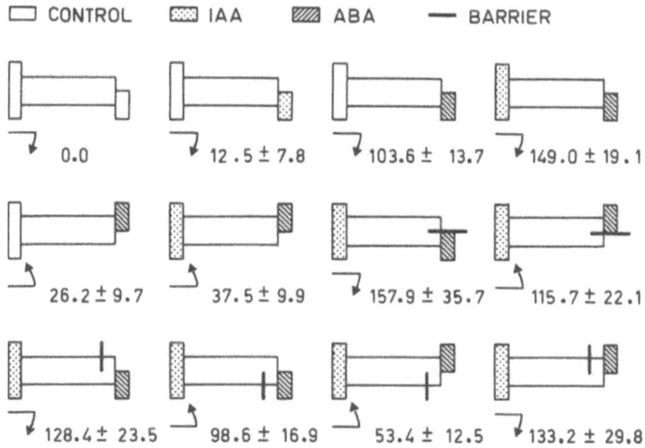

Fig. 12. Curvature (relative value with ± standard error) after 7 ± 0.5 h of horizontal maize (cv. Kelvedon) root segments decapitated at 0.6 ± 0.1 mm with agar block on basal and apical cut surfaces. IAA and ABA concentration: 10^{-6} M. Barrier: mica

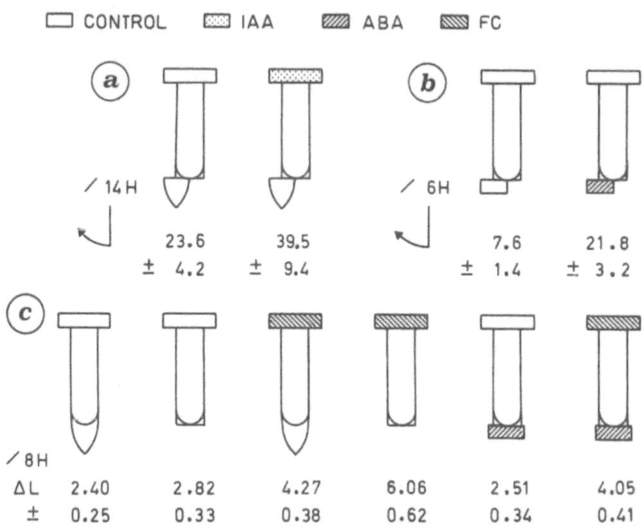

Fig. 13 a–c. Interaction between IAA (10^{-6} M), ABA (5×10^{-6} M), and FC (10^{-5} M) applied (in agar blocks) on basal or apical cut surfaces of decapped maize (cv. Kelvedon) root segments (10 ± 0.2 mm). (a,b) curvature (in degrees ± standard error); (c) elongation (in mm ± standard error)

fusicoccin (FC) - which antagonizes ABA (44, 45) - strongly increased the root elongation when it was applied on the basal cut section. Such stimulation of growth was found to be greater when the root was first decapped (44). Here again, exogenous ABA may act in a way similar to that of the inhibiting substances which are formed by the cap cells and which move basipetally through the apex into the extending zone of the root.

VIII. Physiologic and Biochemical Interactions[1]

It is clearly stated that the transport and the accumulation of these inhibiting substances, produced by the root cap, can influence several "systems" which are themselves - more or less directly - involved in growth regulation. In our Institute, and for some years, a number of problems of this type have been studied in cooperation with several of our co-workers. Some of them - those that have progressed sufficiently - will be briefly presented, thus constituting the final section of this paper.

As discussed previously, the implication of the transport of both IAA and ABA for the growth and georeaction of maize roots has been clearly established. Now some regulators, such as fusicoccin - which acts strongly on root elongation -, may also change the transport properties of IAA (PD, 130). The role of these endogenous growth factors could be better understood by analysing their translocation as modified by some growth effectors.

Careful and sure determinations of the nature and level of some growth hormones in different parts of growing and geostimulated maize roots are now possible using the GC-MS technique (PD, 139).

The redistribution of some enzymes (lipase: PD,64; phosphomonoesterase: PD,110) during the elongation and georeaction of maize root is of interest, and some correlations between growth and metabolism have to be proposed. The metabolic implications were also observed when measuring the oxygen uptake in different parts of the geostimulated roots of maize (PD,10).

The different types of microsurgical assays (discussed previously in the present paper), related to the maize root tips, may sharply change the nucleus reactivity of some kinds of cells (quiescent center, calyptrogen, meristems...). The regeneration of a few tissues (root cap and apex), as well as their mitotic activity, was analysed in terms of DNA content (PD,133).

All of these reactions that allow growth and its regulation require energy, at least a part of which was tested precisely when analysing changes of ATP, ADP...content (PD,162) in the tips of elongating and georeacting maize roots.

On the other hand, growth localization throughout the elongating axis of maize roots - in relation to accumulation of inhibitor - can also be analysed by several other methods. Scanning studies of root surfaces reveal some typical morphological characteristics of the cuticle from root protoderm (PD,69). The hydroxyproline content of the cell wall fractions may also yield information about the role played by wall constituents in controlling cell extension. In maize roots, hydroxyproline clearly regulates the growth potential: when its level increases, elongation decreases and vice versa (PD,81).

D. Conclusions

Root growth and geotropism of the root seem to be clearly controlled by opposite flows of at least two endogenous factors: basipetal trans-

---

[1] All references quoted in this section correspond to the paper demonstrations (PD) quoted at the end of this book.

port of inhibiting regulators (namely ABA) produced by the root cap and acropetal transport of auxins (namely IAA), which may accumulate inside the root cap. When the roots were maintained in a horizontal position (geostimulation followed by the georeaction), two series of data were reported about the growth-inhibiting substances: their production was found to be greater in the lower part of the root cap, and their downward lateral movement was observed but only inside the root apex. These last findings may explain an increase in the growth-inhibitor levels in the lower part of the extending zone of the growing roots, horizontally maintained, causing a positive georeaction. Finally, it now seems clear that IAA may act on the distribution of these growth inhibitors, since it may increase the georeaction controlled by the inhibiting compounds produced by the cap.

References

1. Audus, L.J., Brownbridge, M.E.: J. Exp. Botany 8, 105 (1957)
2. Audus, L.J., Gunning, B.E.S.: Physiol. Plantarum 11, 685 (1958)
3. Audus, L.J., Tresch, R.: Physiol. Plantarum 6, 451 (1953)
4. Barlow, P.W.: J. Theoret. Biol. 57, 433 (1976)
5. Bennet-Clark, J.A., Tambiah, M.S., Kefford, N.P.: Nature (London) 169, 452 (1952)
6. Bennet-Clark, T.A., Younis, A.F., Esnault, R.: J. Exp. Botany 10, 69 (1959)
7. Bowen, M.R., Wilkins, M.B., Cane, A.R., McCorquodale, I.: Planta 105, 273 (1972)
8. Bridges, I.G., Hillman, J.R., Wilkins, M.B.: Planta 115, 189 (1973)
9. Cholodny, N.: Jahrb. Wiss. Botan. 65, 447 (1926)
10. Cornforth, J.W., Milborrow, B.V., Ryback, G., Wareing, P.F.: Nature (London) 205, 1269 (1965)
11. Davies, P.J., Mitchell, E.K.: Planta 105, 139 (1972)
12. Davies, P.J., Doro, J.A., Tarbox, A.W.: Physiol. Plantarum 36, 333 (1976)
13. Elliott, M.C., Greenwood, M.S.: Phytochemistry 13, 239 (1974)
14. Gibbons, G.S.B., Wilkins, M.B.: Nature (London) 226, 558 (1970)
15. Juniper, B.E., French, A.: Planta 95, 314 (1970)
16. Juniper, B.E., Groves, S., Landau-Schachar, B., Audus, L.J.: Nature (London) 209, 93 (1966)
17. Konings, H.: Acta Botan. Neerl. 13, 566 (1964)
18. Konings, H.: Acta Botan. Neerl. 16, 161 (1967)
19. Kundu, K.K., Audus, L.J.: Planta 117, 183 (1974)
20. Kundu, K.K., Audus, L.J.: J. Exp. Botany 25, 479 (1974)
21. Larsen, P.: Physiol. Plantarum 6, 735 (1953)
22. Lexander, K.: Physiol. Plantarum 6, 406 (1953)
23. Masuda, Y.: Physiol. Plantarum 15, 780 (1962)
24. Milborrow, B.V.: Planta 76, 93 (1967)
25. Milborrow, B.V.: Ann. Rev. Plant Physiol. 25, 259 (1974)
26. Mitchell, E.K., Davies, P.J.: Plant Cell Physiol. (Tokyo) 13, 1135 (1972)
27. Mitchell, E.K., Davies, J.P.: Physiol. Plantarum 33, 290 (1975)
28. Nougarède, A., Pilet, P.E.: C. R. Acad. Sci. 279, 477 (1974)
29. Pernet, J.J., Pilet, P.E.: Planta 128, 183 (1976)
30. Pernet, J.J., Cherix, M.C., Pilet, P.E., (in prep.)
31. Pilet, P.E.: Bull. Soc. Botan. Suisse 61, 410 (1951)
32. Pilet, P.E.: Physiol. Vég. 1, 171 (1963)
33. Pilet, P.E.: Nature (London) 204, 561 (1964)
34. Pilet, P.E.: J. Exp. Botany 21, 446 (1970)
35. Pilet, P.E.: Nature (London) 233, 115 (1971)
36. Pilet, P.E.: Bull. Soc. Botan. Suisse 81, 52 (1971)
37. Pilet, P.E.: Physiol. Vég. 10, 347 (1972)

38. Pilet, P.E.: Planta 106, 169 (1972)
39. Pilet, P.E.: Planta 111, 275 (1973)
40. Pilet, P.E.: C. R. Acad. Sci. 276, 2529 (1973)
41. Pilet, P.E.: Plant Sci. Lett. 1, 137 (1973)
42. Pilet, P.E.: Physiol. Plantarum 33, 94 (1975)
43. Pilet, P.E.: Planta 122, 299 (1975)
44. Pilet, P.E.: Plant Sci. Lett. 5, 137 (1975)
45. Pilet, P.E.: Plant Sci. Lett. 7, 81 (1976)
46. Pilet, P.E.: Planta 130, 245 (1976)
47. Pilet, P.E.: Planta 131, 91 (1976)
48. Pilet, P.E., Nougarède, A.: Plant Sci. Lett. 3, 331 (1974)
49. Pilet, P.E., Went, F.W.: Am. J. Botany 43, 190 (1956)
50. Rivier, L., Milon, H., Pilet, P.E., (in prep.)
51. Rivier, L., Pilet, P.E.: Planta 120, 107 (1974)
52. Robinson, E., Wareing, P.F.: Physiol. Plantarum 17, 314 (1964)
53. Sachs, J.: Arb. botan. Inst. Würzburg 1, 385 (1874)
54. Scott, T.K., Wilkins, M.B.: Planta 83, 323 (1968)
55. Shaw, S., Wilkins, M.B.: Planta 109, 11 (1973)
56. Shaw, S., Wilkins, M.B.: J. Exp. Botany 25, 199 (1974)
57. Stuart, N.W., Cathey, H.M.: Ann. Rev. Plant Physiol. 12, 369 (1961)
58. Went, F.W.: Proc. Kon. Akad. Wetensch. Amsterdam 30, 10 (1926)
59. Wilkins, H., Wain, R.L.: Planta 121, 1 (1974)
60. Wilkins, H., Wain, R.L.: Planta 123, 217 (1975)
61. Wilkins, M.B.: Current Adv. Plant Sci. 13, 317 (1975)
62. Yeomans, L.M., Audus, L.J.: Nature (London) 204, 559 (1964)

Paper Demonstration References (see Section IX):

2 - 3 - 11 - 24 - 39 - 61 - 62 - 64 - 66 - 72 - 76 - 90 - 105 - 124 - 127 - 133 - 142 - 146 - 149 - 151 - 155 - 176 - 179

V. Gibberellins and Cytokinins

14. Some Aspects of Gibberellin Metabolism in Higher Plants

J. MacMillan

Abstract

Selected aspects of gibberellin (GA) metabolism in higher plants are reviewed. Technical problems of identifying endogenous GAs, and metabolites from radioactive precursors, are discussed. The use of $[^2H]$-labelling in conjunction with mass spectrometry is suggested as a method of identifying metabolites from applied GAs in the presence of the endogenous compound, and the preparation of some 1,3-$[^xH]$-GAs from GA_3 and GA_4/GA_7 is described. Some recent results on the ring contraction step, and on the conversion of C_{20}- to C_{19}-GAs, in GA biosynthesis are presented. Loss of biological activity (deactivation) of C_{19}-GAs by 2β-hydroxylation is described.

A. Introduction

There are two main approaches to the study of the hormonal function of gibberellins (GAs). One comprises studies of the effects of exogenously applied GAs on specific biochemical processes in the whole plant or at the cellular or sub-cellular level. The other seeks to identify endogenous GAs and to define the factors that determine their endogenous levels at sites of action.

In the first approach it is not always known if the applied GA is endogenous in the plant system or if the biochemical process is directly related to GA-controlled growth. GA_3 is most commonly used because of its availability from fungal sources and because of its broad spectrum of biological activity. The overall strategy is to define primary sites of action and primary modes of action and, ultimately, to apply these findings to the hormonal control of growth and development in whole plants.

The second approach, which complements the first, includes investigations of biosynthetic sites and pathways, translocation, and deactivation of GAs together with an analysis of the factors that control these processes. This paper deals only with this second approach and presents a selective and biased review of three aspects - techniques, biosynthesis, and metabolic deactivation. Although only GAs are discussed, most of the general comments apply also to the other classes of plant hormones.

B. Analytical Procedures

Many methods, reviewed by Crozier (14) in this volume, are available for the fractionation of crude plant extracts as a preliminary to the

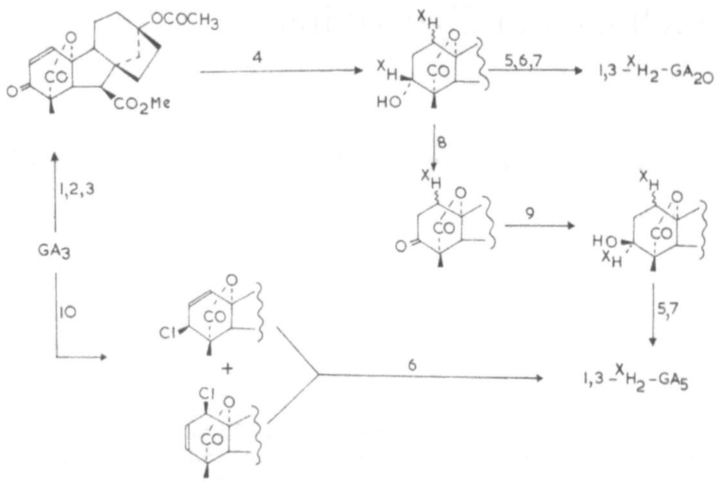

Fig. 1. Preparation of 1,3-$^{X}H_2$-GA$_5$ and -GA$_{20}$ for metabolic studies. 1,3-$^{X}H_2$-GA$_4$ and -GA$_9$ can be prepared in similar fashion from a GA$_{4/7}$ mixture. Reagents: *1*: CH_2N_2; *2*: MnO_2; *3*: Ac_2O; *4*: LiB$^{X}H_4$; *5*: $POCl_3$; *6*: (n-Bu)Sn^{X}H; *7*: OH^-; *8*: CrO_3; *9*: Li^{X}H(O-secBu)$_3$; *10*: Ph_3P, CCl_4

identification of endogenous GAs. The choice of solvents for extraction and fractionation can be important. For example the use of acetone may lead to isopropylidene derivatives as artefacts of GA-diols such as GA$_{32}$ (46). More serious perhaps is the possibility that direct solvent extraction of macerated plant tissue does not extract protein-, or membrane-, associated GAs (18, 32).

Conclusive identification of GAs rests upon a comparison of their physical constants with those of authentic standards. Rigorous proof of identity is not always easy but is essential. The most sensitive method is mass spectrometry (MS). When combined with gas chromatography in GC-MS, an additional and useful fractionating step is introduced so that GC-MS can be applied to mixtures. Sensitive quantitation can also be achieved by MS and GC-MS, especially by selected ion current monitoring in conjunction with a careful choice of internal standard.

Even with GC-MS special problems of identification can arise in metabolic studies. For example, to confirm the metabolic step between the endogenous GA$_X$ and GA$_Y$ by applying ^{3}H-GA$_X$, it would be necessary to identify ^{3}H-GA$_Y$ in the presence of endogenous GA$_Y$. For unequivocal identification, it is not sufficient to show that the radioactivity associated with the metabolite has the same R_f and R_T as GA$_Y$. Usually the specific radioactivity is so low that even GC-MS cannot detect the very small percentage of the molecules of GA$_Y$ that contain [^{3}H] from those that do not. If the mass of the [^{3}H]-metabolite, calculated from radioactivity, is sufficient to give a moderately intense MS, it is then possible to scan continuously throughout the mass peak by GC-MS and provide the negative evidence that the radioactive peak contains only GA$_Y$ (20). It is easier, in principle, to detect ^{14}C-metabolites in the presence of the same unlabelled compound. For [^{14}C]- and [^{3}H]-compounds of the same specific activity, the proportion of labelled to unlabelled molecules is ca. 10^3 greater for ^{14}C. Thus MS can be used to distinguish ^{14}C-labelled metabolites, and this technique has been used both to identify and to determine the specific radioactivity of metabolites in the *Cucurbita maxima* enzyme system (12, 25).

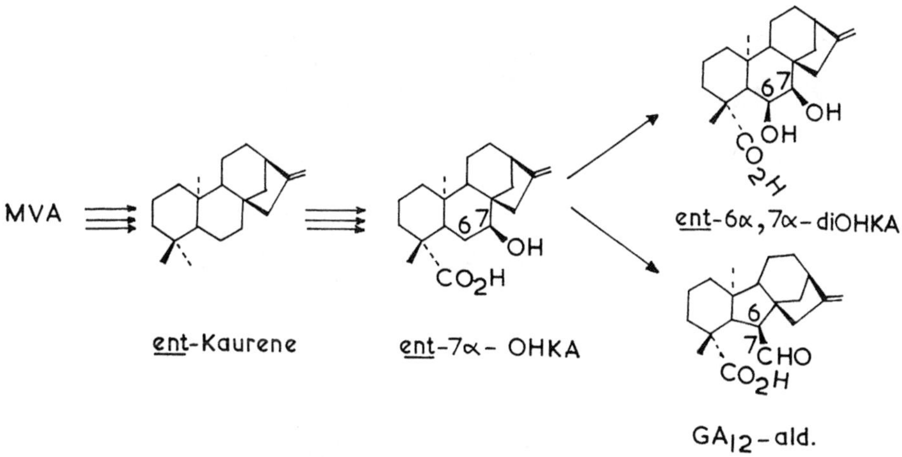

Fig. 2. Gibberellin biosynthesis. Conversion of MVA into GA_{12}-aldehyde and *ent*-6α,7α-dihydroxykaurenoic acid

However, a more convenient method of overcoming this problem is the use of substrates doubly labelled with [^{2}H] and [^{3}H] with a very high [^{2}H]-content. The radioisotope can then be used to follow the process, and the [^{2}H]-metabolite can be distinguished from the endogenous compound by MS. For this purpose methods have recently been developed (2) for the preparation of 1- and 3-[^{1}H], [^{2}H], and [^{3}H]-GAs. Those illustrated in Figure 1 from GA_3 also apply to GA_7 or a commercial GA_4/GA_7 mixture.

Next in importance to unambiguous identification is sensitivity of detection. High sensitivity is necessary if metabolic studies are to be conducted at physiologically meaningful concentrations (see below). High sensitivity of detection, however, requires scrupulously clean techniques in order to avoid accidental contamination, even in trace amounts, from GAs being used in the laboratory as standards. The commonly used GA_3 is particularly troublesome in this context (unpublished observations).

C. Biosynthetic Metabolism

Detailed information of GA biosynthesis in higher plants is so far confined to cell-free enzyme preparations. Most of these studies relate to seeds, although physiological studies (see 34) suggest that there are at least three sites of biosynthesis (root tips, shoot tips, and seeds).

In *Cucurbita maxima* it has been established that developing seeds contain all the enzymes required for the conversion of mevalonic acid (MVA) to GA_4 and GA_{43} (see Figs.2 and 3). Since GA_4 and GA_{43} are known (24) to occur naturally in the seed of *C. maxima*, these conversions presumably occur in vivo.

The steps (Fig.2) from MVA to *ent*-kaurene are catalysed by the high-speed supernatant fraction and require Mn^{2+} (or Mg^{2+}) and ATP. They were elucidated for an enzyme preparation from *Echinocystis (Marah)* nucellus-endosperm (28, 45) and subsequently for enzyme preparations from seed of *Pisum* (1, 21) and *Cucurbita* (22), from shoots of *Ricinus* (43), and from shoot tips of *Pisum* (13).

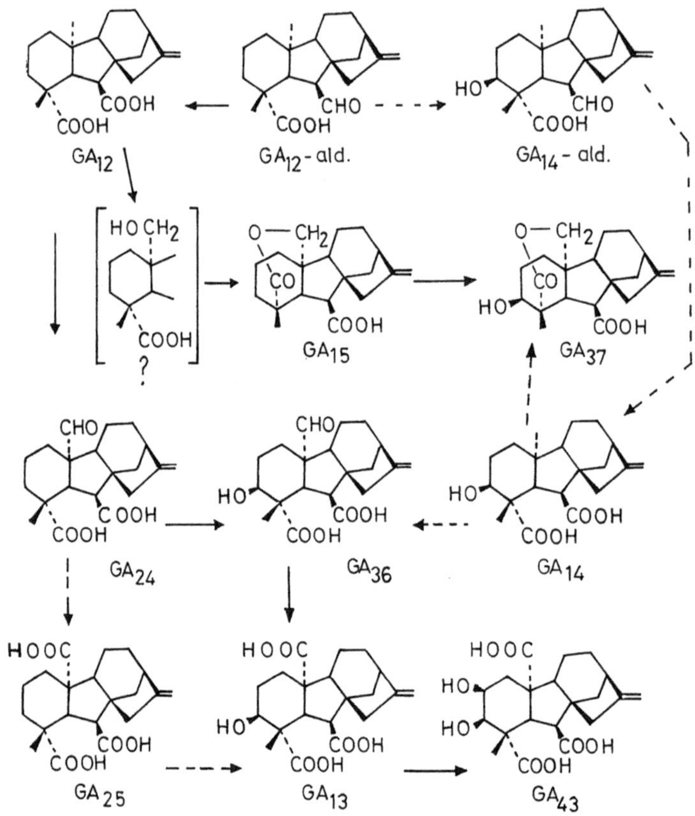

Fig. 3. Conversion of GA_{12}-aldehyde to other C_{20}-GAs, catalysed by high-speed supernatant fraction of *C. maxima* system [after Graebe and Hedden (24)]

The conversions from *ent*-kaurene via the 19-ol, 19-al, 19-oic acid to *ent*-7α-hydroxykaurenoic acid were first established in the *Echinocystis* system (45). They are catalysed by the 200,000 *g* particulate fraction and require oxygen and NADPH. Under the same conditions, the further conversion of *ent*-7α-hydroxykaurenoic acid to GA_{12}-aldehyde was first demonstrated in the *C. maxima* system (23). This ring contraction step is of particular interest since it represents a branch point to hormonal and non-hormonal metabolism of tetracyclic diterpenes and may thus be a point of control of GA biosynthesis. The following recent results provide more details of this step.

First, there is a branch point (Fig.2) from *ent*-7α-hydroxykaurenoic acid to *ent*-6α,7α-dihydroxykaurenoic acid and to GA_{12}-aldehyde. Both products are formed simultaneously at nearly the same rate with the same cofactor requirements and the same pH and temperature optima (24). Furthermore the Lineweaver-Burk and Hill plots for the rates of formation of both products are linear with slopes close to unity, suggesting first order reactions. These facts strongly suggest that both products are formed from the same intermediate whose formation is rate-limiting (24). Secondly, it has been established that both reactions incur the loss of one hydrogen from the 6-position in *ent*-7α-hydroxykaurenoic acid (27). A large (primary) isotope effect was observed (24, 27), consistent with the breaking of a C_6-H bond in the rate-determining step. It is probably the *ent*-6α-hydrogen that is lost, as has been reported for the fungus *Gibberella fujikuroi* (30). However, attempts to confirm the loss of the *ent*-6α-hydrogen directly from *ent*-6α-[^{3}H]-7α-hydroxykaurenoic acid have so far been thwarted by the large observed isotope effect operating on a substrate of low specific radioactivity. Under these circumstances, although the products, GA_{12}-aldehyde and ent-6α, 7α-dihydroxykaurenoic acid, lost 95-98% of the [^{3}H]-label (3), it is not easy to eliminate the possibility that the unlabelled products are formed entirely from the unlabelled molecules of the substrate. However, the results do suggest that it is not the ent-6β-hydrogen that is lost; loss of this hydrogen from the ent-6α-[^{3}H]-substrate would not be subject to a primary isotope effect and the products should be labelled. Thirdly *ent*-7β-[^{3}H]-7α-hydroxykaurenoic acid is converted into GA_{12}-aldehyde and *ent*-6α,7α-dihydroxykaurenoic acid without loss of label, showing that the ent-7α-hydrogen is retained.

The pathways from GA_{12}-aldehyde via GA_{12} to GA_{37} and GA_{43} are catalysed by the 200,000 g supernatant with an absolute requirement for $Fe^{2+/3+}$. [A 2β-hydroxylating enzyme with similar requirements has been obtained from germinating seed of *Phaseolus vulgaris* (39)]. These conversions are inhibited by Mn^{2+}. The steps shown in Figure 3 with unbroken arrows are considered to be the normal pathways observed at low substrate concentrations (0.1 μg/100 μl), while those with broken arrows are observed only at higher concentrations and may therefore be induced (24).

GA_4 was obtained together with GA_{43}, GA_{13}, and GA_{37} from GA_{12}-aldehyde using a 15,000 g supernatant fraction (25). GA_9, which has never been detected in the *C. maxima* system, is quantitatively converted by this system into GA_4, but the intermediates between GA_{12} and GA_4 are not known. Indeed the way in which the carbon atom-20 is lost in the conversion of the C_{20}-GAs into the C_{19}-GAs is the one main gap in our knowledge of GA biosynthesis. Since this step converts relatively inactive GAs into the more biologically active GAs, the conversion is of obvious importance. However, in the *C. maxima* system (26), as in *G. fujikuroi* (9), the two obvious series of intermediates - GA_{15}, GA_{24}, GA_{25} and GA_{37}, GA_{36}, and GA_{13} - do not act as precursors of GA_9 and GA_4, respectively. The possibilities that the hydroxy-acid, shown in brackets in Figure 3 and inferred as an intermediate by Graebe et al. (26), or that the carbon-7 should be at the aldehyde stage of oxidation as suggested by MacMillan (35), have not been tested directly. However, one recent piece of information has been provided by Bearder and MacMillan (6). They prepared GA_{12}-alcohol, previously shown (9, 29) to be an excellent precursor of C_{19}-GAs, with ^{18}O-labels in the carboxyl group (Fig.4) and fed this alcohol to mutant Bl-41a (8) of the fungus, *G. fujikuroi*. The ^{18}O-content of the substrate and products, determined by GC-MS, are shown in Figure 4. These results show that both oxygens in the lactone ring of C_{19}-GAs come from the 19-carboxyl group. Although GA_{13} does not act as a precursor of C_{19}-GAs, the chemical oxidation of GA_{13} with lead tetra-acetate has recently been shown (7) to give GA_4 and its 20,4-lactone (Fig.5), thus providing the first structural correlation between the C_{20}- and C_{19}-GAs. Interestingly, the 20,4-lactone isomer of GA_4 was almost as active as GA_4 in several bioassays (10).

A beginning has been made to correlate GA biosynthesis in higher plants with growth. For example Ecklund and Moore (17) have investigated ent-kaurene synthesis in cell-free preparations from light- and dark-grown seedlings of *P. sativum* cv. Alaska and cv. Progress No. 9. They observed greater activity from light-grown than dark-grown plants, and this does not correlate either with the well-established light inhibi-

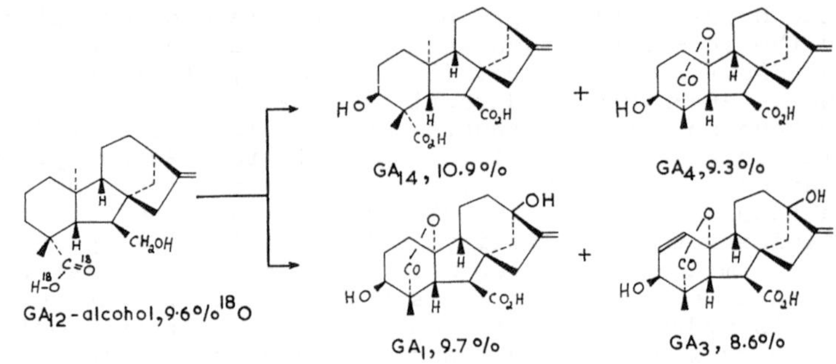

Fig. 4. Conversion of ^{18}O-GA_{12}-alcohol into GA_{14}, GA_4, GA_1, and GA_{13} by cultures of *G. fujikuroi*, mutant B1-41a. Results show that both O-atoms in lactone bridge of C_{19}-GAs are derived from 19-oic acid of C_{20}-GAs

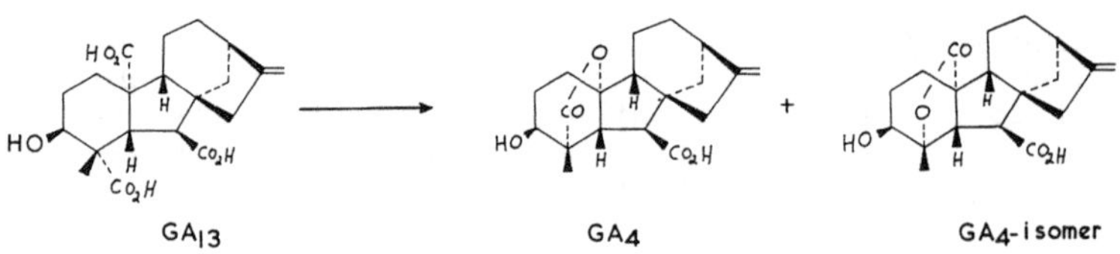

Fig. 5. Lead tetra-acetate oxidation of GA_{13} to GA_4 and the 20,4-lactonic isomer of GA_4, providing first chemical conversion of C_{20}-GA to C_{19}-GA

tion of growth or the greater growth response of light-grown seedlings to applied GAs. However, they did find a lower activity in the dwarf cultivar than in the tall one. In a related study of dwarfism in *Zea mays*, Hedden and Phinney presented data in a demonstration at this Conference, showing that the genetic lesion in the d-5 mutant controls the cyclisation step from copalyl pyrophosphate to ent-kaur-16-ene; as a result cyclisation in the dwarf mutant is diverted to the isomeric ent-kaur-15-ene.

In biosynthetic studies in whole plants, difficulties are sometimes experienced in obtaining incorporation of substrates known from the cell-free studies to be intermediates. A recent example is provided by Faull et al. (19) who found that barley seedlings incorporated $^{14}CO_2$ into a GA_1-like compound but not ^{14}C-MVA or ent-17-^{14}C-kaurenoic acid. Such non-incorporations may be caused by a failure to reach the active site of biosynthesis because a specific carrier protein is required (e.g., 37) or because of compartmentalisation within the plant.

D. Deactivating Processes

The processes by which active endogenous GAs are removed or deactivated are even less well understood than their biosynthesis. For example the effect of light or temperature on the metabolism of GAs, reviewed in 1973 (33), is particularly confusing, and more recent work has not resolved the contradictions. This ignorance is due partly to the technical difficulties, already referred to, in identifying metabolites, but it is also due to the fact that most studies, with a few notable exceptions, of the metabolism of applied GAs are planned with insufficient regard to the GA status in the plant (see below).

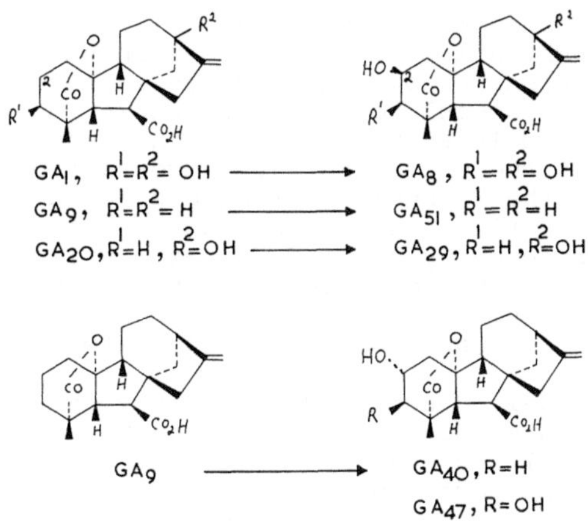

Fig. 6. Examples of 2β-hydroxylation of C_{19}-GAs by higher plants and of 2α-hydroxylation by *G. fujikuroi*

2β-Hydroxylation of applied C_{19}-GAs is a commonly observed process (Fig.6). The best documented example is the conversion of GA_1 to GA_8, which has been observed in many plant tissues (e.g., 41). Other examples include: GA_{20} to GA_{29} in dark-grown seedlings (40), germinating seed (40), and developing seed (20) of *P. sativum*, and in *Bryophyllum daigremontianum* (16); and GA_9 to 2β-hydroxy GA_9 (GA_{51}) in developing seed of *P. sativum*. It is of interest that 2β-hydroxylation does not occur in the fungus *G. fujikuroi*, although 2α-hydroxylation occurs to a minor extent. Thus (Fig.6) GA_9 gives GA_{40} in the mutant B1-41a (4), and GA_{47} occurs in the wild-type strain GF-1a (11, 40). The 2β-hydroxy GAs, shown in Figure 6, possess zero or very low biological activity (10, 42) (in contrast, the 2α-hydroxy GAs are quite active). It is tempting to conclude that 2β-hydroxylation is a deactivating process of regulatory significance in plants.

Caution, however, must be exercised in interpreting results from observed metabolic conversions of radiolabelled GAs applied to plants (see 15, 38). It is necessary to ensure that the observed conversions are normal endogenous processes. Frydman and MacMillan (20) have suggested that the following criteria should be met: (1) the applied GA and its metabolite must be shown to be endogenous; (2) the GA should be applied at levels close to the previously determined endogenous levels; (3) the GA should be applied at the stage of development at wich the endogenous level is maximal; and (4) the period of metabolism should correspond to the time interval between the maximal endogenous levels of the GA and its metabolite. To meet these requirements, time course studies on the GA levels are necessary.

Applying these criteria, Frydman and MacMillan (20) showed that GA_{20} is metabolised to GA_{29} in 50-90% yield in cultured seed of *P. sativum* cv Progress No. 9. The endogenous levels of GA_9, however, were so low that metabolites could not be identified when criterion (2) was complied with. But when higher levels were applied, GA_9 was metabolised to a monohydroxy compound, presumed to be 2,3-dihydro-GA_{31}, and its conjugate together with either 2-epi GA_{40} or GA_{20}, depending upon the conditions under which the plants were grown. More recent data, presented by the same authors in a demonstration at this Conference, showed that 2-epi GA_{40} is endogenous (GA_{51}). Thus 2β-hydroxylation of GA_9 to GA_{51} occurs under conditions that satisfy criteria (1), (3), and (4) and may therefore be a normal endogenous process. The presumed 2,3-dihydro GA_{31} cannot be detected as an endogenous GA and is probably an artefact, possibly as a result of feeding [^3H]-GA_9 at levels in excess of the endogenous levels. GA_{20} is formed from GA_9 only in seed that are at a much earlier stage of development, and the conversion is probably not a normal endogenous process in seed of Progress No. 9. It is interesting to speculate that the enzyme for 2β-hydroxylation in the developing seed is active when the endogenous GA_9 and GA_{20} levels are high but that the enzyme for 13-hydroxylation peaks much earlier. The function of the 13-hydroxylating enzyme may be to 13-hydroxylate an intermediate early in the biosynthetic pathway to GA_{20}, yet when presented with exogenously applied GA_9 at a stage when GA_9 is not normally present, the 13-hydroxylase can convert the "unnatural" GA_9 to GA_{20}. In other words, these hydroxylating enzymes may possess low substrate specificity, as has been demonstrated for the fungal enzymes (e.g., 5).

Another well-documented de-activating process is glucosylation, particularly in seeds where many glucosyl ethers and esters of GAs occur (e.g., 31). In seed the levels of free GA acids decrease during maturation. This fact, together with the initial discovery of large amounts of GA-conjugates in mature seed (44), has led to the belief that conjugation of free GAs is a terminal process in seed maturation and that these biologically inactive conjugates are hydrolysed to free GAs, required during germination. However, there is no convincing evidence for this function for glucosides. Indeed when 2β-hydroxylation is followed by 2β-O-glucosylation, hydrolysis of the conjugate would yield an inactive free GA, and there is no evidence that 2β-hydroxylated GAs are reconverted to active GAs. Detailed investigations of the conjugation of GAs under conditions that satisfy the foregoing criteria are sorely needed.

E. Concluding Remarks

To many plant physiologists, gibberellinology must seem a very complex subject. The problem of defining the physiologic significance of such a large number of GAs (to date 51, of which 43 are known to occur in higher plants) must seem a daunting prospect. Added to this are the technical difficulties of identifying endogenous GAs, their precursors, and metabolites. Without wishing to be discouraging, there is no escaping the fact that conclusive identification of these compounds is essential. Nevertheless metabolic studies are gradually providing information that suggests that over half of the biologically inactive GAs are probably not on the biosynthetic pathway to biologically active GAs. If 2β-hydroxylation is an irreversible deactivating process, the 2β-hydroxy GAs can be eliminated from the list of GAs directly involved in GA-controlled growth. Additional simplification may come from future studies on the significance of hydroxylation in ring C, and of 18-oxygenation, of the C_{19}-GAs. Thus one can begin to see a picture emerging in which a relatively small number of GAs are recognised as primary hormones, directly controlling GA-induced growth.

Thus metabolic studies may ultimately simplify and rationalise a subject which, at first sight, seems to be very complex.

References

1. Anderson, J.D., Moore, T.C.: Plant Physiol. 42, 1527 (1967)
2. Beale, M., Bearder, J.R., MacMillan, J.: unpublished results
3. Beale, M., Bearder, J.R., Graebe, J.E., MacMillan, J.: unpublished results
4. Bearder, J.R., Frydman, V.M., Gaskin, P., Hatton, I.K., Harvey, W.E., MacMillan, J., Phinney, B.O.: J. Chem. Soc. Perkin I, 178 (1976)

5. Bearder, J.R., Frydman, V.M., Gaskin, P., MacMillan, J., Wels, C.M., Phinney, B.O.: J. Chem. Soc. Perkin I, 173 (1975)
6. Bearder, J.R., MacMillan, J.: unpublished data
7. Bearder, J.R., MacMillan, J.: J. Chem. Soc., Chem. Comm. 1976, p. 421
8. Bearder, J.R., MacMillan, J., Chaffey, M.B., Phinney, B.O.: Phytochemistry 13, 911 (1974)
9. Bearder, J.R., MacMillan, J., Phinney, B.O.: J. Chem. Soc. Perkin I, 721 (1975)
10. Beeley, L.J., Hoad, G.V., Sponsel, V.M.: unpublished data
11. Beeley, L.J., MacMillan, J.: J. Chem. Soc. Perkin I, 1002 (1976)
12. Bowen, D.H., MacMillan, J., Graebe, J.E.: Phytochemistry 11, 2253 (1972)
13. Coolbaugh, R.C., Moore, T.C., Barlow, S.A., Ecklund, P.R.: Phytochemistry 12, 1613 (1973)
14. Crozier, A.: This Volume, p. -.
15. Davies, L.J., Rappaport, L.: Plant Physiol. 56, 60 (1975)
16. Durley, R.C., Pharis, R.P., Zeevaart, J.A.D.: Planta 126, 139 (1975)
17. Ecklund, P.R., Moore, T.C.: Plant Physiol. 53, 5 (1974)
18. Evans, A., Smith, H.: Proc. Natl. Acad. Sci. U.S. 73, 138 (1976)
19. Faull, K.F., Coombe, B.G., Paleg, L.G.: Australian J. Plant Physiol. 1, 199 (1974)
20. Frydman, V.M., MacMillan, J.: Planta 125, 181 (1975)
21. Graebe, J.E.: Phytochemistry 7, 2003 (1967)
22. Graebe, J.E.: In: Plant Growth Substances. Carr, D.J. (ed.). Berlin: Springer, 1970, p. 151
23. Graebe, J.E., Bowen, D.H., MacMillan, J.: Planta 102, 261 (1972)
24. Graebe, J.E., Hedden, P.: In: Biochemistry and Chemistry of Plant Growth Regulators, Schrieber, K., Schütte, H.R., Sembdner, G. (eds.). GDR: Halle (Saale) 1974
25. Graebe, J.E., Hedden, P., Gaskin, P., MacMillan, J.: Planta 120, 307 (1974)
26. Graebe, J.E., Hedden, P., MacMillan, J.: In: Plant Growth Substances 1973, Tokyo: Hirokawa Publ. Co. 1974
27. Graebe, J.E., Hedden, P., MacMillan, J.: J. Chem. Comm., 1975, p. 161
28. Graebe, J.E., Dennis, D.T., Upper, C.D., West, C.A.: J. Biol. Chem. 240, 1847 (1965)
29. Hanson, J.R., Hawker, J.: Phytochemistry 12, 1073 (1933)
30. Hanson, J.R., Hawker, J., White, A.F.: J. Chem. Soc. Perkin I, 1892 (1972)
31. Hiraga, K., Yokota, T., Murofushi, M., Takahashi, N.: In: Plant Growth Substances 1973, Tokyo: Hirokawa Publ. Co., 1974
32. Jones, R.L.: Planta 81, 1612 (1968)
33. Jones, R.L.: Ann. Rev. Plant Physiol. 24, 571 (1973)
34. Lang, A.: Ann. Rev. Plant Physiol. 21, 537 (1970)
35. MacMillan, J.: In: Aspects of Terpenoid Chemistry and Biochemistry. Goodwin, T.W. (ed.). London: Academic Press 1971
36. MacMillan, J., Wels, C.M.: Phytochemistry 13, 1413 (1974)
37. Moore, T.C., Barlow, S.A., Coolbaugh, R.C.: Phytochemistry 11, 3225 (1972)
38. Patterson, R.J., Rappaport, L.: Planta 119, 183 (1974)
39. Patterson, R.J., Pappaport, L., Breidenbach, R.W.: Phytochemistry 14, 363 (1975)
40. Railton, I.D., Murofushi, N., Durley, R.C., Pharis, R.P.: Phytochemistry 13, 793 (1974)
41. Rappaport, L., Davies, L., Lavee, S., Nadeau, R., Patterson, R., Stolp, C.F.: In: Plant Growth Substances 1973, Tokyo: Hirokawa Publ. Co. 1974
42. Reeve, D.R., Crozier, A.: In: Gibberellins and Plant Growth, Kishnamoorthy, H.N. (ed.). New Delhi: Wiley Eastern Ltd. 1975

43. Robinson, D.R., West, C.A.: Biochemistry 9, 80 (1970)
44. Schrieber, K., Weiland, J., Sembner, G.: Phytochemistry 9, 189 (1970)
45. West, C.A.: In: Biosynthesis and its Control in Plants, Milborrow, B.V. (ed.), London: Academic Press, 1973, p. 143
46. Yamaguchi, I., Yokota, T., Murofushi, N., Ogawa, Y., Takahashi, N.: Agr. Biol. Chem. (Tokyo) 34, 1439 (1970)

Paper Demonstration References (see Section IX):

8 - 28 - 34 - 60 - 110 - 148 - 153 - 160 - 173 - 180

15. Cytokinin-Binding Proteins in Higher Plants

J. E. Fox and J. Erion

A. Introduction

In 1970 Matthysee and Abrams (13) described a cytokinin receptor protein from pea bud chromatin that, in the presence of an active cytokinin, increased RNA synthesis in an in vitro system (pea bud chromatin or DNA plus *E. coli* RNA polymerase). This stimulation of RNA synthesis could not be repeated by Berridge et al. (1, 2) who, however, described reversible, nonenzymatic binding of cytokinins to 80 S ribosomes from *Brassica pekinensis* and concluded that several binding sites, which could not be washed off with 0.5 M KCl, exist on the ribosome. These initial reports were not followed up by further studies until our own recent preliminary report (6) of a cytokinin-binding protein from higher plant ribosomes. Although we were able to confirm that cytokinins do indeed bind to higher plant ribosomes, the binding is much more complex than previously described. In addition to nuclear and ribosomal components, cytokinins have also been reported to interact with membrane-associated proteins. An allosteric effect of cytokinins on the binding of calcium to a glycoprótein on the membrane surface of fungal cells has been claimed (9, 10), and LeJohn recently reported (11) that this glycoprotein binds not only calcium, but also cytokinins, auxins, and the auxin analog tryptophane, and that the uptake of both auxins and tryptophane is regulated quite specifically by cytokinins. In addition there have been several preliminary reports (8, 19, 20) of cytokinin binding to particulate fractions from tobacco and moss protonemata cells, and at least two types of sturable sites have been described. However, since the experimental details of these reports have not yet appeared in the literature, they cannot be evaluated. Finally, the isolation of a cytokinin-binding protein with an apparent molecular weight of 4000 daltons has been recently described from tobacco leaves (21). In the present study we describe the isolation, purification, and characteristics of at least three cytokinin-binding proteins from wheat germ.

B. Materials and Methods

Raw wheat germ was purchased from Niblack Foods (Rochester, N.Y.). Ribosomes were prepared by previously described procedures (6). [Methylene-^{14}C]-bzl^{6}Ade (N^6-benzylaminopurine) (5.6 mCi/mmol) was synthesized as described previously (4). [8-^{14}C]-bzl^{6}Ade (12.8 mCi/mmol) was purchased from Amersham-Searle. [p-^{3}H-benzyl]-bzl^{6}Ade (9 Ci/mmol) (18) was a gift from M. Sussman and H. Kende, (Michigan State University).

Protein concentrations were determined by the method of Lowry et al. (12) as modified by Shatkin (17). Cytokinin-binding activity was measured by equilibrium dialysis as described previously (6). In routine assays of column eluates, a 0.5 ml aliquot of a column fraction was dialyzed for 12 to 18 h against buffer containing radioactive bzl^{6}Ade. Aliquots (0.4 ml) were removed from inside and outside the dialysis bag, and radioactivity was measured in Hydromix (Yorktown Research). In standard procedures, all fractions assayed from a chromatography run

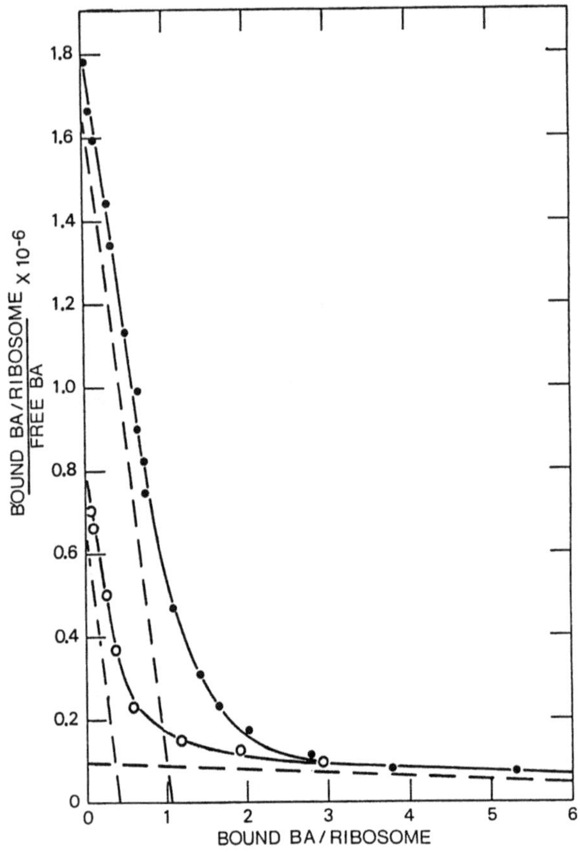

Fig. 1 A,B. Scatchard plot of concentration-dependent binding of bzl[6]Ade to native ribosomes (A) and 0.5 M KCl-washed ribosomes (B). Experimental points were obtained and calculations made as described previously (6)

(30-50 dialysis bags) were dialyzed together in a single stirred container so that the final concentration of free ligand was identical for all fractions.

C. Results and Discussion

I. Cytokinin Binding to Native and Salt-Washed Ribosomes

We reported previously that cytokinin binding to ribosomes isolated from higher plants is a complex phenomenon involving multiple binding sites having at least two, quite different, binding characteristics. We have now carefully repeated that work in a series of equilibrium dialysis studies, the results of which, plotted by the method of Scatchard (16), yield a biphasic curve (Fig.1A) generally interpreted (e.g., 3, 14) to indicate a limited number of binding sites of high affinity together with a larger number of lower affinity sites. The high affinity site on wheat germ ribosomes has a Kd 5×10^{-7} M for the cytokinin bzl[6]Ade and would thus bind significant levels of this cytokinin at levels 1×10^{-8} M to 1×10^{-6} M, which are physiologically active in tissue cultures (e.g., 5). The Scatchard plot (Fig.1A) further indicates that wheat germ ribosomes isolated as described previously (6) have one high affinity site per ribosome. Similar studies with ribosomes from other higher plant sources indicate a considerable

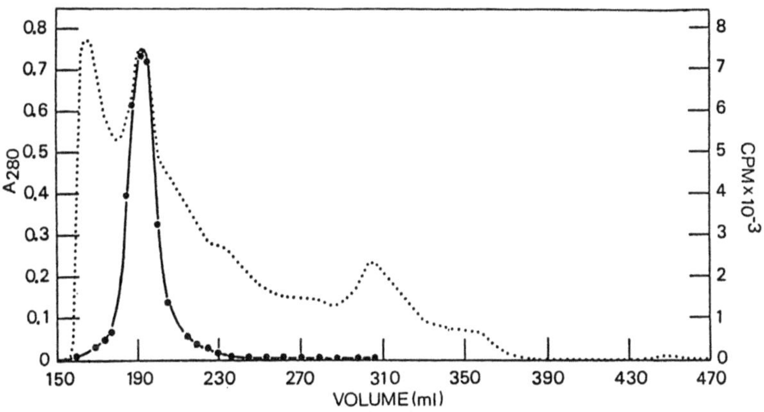

Fig. 2. Sephadex G-100 chromotography of partially purified preparation (see text) of CBF-1 obtained from ribosomes. 100 mg of material dissolved in 5 ml of buffer (25 mM tris acetate pH 7.6, 25 mM potassium acetate, 5 mM magnesium acetate, 5 mM 2-mercaptoethanol) were added to top of column 2.5 × 90 cm and eluted with same buffer at rate of 1 ml/min. Fractions of 1 ml were collected and assayed for cytokinin-binding activity as described in Materials and Methods. For the estimation of molecular weight, column was calibrated with known protein standards

variation in the ratio of high-affinity cytokinin binding sites per ribosome with one as the upper limit so far detected. The lower affinity (Kd 1×10^{-4} M) sites by contrast require high concentrations of cytokinins for a significant amount of binding to occur with 10 or more available sites per ribosome, and it would appear as though this is the binding site described previously in the literature (1, 2). That the two types of binding are indeed distinct, is shown in Figure 1B. Here ribosomes were washed with 0.5 M KCl before the assay for concentration-dependent binding, and it may be seen that the high-affinity site is preferentially removed by this treatment. The inability of previous workers to detect the loss of cytokinin binding sites in salt-washed ribosomes (1, 2) is explained by their use of relatively high cytokinin levels in the binding assays, in which case most of the binding is to the low-affinity site unaffected by the salt treatment.

II. Purification of the Cytokinin-Binding Moieties

We demonstrated previously the proteinaceous nature of the cytokinin binding factor (CBF-1) isolated from wheat germ ribosomes. Although its purification will be described in detail elsewhere, a general statement of the procedure is given here. From 250 g wheat germ, 66,000 A_{260} units of ribosomes were isolated as described previously (6) and suspended at a concentration of 500 A_{260} units/ml in a high salt buffer (25 mM Tris-acetate pH 7.6, 0.75 M potassium acetate, 5 mM magnesium acetate, 5 mM 2-mercaptoethanol). The salt-washed ribosomes were removed by centrifugation, and the supernatant after dialysis against the same buffer with 50 mM potassium acetate was precipitated with $(NH_4)_2SO_4$ and the 20-80% saturated fraction, which proved to have all the binding capacity, was dialyzed against buffer for further use. The dialysate was then heated at 60°C for 10 min and centrifuged to remove coagulated material. This treatment removed 60-75% of the protein without reducing binding capacity of the supernatant. The protein was then

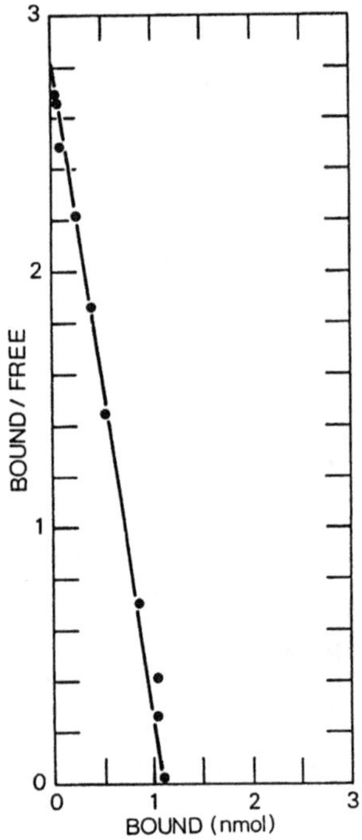

Fig. 3. Scatchard plot of concentration-dependent binding of bzl^{6}Ade to purified preparation of CBF-1. Aliquots of CBF-1 (see text for details of purification) were subjected to equilibrium dialysis as described previously (6), and calculations were made on the assumption of a binding moiety, having a molecular weight of 93,000 daltons

concentrated by a second $(NH_4)_2SO_4$ precipitation, and the subsequent dialysate in a small volume was subjected to DEAE-cellulose chromatography. At pH 7.0 all of the cytokinin-binding material may be eluted with a buffer containing 0.1 M KCl. This fraction is dialyzed against a pH 7.6 buffer containing 0.1 mM EDTA, but no KCl is applied to the top of a second DEAE-cellulose column; the column is washed with the same buffer containing 30 mM KCl, and the binding protein is then eluted as a sharp peak with this buffer containing 150 mM KCl. The binding protein is then further purified on a phosphocellulose column by gradient elution (0.05 M KCl to 0.75 M KCl) in a Tris-acetate pH 7.6 buffer, and the active fractions that elute at 0.25 M KCl at this pH are concentrated as before. Sephadex G-100 chromatography of the partially purified material (Fig.2) indicates that the cytokinin-binding fraction has a molecular weight between 90,000 and 100,000 daltons. Using the procedure described by Maizel (15) for electrophoresis of native, basic proteins, we determined that our most highly purified fraction migrated on 7.5% acrylamide gels as a single band. However, SDS gel electrophoresis on 10% acrylamide by the method of Weber and Osborne (22) resolves two subunits with apparent molecular weights of 56,000 and 37,000 daltons. The intensity of staining as determined by absorbance at 660 nm, gave a ratio of 6:4 for the peak absorbances of the large subunit to the small. This is in good agreement for the expected ratios of a 93,000 molecular weight protein with two subunits of the foregoing size. The molecular weight calculated here, for the protomer, is also in good agreement with that determined by G-200 Sephadex chromatography. Highly purified preparations of CBF-1 have been subjected to a concentration-dependent cytokinin binding study by equilibrium dialysis as described previously for the intact ribosomes (6). On the basis of the calculated molecular weight a Scatchard plot was constructed from the results of this binding study (Fig.3), and this matches very closely the characteristics of the high affinity site inferred from the Scatchard plot of cytokinin binding to native wheat germ ribosomes (Fig.1A).

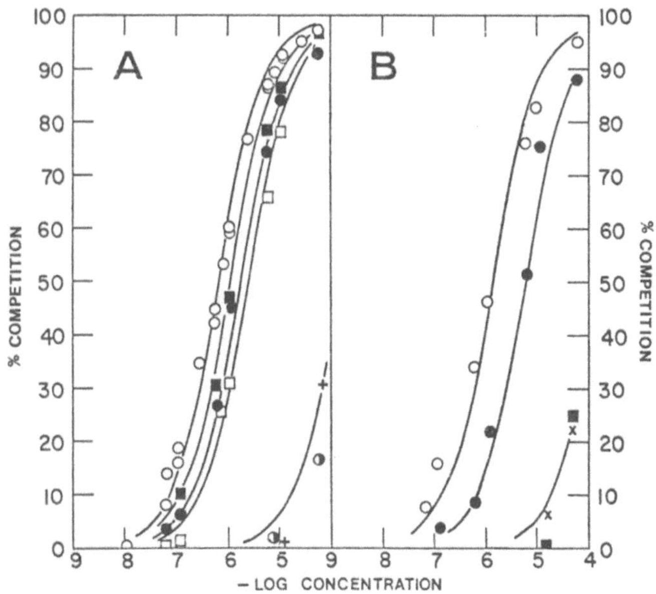

Fig. 4 A,B. Competition binding by cytokinins and related substances against bzl^{6}Ade (see text for details of experimental procedure). In this study, concentration of radioactive ligand was 2 × 10^{-9} M ^{3}H-bzl^{6}Ade. Unlabelled competing ligands were: (A), (o), bzl^{6}Ade; (■): 6-n-butylaminopurine; (□): 6-furfurylaminopurine; (●): 6-(3-methyl-2-butenylamino) purine; (+): 6-methylaminopurine; (◑): adenine. (B), (o): 6-benzylaminopurine riboside; (●): 6-(methyl-2-butenylamino)-purine riboside; (x): 6-methylaminopurine riboside; (■): adenosinc

III. Specificity of the Cytokinin-Binding Protein

A large number of cytokinins, cytokinin nucleotides, analogues and related purines, and plant growth substances were tested for their ability to compete with 6-benzylaminopurine for the binding site. Native ribosomes or partially purified preparations of CBF-1 were dialyzed against a singlc, low concentration of the labelled cytokinin together with increasing levels of unlabelled antagonists. A cytokinin concentration well below the level needed to saturate the high affinity site was chosen in order to ensure that most of the competition observed would be for this site rather than for the multiple low-affinity sites. Thus in most experiments the concentration of the labelled component was either 1.75 × 10^{-7} M or 2 × 10^{-9} M (in the case of ^{3}H-BA), at which levels greater than 95% of the binding is to the high-affinity site. The results (some of which are shown in Fig.4) indicate that binding, as judged by the ability of a substance to compete with bzl^6 Ade and thus reduce bound radioactivity, is in general directly related to cytokinin activity. The presence of a 9-ribosyl substituent, however, considerably reduces the apparent binding. Thus i^{6}Ade [N^6-(3-methyl-2-butenylamino) purine] binds about ten times more effectively than i^{6}A [9-ribofuranosyl-N^6-(3-methyl-2-butenylamino) purine], and adenosine hardly binds at all, while adenine or 6-methylaminopurine exhibits slight binding but only at very high concentrations (about 500 times less effective than i^{6}Ade in this regard). Substituents at the 7 position either abolish binding completely (BA-7-glucopyranoside) (7) or reduce it greatly (BA-7-xylopyranoside) (7). Inactive analogues,

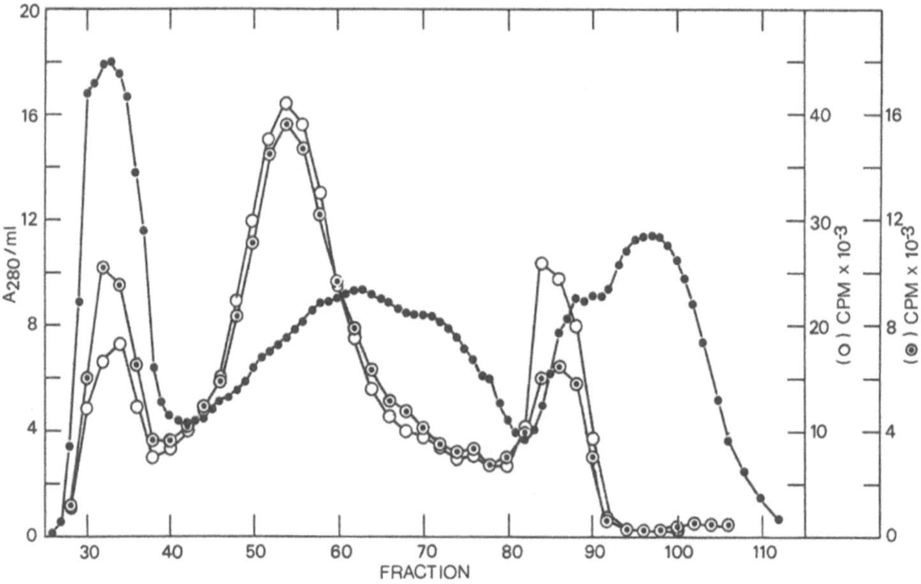

Fig. 5. Sephadex G-200 chromatography of S100 supernatant, cytokinin-binding proteins. S100 protein, concentrated with ammonium sulfate and solubilized in buffer (see legend to Fig.2) was chromatographed on Sephadex G-200 column (2.5 × 90 cm). Aliquots (0.5 ml) were assayed for binding by equilibrium dialysis using: in one assay series, (o) ^{3}H bzl^{6}Ade (sp. act. 9 Ci/mmol), at free concentration of 2×10^{-9} M, and in second assay series, (⊙) ^{14}C bzl^{6}Ade (sp. act. 5.6 mCi/mmol), at free concentration of 1.2×10^{-6} M. *CPM*: amount of bound radioactivity per 0.4 ml of each fraction assayed (4.9 ml/fraction)

such as bzl^{3}Ade, other growth promoters, such as gibberellic acid, abscisic acid, or indole-3-acetic acid, and preparations of transfer RNA either show no detectable interaction with the cytokinin-binding site or binding only at such high concentrations as to cast doubt on its physiologic significance.

IV. Cytokinin-Binding Factors Not Associated with the Ribosome

The high-speed post ribosomal supernatant can also be shown by equilibrium dialysis to contain substances that bind cytokinins (6). Although at no time during the purification of CBF-1 from the ribosomal 0.5 M KCl wash have we ever obtained evidence for more than a single cytokinin-binding substance, a 20-80% $(NH_4)_2SO_4$ precipitate of the ribosome-free supernatant can be resolved into at least three fractions of cytokinin binding activity by chromatography on Sephadex G-200 (Fig.5). One of these can be purified concomitantly with CBF-1 and has apparently identical properties and binding characteristics with those of CBF-1. If this cytokinin-binding factor is indeed identical with CBF-1, our data suggest that 2-3 molecules of CBF-1 exist free in the cytosol for each molecule associated with the ribosome. A second binding fraction, designated CBF-2, binds cytokinins at least as well as CBF-1, but has a much lower molecular weight in the region of 30,000. Whether or not it is related to CBF-1 has not yet been determined, but CBF-2 may be similar to the relatively low molecular weight, cytokinin-binding substance described from tobacco leaves (21). A third fraction of binding

Table 1. Cytokinin binding to higher plant ribosomes

Ribosome preparation	Final ligandconc. at equilibrium $\times M^{-1}$	Moles cytokinin bound per mol ribosomes
5-Day corn root ribosome (176 A260/ml)	1.75×10^6 4.55×10^7	1.6 0.96
5-Day dark-grown corn shoot ribosomes (352 A260/ml)	1.75×10^6 4.55×10^7	0.18 0.15
Wheat germ ribosome (206 A260/ml)	1.75×10^6 4.55×10^7	1.4 0.86

Ribosomes were extracted as described previously (6). 0.5 ml of each preparation was dialyzed at 4°C with constant shaking for 40 h against two concentrations noted of (methylene-^{14}C) bzl^{6}Ade sp. act. 5.6 mCi/mmol. 200 ml aliquots were assayed for radioactivity by liquid scintillation spectrmetry in standard fluid containing 25% triton X-100. Calculations were made as described previously (6)

activity (CBF-3) has a molecular weight greater than 250,000 but considerably less affinity for cytokinins than CBF-1, similar to the low-affinity ribosomal binding sites.

V. Cytokinin Binding to Ribosomes from Seedlings

A series of studies has been carried out with ribosomes collected from various organs of light- and dark-grown seedlings in several plants. Although the results of these studies will be reported in detail elsewhere, Table 1 contains a summary of one such study with dark-grown corn seedlings. In general our data indicate that root ribosomes show high cytokinin-binding capacity, at least as good as wheat germ ribosomes, while shoot ribosomes are usually considerably less effective in binding cytokinins at low levels. This result may indicate a shift in the equilibrium between CBF-1 in the cytosol and on the ribosome during seedling development, which could be related to the physiologic role of the cytokinins. It should be emphasized, however, that we have no evidence for any biological role of the particulate and soluble cytokinin-binding substances described here.

Acknowledgement

This work was supported by National Science Foundation grant BMS 74-07386.

D. Abstract

Wheat germ ribosomes exhibit both high and low affinity cytokinin binding sites that can be readily discriminated. The high affinity site is removed in 0.5 M KCl, and the active component has been purified to near homogeneity yielding a protein (CBF-1) with a molecular weight of about 93,000 daltons. Wheat germ ribosomes contain one copy of CBF-1 per ribosome, which binds active cytokinins with a Kd in the region 5×10^{-7} M - 1×10^{-6} M. Competition studies indicate a high degree of specificity for cytokinin-active moieties.

CBF-1 also exists free in the cytosol together with at least two other cytokinin-binding components, CBF-2, a 30,000 dalton moiety with high affinity for cytonins and a low affinity substance with a molecular weight in excess of 250,000 daltons. In studies with ribosomes from intact seedlings it was shown that root ribosomes contain more of the high affinity cytokinin-binding moiety than dark-grown shoot ribosomes. The biological role of cytokinin-binding proteins is as yet unknown.

References

1. Berridge, M., Ralph, R., Letham, D.: Biochem. J. 119, 75 (1970)
2. Berridge, M., Ralph, R., Letham, D.: Plant Growth Substances. Carr, D.S. (ed.). Berlin: Springer-Verlag, 1972, p. 248
3. Davies, I.J.: In: Molecular Techniques and Approaches in Developmental Biology. Chrispeels, M. (ed.). New York: Wiley Interscience, 1973
4. Dyson, W., Fox, J.E., McChesney, J.: Short term metabolism of urea and purine cytokinins. Plant Physiol. 49, 506 (1972)
5. Fox, J.E.: 1964 IAA-Kinetin antagonism in certain tissue culture systems. Plant Cell Physiol. (Tokyo) 5, 251 (1964)
6. Fox, J.E., Erion, J.: Biochem. Biophys. Res. Commun. 64, 694 (1975)
7. Fox, J.E., McChesney, J.D.: Proc. VIII Int. Conf. Plant Growth Substances, Tokyo, August 1973. (1975)
8. Gardiner, G., Sussman, M.R., Kende, H.: 1975 Plant Physiol. 56, (Suppl), 28 (1975)
9. LeJohn, H.B., Cameron, L.E.: Biochem. Biophys. Res. Commun. 54, 1053 (1973)
10. LeJohn, H.B., Stevenson, R.M.: Biochem. Biophys. Res. Commun. 54, 1061 (1973)
11. LeJohn, H.B.: Can. J. Biochem. 53, 768 (1975)
12. Lowry, O.H., Rosebrough, N.J., Farr, A.L., Randall, R.J.: 1951 J. Biol. Chem. 193, 265 (1951)
13. Matthysee, A.G., Abrams, M.: Biochim. Biophys. Acta 199, 511 (1970)
14. McDaniel, C., Kirtley, M., Tanner, M.: J. Biol. Chem. 249, 6478 (1974)
15. Maziel, J.V., Jr.: In: Methods in Virology. Maramorosch, K., Koprowski, H. (eds.). New York: Academic Press 1971
16. Scatchard, G.: Ann. N. Y. Acad. Sci. 51, 660 (1949)
17. Shatkin, A.J.: 1969 In: Fundamental Techniques in Virology. Habel, K., Salzman, N.P. (eds.). New York: Academic Press, 1969, p. 231
18. Sussman, M., Firn, R.: 1976 Phytochemistry 15, 153 (1976)
19. Sussman, M., Firn, R.D., Kende, H.: Plant Physiol. 51 (Suppl), 4 (1973)
20. Sussman, M., Kende, H.: Plant Physiol. 56 (Suppl), 28 (1975)
21. Takegami, T., Yoshida, K.: 1975 Biochem. Biophys. Res. Commun. 67, 782 (1975)
22. Weber, K., Osborn, M.J.: J. Biol. Chem. 244, 4406 (1969)

Paper Demonstration References (see Section IX):

37 - 48 - 55 - 77

16. Cytokinin Relations in the Whole Plant

P. F. WAREING, R. HORGAN, I. E. HENSON, and W. DAVIS

Abstract

Cytokinin levels in *Xanthium strumarium* decline rapidly throughout the plant after only one short-day cycle. This effect appears to be due to reduced cytokinin production by the roots, in response to a signal produced in the leaves during long dark periods.

Disbudding of seedlings of *Phaseolus vulgaris* leads to the accumulation of high levels of dihydrozeatin-O-β-D glucoside in the leaves, where it appears to be formed from an intermediate supplied from the roots. In short-term experiments 8-^{14}C zeatin is rapidly converted in the leaves to zeatin-O-β-D glucoside, which is later degraded to adenine and related compounds. It is suggested that dihydrozeatin-O-β-D glucoside accumulates in the leaves of disbudded plants because it is resistant to "cytokinin-oxidase" type enzymes. In roots and buds of *P. vulgaris*, zeatin is converted to zeatin ribotide rather than to glucosides. Cytokinin interrelations in the whole plant are discussed in relation to biosynthesis, transport, and metabolism.

A. Introduction

The possibility that cytokinins may function as regulatory hormones in the plant as a whole has been recognized for some years, especially in relation to such processes as leaf senescence (16), apical dominance (26), and stolon development in potato (11). Studies on senescence in detached leaves led to the idea that in some species leaves may depend upon the supply of cytokinins produced in the roots, suggesting that the capacity for cytokinin biosynthesis is localized in the plant. There is, indeed, evidence that cytokinins may be produced in the root tips and released into the xylem sap, whence they are transported in the transpiration stream to the shoot (10). These observations have led to the widely held hypothesis that cytokinins produced in the roots are transported to the shoot and that mature leaves, at least, depend upon such cytokinins for their normal functioning.

Young developing seeds contain high levels of cytokinins, but it is not known whether they are produced in the seeds themselves or derived from other parts of the plant. It is also uncertain whether shoot buds are sites of cytokinin synthesis in the intact plant (p. x).

Apart from the question as to whether cytokinins have a role as regulators within the plant itself, other evidence has been accumulating which suggests that endogenous cytokinins may be profoundly affected by various environmental factors, such as photoperiod (20), short light exposures involving phytochrome (7, 19), temperature (1), water stress (8), and mineral nutrition (17), and that plant responses to such factors may be partly mediated by quantitative and qualitative variations in the endogenous cytokinins.

Thus there is extensive evidence that endogenous cytokinins probably play an important role as hormones in various aspects of growth and development. In the present paper we shall present the results of some recent studies that have a bearing on the possible sites of cytokinin production in the plant and on the interrelations of root and shoot cytokinins.

B. Effects of Photoperiod on Cytokinin Relations in *Xanthium strumarium*

Recent work on photoperiodic control of cytokinin levels in *Xanthium strumarium* (5, 20) casts some light on the interrelations between root and shoot cytokinins and on their regulation in the whole plant. Transfer of plants of *X. strumarium* raised under long days (LD) to short days (SD) results in a marked decline in the cytokinin levels in the leaves, buds, and root exudates after only one SD cycle (5). This appears to be a true photoperiodic effect, since it is at least partially nullified by interrupting the long dark period by a 'night-break', although the involvement of phytochrome has not yet been established by demonstration of R/FR reversibility.

Further work has been carried out to determine the interrelations of shoot and root cytokinins during this decline in their levels (6). It is well established that the induction of flowering in *Xanthium* is controlled by the daylength conditions to which the leaves are exposed. It might be expected, therefore, that the observed cytokinin changes would be initiated in the leaves. However, two pieces of experimental evidence are inconsistent with this hypothesis, i.e.: (1) When only three leaves on each plant were exposed to SD while the remainder of the plant was maintained under LD, the cytokinin levels in the SD leaves were not significantly lower than those in the LD leaves, although the levels in the root exudates of the treated plants were intermediate between those in plants wholly exposed to LD or to SD. (2) When detached leaves or whole shoot systems cut off from the roots were exposed to LD or SD, there was a rapid decline with both treatments but no significant difference in cytokinin levels.

From the foregoing evidence it would seem that the effect of SD on cytokinin activity in *Xanthium* leaves occurs only in intact plants and that the decline is not a localised and direct effect of photoperiod on the leaf itself. The rapid decline in cytokinin activity in detached leaves is consistent with the hypothesis that in the intact plant the leaves are dependent on imported cytokinins, which are rapidly metabolised so that the levels at any given time reflect the turnover rate. Rapid metabolism of the unnatural cytokinin, benzyladenine, has been demonstrated in *Xanthium* leaves (12), and preliminary experiments in our laboratory indicate that ^{14}C-zeatin is similarly rapidly degraded in this material. It seemed possible, therefore, that the lower levels observed in SD leaves might reflect a greater rate of degradation or inactivation of cytokinins imported from the roots. To test this hypothesis, experiments were carried out in which unlabelled zeatin was supplied to detached leaves previously maintained under LD or SD, after which the leaves were extracted and cytokinin activity was determined by bioassay. It was found again that the cytokinin levels declined markedly over a 25-h period but that there were no significant differences between LD and SD leaves in this respect, thus indicating that the decline in the levels in intact plants under SD cannot be due to a higher rate of cytokinin inactivation under SD than LD. It would seem, therefore, that the decline in cytokinin levels in SD leaves results from the reduced supply of cytokinins from the roots in association with a constant rate of inactivation. Indeed, it would seem that the decline in cytokinin levels throughout the plant under SD reflects reduced production by the roots.

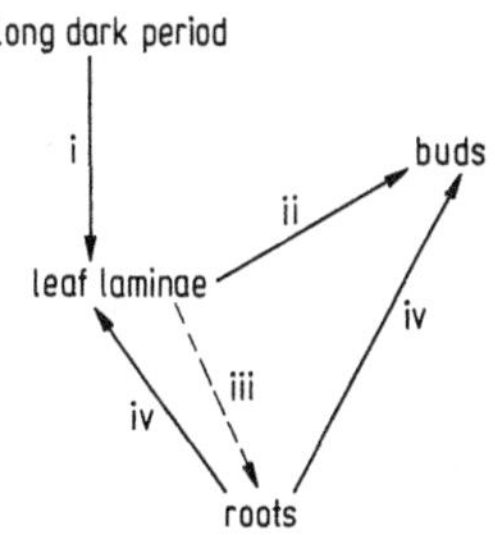

Fig. 1. Scheme outlining possible sequence of events in photoperiodic control of cytokinin levels in *Xanthium*. *i*: environmental stimulus perceived by the leaf; *ii*: transmissible stimulus generated in leaf affecting flower evocation at apex; *iii*: transmissible stimulus generated in leaf affecting cytokinin synthesis and/or release in roots; *iv*: cytokinin supply from root to shoot via transpiration stream

On the other hand, the primary site of detection of changes in daylength must be the leaves, and hence it would seem necessary to postulate the origin of a signal in the leaves which is transmitted to the roots and causes reduced cytokinin production or release there. The nature of this signal remains unknown, but evidently it is transmitted in the bark, since the removal of a bark ring from the stem, between the leaves and the root system, prevented the decline in cytokinin levels in the leaves and root exudates when the leaves were exposed to SD.

Experiments were carried out to determine more precisely the time relationships of the decline in cytokinins resulting from exposure to a single, long dark period. It was found that if the plants were harvested and extracted at the end of single dark periods of various durations, some decline in leaf extracts and root exudates was detected at the end of a 12-h dark period; this decline was marked at the end of a 16-h dark period, compared with the levels present at the end of the normal 6-h dark period associated with the experimental long-day conditions.

Thus, if we are correct in assuming that the decline in cytokinin levels throughout the plant is brought about primarily by reduced cytokinin production in the roots and that the decline in the leaves is the result of a reduced cytokinin supply from the roots, then the 'signal' transmitted from the leaves when the critical dark period of 8 1/2 h is exceeded must also be transmitted to the roots very rapidly. The postulated interrelations between the cytokinins of shoot and root in *Xanthium* in response to long dark periods are summarized in Figure 1.

Although the endogenous cytokinins of plants of *X. strumarium* have not yet been positively identified, the physical properties of two major active fractions present throughout the plant strongly suggest that these are zeatin and zeatin riboside, respectively. A further active fraction is present in mature leaves, which appears to be a glucoside of zeatin, since treatment with β-glucosidase released an active fraction with the properties of zeatin. However, this fraction did not appear to be the 7- or 9-glucoside of zeatin. The main cytokinin present in the root exudates appears to be zeatin riboside. Thus, the unidentified glucoside of the leaf may be derived from zeatin riboside imported into the leaf via the transpiration stream. On the other hand, the main cytokinins in the buds appear to be zeatin and zeatin riboside, and the putative glucoside is not present. It would appear that, as the leaf matures, it acquires the capacity to convert the imported zeatin riboside into a more polar glucoside.

C. Cytokinin Metabolism in *Phaseolus vulgaris*

Any assessment of the role of cytokinins in root/shoot interrelations must take into account the turnover of these compounds in the plant as a whole and in the various plant organs. Unfortunately it is not yet possible to give a detailed account of the two components of cytokinin turnover, i.e., biosynthesis and metabolism, in any single plant system. We can, however, examine some data on cytokinin metabolism to help us towards a clearer understanding of the dynamic aspects of cytokinins in plants.

We recently started an investigation into the cytokinin status of rooted and derooted cuttings of *Phaseolus vulgaris*, with and without apical and axial buds. The original observations of Engelbrecht (2) that bean leaves allowed to root on the petiole accumulated large quantities of cytokinin in the laminae and that the appearance of extractable cytokinin is closely co-ordinated with the appearance of roots offers further evidence for the hypothesis that roots are the site of cytokinin biosynthesis. We have chosen a slightly modified version of this system for study, in which cuttings from young *P. vulgaris* plants, consisting of the two primary leaves and a 10-cm length of stem below the leaves, are allowed to form roots on the bottom 5 cm of the stem. The relatively high levels of cytokinin produced and the ease of feeding the system for tracer studies make it a very suitable one to use to attempt to study the dynamic aspects of cytokinin interrelations in roots and shoots.

Using our modified system, we have confirmed Engelbrecht's original observations and have positively identified the cytokinin that accumulates in the leaves as dihydrozeatin-O-β-D-glucoside (23). Initial attempts to demonstrate biosynthesis of this compound by supplying the roots with radioactively labelled adenine have not been successful, so the unequivocal demonstration of cytokinin biosynthesis in roots must await further experimentation. Dihydrozeatin-O-β-D-glucoside is not detected in the roots or root exudate, so presumably it is not produced by the roots or transported to the leaves as such, but is produced by metabolism in the leaves of a root-produced intermediate. We have not yet positively identified this intermediate, but chromatographic evidence suggests that zeatin riboside or dihydrozeatin riboside is present in the root exudate.

The fact that dihydrozeatin-O-β-D-glucoside is produced by metabolism in the leaf has prompted us to investigate the metabolism of cytokinins in this system in some detail. 8-^{14}C-zeatin is metabolised rapidly by leaves of de-rooted and debudded bean seedlings of the type previously described, to zeatin-O-β-D-glucoside. This compound has been positively identified by a combination of UV and mass spectrometry (MS) and enzymatic methods. However, this compound is not as stable as the endogenous dihydrozeatin-O-β-D-glucoside and is eventually degraded to adenine and related compounds. When ^{14}C-zeatin is fed to *rooted* debudded bean seedlings, the major root-produced metabolite is zeatin ribotide, identified by enzymatic degradation to zeatin riboside and identification of this compound by gas chromatography (GC)-MS. PEI chromatography has also revealed the presence of small quantities of the di- and tri-phosphates. The major metabolite in the stem and petiole of the rooted system is dihydrozeatin riboside, identified by UV, MS, and GC radiocounting. Other minor metabolites found in the leaves of both systems and tentatively identified by enzymatic and chromatographic methods include dihydrozeatin-O-glucoside and glucosides of zeatin riboside and dihydrozeatin riboside. Extrapolating from these results to the situation in the normal plant, it is possible that the very high glucosylating capacity of the leaf may cause it to act as an effective sink for root-produced cytokinins, while the longevity of the endogenous dihydrozeatin-O-β-D-glucoside relative to zeatin-O-β-D-glucoside is probably due to its resistance to attack by enzymes of the "cytokinin oxidase" type. Whitty and Hall (25) have shown that a purified enzyme of this type, from corn endosperm, will readily cleave cytokinin side chains of the isopentenyl type while saturated side chains are not attacked.

Another interesting aspect of our experimental system is the possibility of using it to investigate the controlling role of apical and axial buds on cytokinin production and distribution. Intact bean plants do not accumulate cytokinins in the leaves to any great extent whereas if the apical and axial buds are removed, accumulation is observed. It is tempting to speculate on the role of the buds in this phenomenon. Two possibilities are that the buds are a more efficient sink for root-produced cytokinins than leaves or that buds are able in some way to control cytokinin production by the roots. Initial results on the metabolism of zeatin in buds are very similar to those obtained with roots, in that zeatin ribotide is the major metabolite, with very little production of glucosides. While this result does not enable us to distinguish clearly between the two possibilities, it does raise the interesting point that since the buds eventually develop into mature leaves the metabolic capacity of the system must change with age from production of ribotide to production of glucoside. However, an autoradiographic study of the distribution of labelled zeatin in bean plants has shown a high accumulation of label in the buds, thus supporting the sink theory. However, the previously discussed observation that photoperiodically induced *Xanthium* leaves can influence the levels of cytokinins in xylem sap suggests the possibility that leaves can also control the production of root-derived cytokinins.

III. Sites of Production and Transport of Cytokinins

The results obtained with *P. vulgaris* and *X. strumarium* described above strongly suggest that cytokinins produced in the root are exported to the shoot and that they undergo rather rapid degradation or conversion in the leaves. The results are also consistent with the generally accepted hypothesis that leaves are dependent on import of root-produced cytokinins for the maintenance of a normal, functional condition.

The question arises as to whether the apical and axillary buds of the shoot also depend upon root-produced cytokinins or whether they have the capacity for cytokinin biosynthesis and are self-sufficient in this respect. Endogenous cytokinins have been extracted from buds of *Dahlia* (9) and other species. In an experiment with *X. strumarium* (6), growing lateral buds were excised and placed with their bases in agar. After 24 h the buds and the agar were extracted and assayed for cytokinins. It was found that there had been a loss of 70-80% of the total cytokinin activity in the buds and diffusates over a 24-h period, suggesting that rapid cytokinin degradation occurs in buds of this species, and that the buds depend upon the import of cytokinins. However, experiments with de-rooted shoots of the potato species, *Solanum andigena*, supplied with a mineral nutrient solution, indicate that bud growth can continue for several weeks even when all root development was excluded by repeatedly cutting off basal stem sections before any roots could be initiated (24). Moreover, decapitation of such de-rooted shoots resulted in outgrowth of the uppermost buds even after 30 days without roots, and extracts of such shoots at this time indicated a small *increase* in cytokinin levels. These experiments seem to indicate unequivocally that, in *S. andigena*, bud growth can continue in the absence of roots, and that the buds have the capacity for cytokinin production if supplied with inorganic nutrients. Whether similar results could be obtained with other species is not known, but it is possible that *S. andigena* is exceptional in this respect. These results appear to be at variance with those of earlier experiments in which it was found that the buds of isolated 1- or 2-node stem sections of *S. andigena* required an exogenous source of cytokinin for growth (27), but the pretreatment of the parent shoot was different in the two series of experiments.

We have also obtained results with detached leaves of sunflower that appear to be inconsistent with the view that they are incapable of cytokinin synthesis. When leaves of sunflower are detached and supplied only with water, there is a rapid decline in the cytokinin levels; this is soon accompanied by signs of senescence, but if, after 2-3 days in water, they are transferred to a complete mineral nutrient solution, their cytokinin content increases significantly and senescence is retarded (17). Thus, the view that leaves and buds lack the capacity for cytokinin production may need to be modified, and their dependence upon root-produced cytokinins may apply only to the conditions in the intact plant.

It is clear that studies are required to establish unequivocally the sites of biosynthesis in the intact plant, and this problem is intimately associated with the elucidation of the biosynthetic pathways. The problem is one of particular difficulty, not only because of the low levels of endogenous cytokinins in plant tissues, but also because of the central role of adenine in metabolism, so that in experiments involving the feeding of labelled adenine as a precursor only a very small fraction is likely to be incorporated into endogenous cytokinin. The problem is further complicated by the fact that the cytokinin groups present in certain species of transfer-RNA are liable to be released during RNA degradation. Indeed, if the latter is the case it is difficult to understand why leaves should require a supply of cytokinin from other parts of the plant, unless the amounts of cytokinin released by t-RNA degradation are insufficient.

The cytokinins most commonly reported to be present in root exudate are zeatin and zeatin riboside, although dihydrozeatin has been found in the xylem sap of *Acer pseudoplatanus* (15). In radish seedlings, zeatin riboside appears to be the translocated form of cytokinin (3). It is generally assumed that the presence of cytokinins in the xylem sap indicates that this is the primary path of transport of cytokinins within the plant, but it is well known that there can be rapid exchange of solutes between xylem and phloem. Moreover, there have been a number of reports of the occurrence of cytokinins in phloem sap (4, 14, 21). Vonk (22) has recently reported that zeatin-8-^{14}C is converted to the mono-ribonucleotide when added to phloem sap of Yucca. Thus, the assumption that the transport of cytokinins in the plant is confined to the movement of zeatin and zeatin riboside in the transpiration stream is probably not valid.

Studies on the hormonal control of stolon development in *S. andigena* (11), suggest that the distribution of root-produced cytokinins is not uniform throughout the shoot, but may be rather tightly controlled; there is, indeed, some evidence that the movement and distribution of cytokinins in the shoot may be affected by auxins (18). It is difficult to envisage how such controlled distribution could be effected if cytokinins were restricted to the xylem.

References

1. Atkin, R.K., Barton, G.E., Robinson, D.K.: J. Exp. Botany 24, 475 (1973)
2. Engelbrecht, L.: Biochem. Physiol. Pflanzen 163, 335 (1972)
3. Gordon, E.M., Letham, D.S., Parker, C.W.: Ann. Botany N. S. (London) 38, 309 (1974)
4. Hall, S.M., Barker, D.A.: Planta 106, 131 (1972)
5. Henson, I.E., Wareing, P.F.: Physiol. Plantarum 32, 185 (1974)
6. Henson, I.E., Wareing, P.F.: J. Exp. Botany (in press)
7. Hewett, E.W., Wareing, P.F.: Planta 114, 119 (1973)
8. Itai, C., Vaadia, Y.: Physiol. Plantarum 18, 941 (1965)
9. Kannangara, T., Booth, A.: J. Exp. Botany 25, 459 (1974)
10. Kende, H., Sitton, D.: Proc. N.Y. Acad. Sci. 1, 235 (1967)
11. Kumar, D., Wareing, P.F.: New Phytologist 71, 639 (1972)
12. McCalla, D.R., Moore, D.J., Osborne, D.J.: Biochim. Biophys. Acta 55, 522 (1962)
13. Michael, G., Mounla, M.A.Kh., Goldbach, H.: Proc. 7th Int. Coll. On Plant Analysis & Fertilizer Production, Hanover, 1974, p. 307
14. Phillips, D.A., Cleland, C.F.: Planta 102, 173 (1972)
15. Purse, J.: PhD Thesis, University of Wales (1975)
16. Richmond, A.E., Lang, A.: Science 125, 650 (1957)
17. Salaama, A.M.S.A., Wareing, P.F.: (in press)
18. Seth, A.K., Davies, C.R., Wareing, P.F.: Science 151, 587 (1966)
19. Van Staden, J., Wareing, P.F.: Planta 104, 126 (1972)
20. Van Staden, J., Wareing, P.F.: Physiol. Plantarum 27, 331 (1972)
21. Vonk, C.R.: Acta Botan. Neerl. 23, 541 (1974)
22. Vonk, C.R.: Acta Botan. Neerl. 25, 153 (1976)
23. Wang, T.L., Horgan, R., Thompson, A.G.: Planta (in press)
24. Wang, T.L., Wareing, P.F.: (unpubl.)
25. Whitty, C.D., Hall, R.H.: Can. J. Biochem. 52, 789 (1974)
26. Wickson, M., Thimann, K.V.: Physiol. Plantarum 11, 62 (1958)
27. Woolley, D.J., Wareing, P.F.: New Phytologist 71, 781 (1972)

Paper Demonstration References (see Section IX):

3 - 20 - 65 - 108 - 160 - 164 - 166 - 167 - 175 - 182

17. Cytokinin and Cell Metabolism

D. Klämbt

Abstract

Cell metabolism is regulated by cell membrane functions due to the control of influx and efflux of the different molecules and ions found in cells. Cell metabolism depends also essentially on gene activation and protein synthesis. It has been shown that cytokinins possess an affinity site within the cell membrane of *Achlya*, related to Ca^{2+}-, amino acid- and auxin-transport (18) and an affinity site on the ribosomes of higher plants (5). Moreover cytokinins influence protein synthesis in vivo (7, 16, 30) and in vitro (17). Cytokinins also influence in vitro protein synthesis in systems prepared from *Lactobacillus acidophilus* and wheat germ. Affinity chromatography of extracts of maize shoots (10) and wheat germ on N^6-(Δ^2-isopentenyl) adenosine-sepharose 6B results in four different cytokinin-binding proteins. Structures and functions of these proteins have to be analyzed.

A. Introduction

In general all growth phenomena can be produced by one of several growth substances, depending only on different plant species. Therefore studying long-term reactions of plants to treatment with growth hormones would never result in special insight into the reaction mechanisms.

Growth and development are mediated by cell metabolism. Cell metabolism is as complex as any living system, but we have made significant progress in the ability to separate and analyze single metabolic processes and even whole processing systems, such as those of RNA synthesis, protein synthesis, lipid synthesis, etc.

In order to carry out a detailed analysis of the action mechanism of growth hormones, we have to simplify our test systems, i.e., we can no longer use intact cells, tissues, organs, or whole plants. They represent buffered regulated systems, in which any type of metabolic change will also change neighborreactions. Thus it is impossible to determine the original change caused by the growth hormone. All data - for instance, those reporting control of cytokinins at the gene level - are probably misinterpretations due to the complexity of the best system (15, 26). Only one paper looks more promising (22), but there have been no further reports since 1970.

Other groups of authors have described cytokinin effects upon the action of single enzymes (3, 6, 9, 12-14, 23). There are reports of action on the activities of catalase, isocitrate lyase = enzymes of the glyoxysomes (23, 28), on hydroxypyruvate reductase and glycolate oxidase = enzymes of the leaf peroxisomes (3, 9, 28), on nitrate reductase (12, 14), amylase (6), and protein kinase (24). Within these groups of authors, the experimental details differ considerably. Reaction measurements are recorded over hours, over days, and even over one week. Of course, all of them mention the differential action upon single enzymes within the total proteins. However, the cell conditions are not sufficiently defined, for special qualitative actions of cytokinins to be proved.

During the last few years, an increasing number of papers have reported a quantitative effect of cytokinin on protein synthesis (1, 4, 7, 16, 30). Although the results were found merely by treating whole plants or callus tissues, they seem to have more significance, since they indicate high specific binding of cytokinin to ribosomes (5), and there are preliminary results on cytokinin-dependent acceleration of protein synthesis in vitro (17).

There is another site of cytokinin effect. LeJohn (18) focused attention on membrane function and cytokinin action in the water mould Achlya sp. and found cytokinin-activated uptake of tryptophan or indole-3-acetic acid (IAA) within 2 min and less. Furthermore LeJohn and Cameron (19) reported specific sites within the membrane at which cytokinin, auxin, and Ca^{2+} seemed to interact. These results were confirmed by isolation of a glycopeptide that specifically bind Ca^{2+}, tryptophan, IAA, and the cytokinins N^6-(Δ^2-isopentenyl) adenine (IPA), N^6 (benzyl) adenine (BA), and kinetin (18). Recently Takegami and Yoshida (27) separated a low molecular weight protein from tobacco leaves by affinity chromatography using a poorly defined cytokinin-sepharose 4B adsorbent. It may be related to that of LeJohn and Cameron.

B. Materials and Methods

I. In vivo Experiments

Sterile tobacco tissues (pith-subcultures of *Nicotiana tabacum* var. Wisconsin 38) grown on low concentrations of auxin and cytokinin with an additional hormone starvation for 12 h were used (16). These tissues were incubated for 4 h in 15 µCi of a ^{14}C-amino acid mixture (CFB. 103 Radochemical Centre, Amersham). For the last 30 min of the 4-h treatment, actinomycin D (final concentration 100 µg/ml) and for the last 20 min auxin [4 mg naphthalene-1-acetic acid (NAA)/2], or kinetin (0.4 mg/2), or no growth hormone, was added. At the end of the incubation period tissues were soaked in a suction funnel, quickly washed with water, and frozen in liquid N_2. All further preparations until the registration of the polysome profiles on isokinetic sucrose gradients were done according to previous work (16).

Incorporation of ^{14}C-amino acids into the polysomes was measured by collecting the polysome fractions between the 5th ml from the tube bottom and the monoribosome peak and recording the total UV absorption at 260 and 280 nm and the radioactivity of the TCA precipitate. Separation of the newly synthesized polypeptide chains by puromycin treatment of the polysomes revealed that more than 82% of the total radioactivity was present in the peptide chains.

II. In vitro Experiments

1. Protein synthesizing systems from young corn shoots (17, 25)
2. Protein synthesizing systems from wheat germ (21)
3. Protein synthesizing systems from *Lactobacillus acidophilus*

All these systems were used to search for a cytokinin effect on protein synthesis. IPA was applied at concentrations from 10^{-9} M to 10^{-4} M, and the incubation period was 30 min.

For systems 1 and 3, protein synthesis on endogenous polysomes was measured using ^{14}C-amino acid mixture or L-(^{14}C)-leucine (CFB. 103 or CFB. 67 Radiochemical Centre, Amersham). In system 2, in which no protein synthesis takes place, poly U-dependent polymerization of L-(^{14}C)-phenylalanine (CFB. 70 Radiochemical Centre, Amersham) was tested. ^{14}C-amino acid incorporation was measured by TCA precipitation and 10-min incubation in 5% TCA at 90°C within a glass-fiber filter (GFA Whatman) or in solution, followed by filtration onto membrane filters (SM 11306) (25).

For a detailed analysis, affinity chromatography is necessary. Besides the first trial (27) mentioned in the introduction, we prepared a very stable and well-defined cytokinin-sepharose 6B (10) using Epoxy-activated Sepharose 6B (Pharmacia, Sweden) and IPA-riboside (Sigma) at pH 13.

Based on a given concentration of epoxy groups of 12 to 20 μval/ml Epoxy-activated Sepharose 6B, the maximum yield was about 5% or 0.5 μmol IPA/ml gel. The unreacted epoxy groups were blocked by ethanolamine (1.0 M), and a pure ethanolamine-Sepharose 6B was also prepared.

C. Results and Discussion

I. In vivo Experiments

When looking for an unequivocal experiment to show whether protein synthesis is directly influenced by cytokinin or auxin, it seems best to measure the rate of incorporation of amino acids. Therefore, ^{14}C-radioactivity applied as ^{14}C-amino acids in relation to the amount of polysomes after a short incubation period should be a measurement of the incorporation rate. The data of these experiments (20) clearly show a stimulating effect of kinetin on the rate of protein synthesis. Under the conditions described previously a 35% higher specific activity of the polysomes could be determined after cytokinin treatment. Auxin had no effect at all. Since more than 82% of the total radioactivity found in the polysomes is due to the nascent polypeptide, the difference of 35% more in polysomes of cytokinin tissues has to be discussed.

The total radioactivities found in the supernatants of the different samples after 30 min centrifugation at 11,000 g were very similar and showed differences not larger than 10%. We interpret these data as representing equal pools of ^{14}C-radioactive amino acids within the cells receiving the different treatments. Since there is an increasing "accumulation" of ^{14}C-amino acids with the incubation time, the products of protein synthesis formed at the very end of the incubation period have to be highest in specific activity. We do find a higher specific radioactivity after cytokinin treatment. This means that the polypeptides found on the polysomes of cytokinin-treated cells are synthesized most recently before cell fixation. However, such a situation is only understood in terms of an increased protein synthesis rate, i.e., an overall shortening of the time necessary for protein synthesis. This result is in agreement with other reports (1, 4, 7, 16, 30).

II. In vitro Experiments

As stated in the Introduction, the search for the actual site of action of plant hormones requires in vitro systems that can be considerably simplified and that are reversible. Although our work is still preliminary, we have obtained interesting results, which indicate the trend for further research.

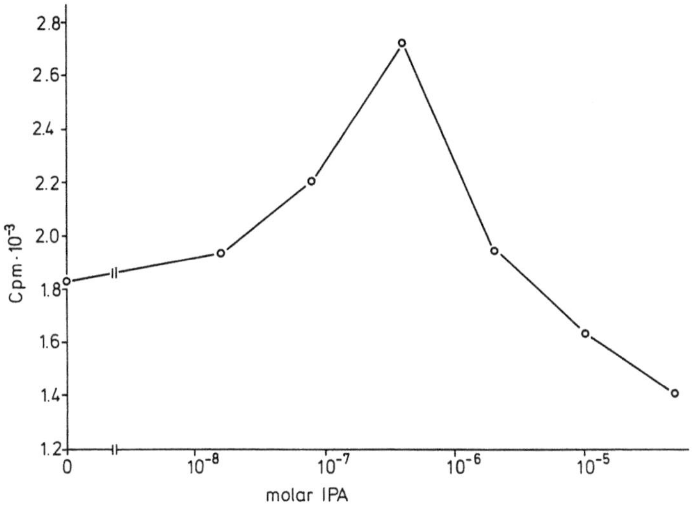

Fig. 1. Cytokinin-dependence of in vitro protein synthesis in wheat germ system prepared according to Marcus et al. (21). Assays contained in 50 µl the modifications: 375 µg S-23-protein, 12.5 µg poly U, 0.13 µCi [^{14}C]-phenylalanine, 2.6 µM Mg^{2+}. IPA, in concentrations indicated in graph. Samples were incubated at 30°C for 30 min

1. *Zea mays* - Young Shoots

Protein synthesis of an in vitro system from young corn shoots (25) can be stimulated by about 20% in the presence of 10 ^{6} M to 10 ^{7} M IPA (17). Since such preparations contain endogenous polysomes and barely show initiation of protein synthesis, stimulation must be related to polypeptide chain elongation.

2. *Triticum sativum* - Wheat Germ

We used the wheat germ system for protein synthesis (21) because it is the best-known plant system. Wheat germ is a nongrowing plant material without intact polysomes. Preparations for in vitro protein synthesis show practically no background amino acid polymerization. Therefore any type of mRNA has to be added. In the presence of poly U without added tRNA, cytokinin stimulates the incorporation of [^{14}C]-phenylalanine by a maximum of 80 to 100%. In the presence of poly U and additional tRNA the stimulation is in the same range (Fig.1). When the experiments are performed with poly U at 2.6 mM Mg^{2+}, initiation of polypeptide synthesis is not specifically controlled; therefore stimulation should be due to peptide chain elongation only.

3. *Lactobacillus acidophilus*

Cytokinins are described as a growth substance in higher plants. Because these compounds are known as normal constituents in the tRNA of all living cells, they could also have special functions in nonplant organisms.

At first we were looking for a stimulating effect of cytokinins on the growth of *Lactobacillus acidophilus* (ATCC 4963), which is partly auxo-

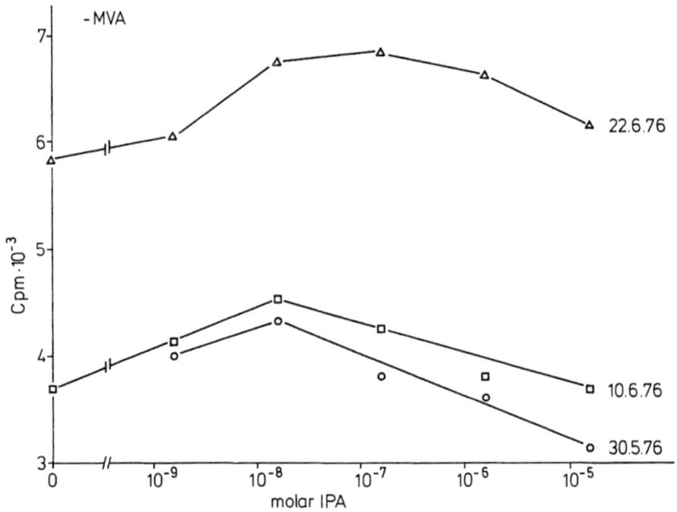

Fig. 2. Cytokinin-dependence of in vitro protein synthesis in *Lactobacillus acidophilus* (ATCC 4963) system grown without addition of mevalonic acid. Assays contained per 200 µl: 20 µmol Tris-Ac, pH 7.8, 1 µmol $MgAc_2$, 10 µmol KCl, 1 µmol 2-mercaptoethanol, 10 nmol ATP, 5 nmol GTP, 20 nmol PEP, 2.5 µg PK, 6.7 A_{260} units dialyzed polysomes, 1.04 mg dialyzed S-100-protein, 0.1 µCi ^{14}C-amino acid mixture, 4.0 nmol each of missing amino acids: L-cysteine, L-glutamine, L-methionine, L-tryptophan. IPA in concentrations indicated in graph. Samples were incubated at 37°C for 30 min

trophic for mevalonic acid. Because mevalonic acid is a precursor of the side chain - Δ^2-isopentenyl-group - of cytokinins, we hoped to control the cytokinin content by application of different mevalonic acid concentrations. Indeed, growth rate and especially the total growth of cultures of *Lactobacillus* were stimulated by IPA and less by kinetin (Leineweber, Klämbt unpublished). These results led us to question whether or not cytokinins stimulate protein synthesis even in this bacterium.

Crude in vitro systems gave no measurable effect. However, when the polysomes were separated from the supernatant by 2-h centrifugation at 110,000 g and both fractions were dialyzed for 12 h, up to 20% stimulation of ^{14}C-amino acid incorporation was measured. Dialyzing only the polysomes showed little effect due to inhibition of protein synthesis by nondialyzed supernatant. There are some differences in the response curves (Figs.2 and 3) due to the fact that the bacteria were cultured with or without 33.7 µM mevalonic acid prior to the preparation. Stimulation in relation to cytokinin concentrations results in an optimum curve. Compared to the in vitro synthesis of higher plant systems, the optimum is reached at lower IPA concentrations.

It seems therefore a general phenomenon that cytokinins to some extent control or stimulate protein synthesis. It is now necessary to search for the actual site of cytokinin action.

Fox and Erion (5) described a specific protein, bound to ribosomes, which specifically interacts with cytokinins with an affinity constant of $1.6 \times 10^6\ M^{-1}$. These findings are in good agreement with the optimal stimulation of in vitro protein synthesis in systems prepared from maize and wheat.

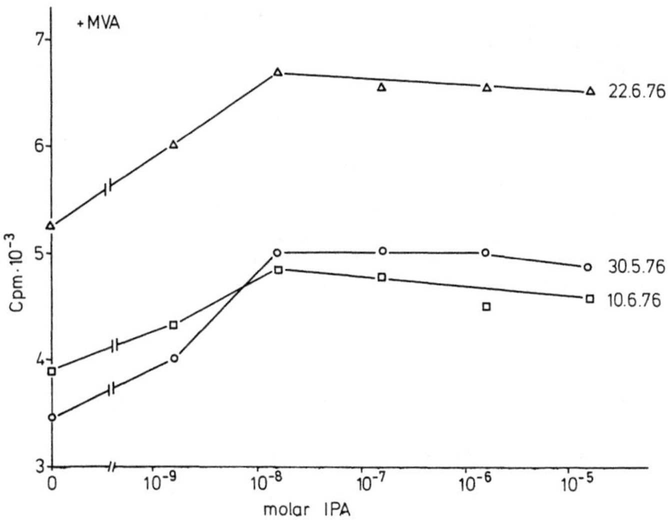

Fig. 3. Cytokinin-dependence of in vitro protein synthesis in *Lactobacillus acidophilus* (ATCC 4963) system grown with optimal addition of mevalonic acid (33.7 µM). Assay conditions are described in Fig.2

4. Affinity Chromatography

Application of different extracts of young shoots of *Zea mays* to the IPA-riboside-sepharose 6B and ethanolamine-sepharose 6B reveals four cytokinin affinity proteins, as shown by acrylamide gel electrophoresis. One protein band seems to be related to ribosomes (10).

Analyzing extracts of wheat germ [S-100, precipitated by $(NH_4)_2SO_4$, 80% saturation] by the same method results in at least three protein fractions closely related to those of corn shoots. With these data we are now able to concentrate on the preparation of the different cytokinin affinity proteins. There may be proteins of IPA metabolism, but probably some IPA affinity proteins may have regulating properties.

Young corn shoots are difficult to collect in large quantities, whereas wheat germ - rich in the desired protein compounds - is easily available. Our intention is, however, to separate and analyze the different cytokinin affinity proteins from wheat germ and to look for their functional relations.

D. Conclusions

Plant growth hormones are necessary for growth. The hypothesis that the different hormones act by changing cell metabolism qualitatively rather than quantitatively has not yet been unequivocally proved. The idea, therefore, that plant hormones act quantitatively within the limitations of special gene activations is much more attractive. The observations and characterizations of the membrane glycopeptide from *Achlya* (18, 19) and the preliminary results of the effect of cytokinin on protein synthesis (17) are in good agreement with such a hypothesis. The same is true for the plant hormone auxin, for which we know binding and regulating sites within the cell membrane (11, 29) and another action site during transcription (8).

Cell metabolism depends strongly upon an exchange of solutes, substrates, etc., through the membrane. Besides the transcription and translation membrane functions are equally important in regulating cell metabolism. As shown for cytokinins (18, 19) and for auxins (11, 29), membranes are easily employed in in vitro systems.

Protein synthesis, on the other hand, is easiest to analyze, and probably every in vitro system can be used to search for cytokinin action and its characterization. By combining these tests with the simple preparative method for separation and enrichment of cytokinin affinity proteins (10, 27), it should be possible to analyze the action of cytokinins upon these two essential phenomena in the very near future.

References

1. Bhattacharyya, J., Roy, S.C.: Biochem. Biophys. Res. Commun. 35, 606 (1969)
2. Burrows, W.J.: Current Adv. Plant Sci. 1975, p. 837
3. De Boer, J., Feierabend, J.: Z. Pflanzenphysiol. 71, 261 (1974)
4. Fankhauser, M., Erismann, K.H.: Planta 88, 332 (1969)
5. Fox, J.E., Erion, J.L.: Biochem. Biophys. Res. Commun. 64, 694 (1975)
6. Gepstein, S., Ilan, I.: Plant Cell Physiol. (Tokyo) 11, 819 (1970)
7. Gordon, M.E., Letham, D.S., Beever, J.E.: Physiol. Plantarum 35, 27 (1975)
8. Guilfoyle, T.J., Lin, C.Y., Chen, Y.M., Nagao, R.T., Key, J.L.: Proc. Natl. Acad. Sci. U.S. 72, 69 (1975)
9. Harvey, B.M., Lu, B.C., Fletcher, R.A.: Can. J. Botany 52, 2581 (1974)
10. Hermwille, E., Klämbt, D.: in press (1976)
11. Hertel, R., Thomson, K.S., Russo, V.E.A.: Planta 107, 325 (1972)
12. Hirschberg, K., Hübner, G., Borris, H.: Planta 108, 333 (1972)
13. Hodge, E.T., Sacher, J.A.: Biochem. Physiol. Pflanzen 168, 433 (1975)
14. Kende, H., Hahn, H., Kays, S.E.: Plant Physiol. 48, 702 (1971)
15. Key, J.L.: Ann. Rev. Plant Physiol. 20, 449 (1969)
16. Klämbt, D.: Planta 118, 7 (1974)
17. Klämbt, D.: Plant Cell Physiol. (Tokyo) 17, 73 (1976)
18. LeJohn, H.B.: Can. J. Biochem. 53, 768 (1975)
19. LeJohn, H.B., Cameron, L.E.: Biochem. Biophys. Res. Commun. 54, 1053 (1973)
20. Maaß, H., Klämbt, D.: in press (1976)
21. Marcus, A., Efron, D., Weeks, D.P.: In: Methods in Enzymology 30, 749 (1974)
22. Matthysse, A.G., Abrams, M.: Biochim. Biophys. Acta 199, 511 (1970)
23. Penner, D., Ashton, F.M.: Biochim. Biophys. Acta 148, 481 (1967)
24. Ralph, R.K., McCombs, P.J.A., Tener, G., Wojcik, S.J.: Biochem. J. 130, 901 (1972)
25. Sim, W.S., Klämbt, D.: Planta 113, 157 (1973)
26. Skoog, F.: Biochem. Soc. Symp. 38, 195 (1973)
27. Takegami, T., Yoshida, K.: Biochem. Biophys. Res. Commun. 67, 782 (1975)
28. Theimer, R.R., Anding, G., Matzner, P.: Planta 128, 41 (1976)
29. Venis, M.A., Batt, S.: Biochem. Soc. Transact. 3, 1148 (1975)
30. Wozny, A., Szeykowska, A.: Biochem. Physiol. Pflanzen 168, 195 (1975)

Paper Demonstration References (see Section IX):

18 - 42 - 45 - 102 - 103 - 129 - 165 - 166 - 167 - 175

VI. Ethylene and Other Regulators

18. Auxin and Ethylene and the Control of Cell Growth. Identification of Three Classes of Target Cells

D. J. Osborne

Abstract

Three classes of target cells have been identified by their growth responses to auxin and ethylene. Type 1 expands in response to auxin but not to ethylene; Type 2 expands in response to ethylene but not to auxin; Type 3 expands in response to either hormone, but the mechanism controlling the growth may be different for the two hormones.

A. Introduction

Since 1901, ethylene has been known for its regulatory effects in reducing the elongation growth of shoots (19). Endogenously produced ethylene is involved in initiating or accelerating fruit ripening (2), and organ abscission (1, 10), and more recently it has been shown to enhance shoot growth in submerged water plants (17, 18). In contrast, auxin (indole-3-acetic acid) IAA will maintain extension growth in segments of the immature shoot tissues of most land plants (28), and will retard fruit ripening (6) and organ abscission (7), even in the presence of ethylene.

The opposing actions of these two hormones are of special interest in the control of plant growth and development, for the extent of ethylene production is, in part, determined by the level of applied or endogenous auxin within the tissue (11), thereby affording a novel regulatory mechanism of cell growth control by a volatile and a non-volatile component (21).

The original concept of a hormone is animal-orientated and envisages the synthesis of small quantities of the chemical in one tissue of the organism and its release and transport to another tissue, where precise regulatory effects are exerted on specific target cells. Such a concept has posed difficulties when applied to plants. The organs, and more particularly the cells involved in hormone synthesis, cannot be readily distinguished from other cells, either structurally or biochemically, and the identification of target cells has, with few exceptions (27), proved difficult. From recent studies of the growth of cells in response to ethylene and auxin, however (21-23), it now seems that the idea of target cells in plants may be generally justified. Because patterns of development and differentiation are determined by a combination of the rate, extent, and orientation of growth of the component cells that comprise a tissue, the recognition of what appear to be specific target cells, is providing a useful and new approach to an understanding of the regulatory role of hormones in plants.

Comparisons of the auxin and ethylene control of expansion growth of shoot cells have revealed three main target types that can be recognised by their responses as follows: Type 1 cells: those that are stimulated to enlarge by auxin but not by ethylene; Type 2 cells: those that expand in response to ethylene but not to auxin; Type 3 cells: those that enlarge when exposed to either ethylene or auxin.

This report describes and compares the regulation of cell expansion by the two hormones in examples of immature cells of all three target types, e.g., cortical cells of the etiolated shoots of a dicotyledonous seedling (Type 1), cells of leaf abscission zones (Type 2), and internode or petiole cells of shoots of several aquatic plants with aerial parts (Type 3).

B. Materials

I. Plants for Type 1 cells

Pisum sativum var. Alaska or Meteor were grown in sand in darkness at 24°C and used at 6-7 days when the third internode was 1.5-2 cm in length (25). All determinations were made from comparable positions just below the hook of the apical bud and marked with charcoal-lanolin at 0 h.

II. Plants for Type 2 cells

1. Dwarf bean, *Phaseolus vulgaris* var. Masterpiece, were raised in glasshouses with supplementary light and used after 21 days when the primary leaves were fully expanded. Explants 1 cm long were cut from the distal abscission zone of these leaves to include the whole of the pulvinus and part of the petiole (29).

2. Leaves of *Sambucus nigra* were collected from a wild stand at different periods during the growing season. After the removal of leaflets, explants 5 mm long were prepared from the rachis to include either (1) the rachis abscission zone that lies close to the point of insertion of the leaflets (zone explants) or (2) segments cut from the rachis midway between the leaflets (rachis explants) (22).

III. Plants for Type 3 cells

1. *Regnellidium diphyllum*. Rhizomes of the water fern were planted in shallow trays of wet mud (18); fronds were produced at intervals of several centimetres and consisted of an erect rachis with two terminal leaflets. Fronds were used when the leaflets were partially or fully expanded and the rachis about 5-10 cm long.

2. *Ranunculus sceleratus* was grown from seed in water-saturated Levington's compost in glasshouses supplemented with extra light. Leaves with petioles were harvested from the rosette plants when the petioles reached 3-10 cm long.

IV. IAA (99% purity) was obtained from Koch Light Ltd., London and was made up as an aqueous solution of the sodium salt. Ethylene 99.9% purity was from Cambrian Chemicals, Croydon, Surrey.

C. Methods

I. Determinations of Cell Size

In most tissues, this was done by measuring cells under a light microscope fitted with a graticule on L.S. or T.S. free-hand sections or on sections fixed in gluteraldehyde and prepared for electron microscopy. In some experiments with water plants, tissues were first macerated in chromic acid and cell numbers per marked segment counted with a haemocytometer. With peas, measurements of cell sizes were made from photographic prints of free-hand sections (21).

II. Measurements of Ethylene Production

Tissues were enclosed in containers of known volume and samples of the gas phase later removed for quantitative ethylene analysis by gas-solid chromatography in a Pye 104 chromatograph (10). Flasks were aerated and resealed after each sampling.

III. Mechanical Properties of Cell Walls

Extensibility of cell walls in Type 1 or Type 3 tissue was determined in marked segments that had elongated on the plants in response to auxin or ethylene by using an Instron tensile tester for stress-relaxation analysis (3, 15).

IV. Measurements of Glycosidase or Glucanase Activity

Enzymes were extracted in acetate or phosphate buffer at optimal pH and molarity, and glucosidase was measured by the liberation of p-nitrophenol from p-nitrophenyl-β-D-glucoside and determined spectrophotometrically (20). The activity of β-1,4-glucanase was determined either by loss of viscosity in a substrate solution of carboxymethylcellulose (8) or by the liberation of ^{14}C-glucose or ^{14}C-cellobiose from a ^{14}C-cellulose preparation derived from cell walls of Canna lily (24).

The presence of 1:3 glucanases was assayed using laminarin as substrate with subsequent determination of reducing groups with 3:5-dinitrosalicylic acid reagent (20). The products were separated chromatographically to visualise free glucose and oligosaccharides.

V. Mercuric percholorate (MP) 0.25 M in 2.5 M perchloric acid was used as an ethylene absorbent in control and IAA-treatment containers.

D. Results and Discussion

I. Type 1 Cells

1. Cell Extension

When etiolated pea seedlings are enclosed with ethylene from 1-50 μl/l for 3 days, cell elongation in the cells of the extending parts of the shoot is clearly depressed (Table 1 for 50 μl/l). Although the shoots are subsequently shorter than the controls in air, the final volume of the enlarging cortical cells of control and ethylene-treated plants do

Table 1. Dimensions of cortical cells (Type 1) in etiolated pea shoots from marked position below the apex after a single spray treatment with IAA or water, or following exposure to ethylene

	Treatment	Length µm	Volume $\mu m^3 \times 10^{-3}$
0 h	Control	24.4	6.3
9 h	Control	31.1	11.7
	IAA 10^{-3} M	48.9[a]	22.4[a]
	C_2H_4 50 µl/l	26.7[a]	11.4
72 h	(Final cell size)		
	Control	82.2	30.2
	IAA 10^{-5} M	79.9	33.9
	IAA 10^{-4} M	75.5[a]	37.9[a]
	IAA 10^{-3} M	73.3[a]	44.2[a]
	C_2H_4 50 µl/l	44.4[a]	24.5

[a] Significant difference from respective control at each time. P = 0.01

Recording at 9 h and at final cell size (21)

not significantly differ at full expansion (72 h, Table 1). This is accounted for by the ethylene-controlled orientation of cell growth to a lateral expansion rather than a longitudinal elongation. In contrast, when a single spray of IAA is supplied to the intact shoot (21), there is initially an enhanced rate of cell enlargement for several hours, and the final cell volume is significantly greater than the control, the higher the concentration of auxin that is supplied. For the first 9 h after treatment with 10^{-3} M IAA, the 100% volume increase above controls is reflected in a 60% greater cell elongation (Table 1); because the cells respond to the auxin-induced ethylene production, which continues after all the auxin taken up is metabolised (at approximately 6 h), volume increments decrease again to the control level, and the subsequent cell expansion is predominantly in the lateral direction. At full expansion (72 h) cortical cells of the IAA-treated plants are then 40% larger than controls, but shorter in length. The IAA-treated plants are therefore taller than controls for only a few hours.

2. Cell Extensibility

When comparisons are made of the extensibility of cell walls using the Instron load/extension technique (3), it is found that compared with controls marked segments from IAA-treated plants exhibit an enhanced extensibility only in the early hours of treatment (3 h, Table 2) corresponding to the period of enhanced cell expansion. No increase in extensibility compared with controls is ever observed with ethylene; instead the extensibility decreases as the reorientation to lateral cell expansion becomes established (18 h, Table 2).

3. Glucanase Activity

When intact peas are sprayed once with 10^{-3} M IAA 1:4-glucanase activity is approximately doubled by 24 h in marked segments taken from the elongating regions of the shoot (Table 3a), but no increases are observed up to 7 h during the time that IAA-induced expansion growth is maximal. Only in long-term experiments (2-5 days) with decapitated peas are large increases in 1:4-glucanase activity always found in the mark-

Table 2. Extensibility (DP) of apical marked segments from intact pea plants after single spray with IAA 10^{-3} M or exposed to ethylene 10 µl/l

	Treatment	Plastic extensibility dynes/cm^2 × 10^{11}
0 h	Control	1.7 ± 0.3
3 h	Control	2.6 ± 0.4
	IAA	3.6 ± 0.3
	C_2H_4	2.4 ± 0.4
18 h	Conrol	4.3 ± 0.3
	IAA	4.3 ± 0.4
	C_2H_4	2.9 ± 0.3

Table 3. 1:4 glucanase activity in marked apical segments of pea seedlings: (a), intact plants sprayed with IAA 10^{-3} M or H_2O (A.V. Atack, unpublished); (b), intact plants exposed continuously to ethylene (I. Ridge, unpublished). Substrate = carboxymethylcellulose

Treatment	Relative 1:4-glucanase activity (per marked segment)	
	Control	IAA
(a) After 3 h	7.0	6.6
" 7 h	8.5	8.4
" 24 h	9.1	12.3[a]
	Relative 1:4-glucanase activity (per mg soluble protein)[b]	
(b) After 24 h		
	Control + MP	12.2 ± 0.2
	C_2H_4 0.5 µl/l	11.6 ± 0.8
	" 5 µl/l	11.5 ± 0.3
	" 50 µl/l	10.8 ± 0.9
	" 500 µl/l	10.8 ± 0.9

[a] Significant difference from other treatments P = 0.01

[b] No increases are observed in experiments in which enzyme activity is expressed per marked segment

ed apical region following continuous auxin treatment (14, 24). With intact peas (Table 3b) ethylene (0.5-500 µl/l) showed either no significant change in 1:4 glucanase activity or a slightly lower activity compared with the controls (24 h) and no increases with extended treatments (3-5 days).

Table 4. Elongation growth of cells of abscission zones of fully expanded, mature leaves of beans. Explants exposed to ethylene 10 µl or treated with IAA 10^{-3} M or water. Cell measurements of zone cells (Type 2) and of cells (Type 1) on either side (29)

	Initial control	Distal cell (pulvinus)	Zone cell	Proximal cell (petiole)
6 h	-	5.7	5.7	7.6
32 h	H_2O	3.8 ± 0.7	5.2 ± 1.2	5.8 ± 1.0
	IAA	6.6 ± 0.5	7.4 ± 0.9	8.5 ± 0.9
	C_2H_4	7.4 ± 1.0	14.1[a] ± 1.1	7.1 ± 0.8
48 h	H_2O	3.6 ± 0.6	3.3 ± 0.5	4.9 ± 0.6
	IAA	4.4 ± 0.6	5.4 ± 0.6	5.2 ± 0.9
	C_2H_4	Abscinded	12.2[a] ± 1.9	9.1[a] ± 1.2

[a] Significant difference from neighbour cells P = 0.01

II. Type 2 Cells

1. Results with *Phaseolus*

In leaf abscission of *Phaseolus vulgaris*, enlargement of a single row of cells in the abscission zone, in response to ethylene, precedes separation of the distal and proximal parts (29). These cells expand only in response to ethylene, and IAA depresses the enlargement (Table 4). They are therefore distinct from, but surrounded by, the Type 1 cells of the rest of the petiole in which cell enlargement and cell elongation are stimulated by auxin (unpublished data). An increase in β-1-4-glucanase activity (29) can be detected in the region of the zone that contains the Type 2 cells before cell expansion is apparent, and the increase is accelerated by ethylene (8, 29).

There are inherent difficulties in detecting changes in cell wall extensibility in the single row of cells in the bean abscission zone, but several investigators (4, 9) have reported a decrease in "break strength" of the abscission zone accompanying increases in glucanase activity and prior to cell separation. In this assay, reduction in breakload could equally well be interpreted as a measure of the cell wall changes and cell extensibility of the Type 2 cells at the zone during expansion growth in ethylene.

2. In *Sambucus nigra*

A band of some 30-40 cells wide of ethylene sensitive Type 2 zone cells are differentiated early in leaf expansion (23). Explants cut from zone regions of immature leaves contain both Type 1 and Type 2 cells and will therefore increase in fresh weight with IAA or ethylene. But different cells are stimulated to enlarge with the two hormones, and only Type 2 cells treated with ethylene show enlargement with subsequent cell separation. Rachis explants from immature leaves (which contain only Type 1 cells) increase in fresh weight and elongate only in response to IAA (Table 5) and, of course, exhibit no cell separation.

When the leaf is fully expanded and mature, neither rachis nor zone explants elongate or increase in fresh weight with IAA, but zone explants will still increase with ethylene, and this can be attributed solely to

Table 5. Percent increase in fresh eight of abscission zone and rachis explants from immature and mature leaves of *Sambucus* on 1% agar + IAA 2×10^{-4} M or on plain agar ± ethylene 5 µl/l (23).

	% Increase in fresh weight of explants Immature		Mature	
	Rachis (48 h)	Zone (45 h)	Rachis (50 h)	Zone
Control	14.1 ± 1.8	27.0 ± 1.6	19.4 ± 2.1	16.9 ± 2.2
IAA	24.9 ± 1.8	38.4 ± 1.5	17.2 ± 1.7	16.1 ± 1.5
C_2H_4	13.2 ± 1.0	60.4 ± 5.7	23.2 ± 1.9	52.9 ± 3.3
	% Abscission			
Control	-	0	-	0
IAA	-	0	-	0
C_2H_4	-	80	-	100

the growth of Type 2 cells in the zone (22). As with the immature leaves, abscission is accelerated by ethylene and inhibited by IAA (Table 5).

In Type 2 cells sheared from zones of abscinding mature or immature leaves an enhanced activity can be demonstrated in 1:3 and 1:4-glucanase and glucosidase in the supernatant solution from these unbroken cells, in response to ethylene but not with IAA (J. Roberts, unpublished). This indicates the secretion of enzymes through the cell wall in response to ethylene. Indeed, electron micrographs of this tissue clearly show enhanced activity of dictyosomes and golgi and visible vesicles associated with degrading middle lamellae (23). These events do not occur in the surrounding Type 1 cells, again indicating the difference between the two classes of cells in the abscission zone.

III. Type 3 Cells

1. Cell Extension

Elongation of cells in the intact rachis or in petioles of the water plants *Regnellidium diphyllum* and *Ranunculus sceleratus* is enhanced when these plants are exposed either to ethylene or to solutions of IAA (Tables 6 and 7) or when the shoot is submerged so that the cells are responding to the endogenous ethylene accumulating within the plant (17, 18). Combining the optimal concentrations of IAA and ethylene can lead to a further enhancement of growth. Growth stimulations, though quantitatively less, are obtained when excised segments are exposed to the two hormones.

2. Cell Extensibility

Determinations of relaxation modulus in *Regnellidium* (Table 6) have shown that both IAA and ethylene increase the extensibility of the cell walls of treated plants compared with controls, and both hormones achieve this when growth is precluded by mannitol.

Table 6. Effect of treatment with ethylene 10 µl/l and IAA 10^{-5} M for 24 h on growth of rachis of submerged, excised fronds of *Regnellidium diphyllum* and on resultant extensibility of 4-mm segments from below frond. (C. Cookson unpublished)

Treatment	Relaxation modulus dynes/cm^2 × 10^7	% elongation
Control (in H_2O)	17.1 ± 1.6	23.6 ± 6.1
IAA	7.7 ± 0.9	90.6 ± 10.2
C_2H_4	7.1 ± 0.9	64.5 ± 10.0

Table 7. (a) Soluble β-glucosidase[a] activity of floating petioles of *Ranunculus sceleratus* after 24 h in IAA 10^{-4} M or in water (± ethylene 10 µl/l) compared with those submerged. (b) Soluble 1:3 glucanase[b] activity in petioles similarly treated with IAA and ethylene (C. Cookson, unpublished)

Treatment		µ-mol p-NP[a] mg protein/h	% elongation
(a)			
Control	0 h	2.41 ± 0.04	-
"	24 h	2.72 ± 0.04	11.2 ± 0.6
IAA	24 h	2.54 ± 0.07	50.8 ± 2.0
C_2H_4	24 h	5.82 ± 0.08	57.0 ± 2.8
Submerged	24 h	4.89 ± 0.23	47.9 ± 1.8
		µg glucose equivalents mg protein/h	
(b)			
Control	0 h	3.23 ± 0.09	-
"	24 h	3.77 ± 0.51	11.7 ± 0.6
IAA	24 h	3.27 ± 0.33	64.7 ± 2.4
C_2H_4	24 h	4.77 ± 0.11	47.2 ± 1.5

Assay at 50 mM acetate pH 4.5
Substrate: [a] p-nitrophenyl-β-D-glucoside
[b] laminarin

3. Glucosidase and Glucanase Activities

Assays of β-glucosidase and 1:3-glucanase in *R. sceleratus* have shown that whereas both IAA and ethylene enhance cell growth of the petiole, only ethylene enhances these enzyme activities. Submergence growth, which is a response to endogenous ethylene, also leads to enhanced glucosidase activity (Table 7). Although these enzyme changes in response to IAA and to ethylene are clearly different, indicating a different mechanism of growth regulation by the two hormones, it could be suggested that cell elongation is mediated through either (1) an enhancement of ethylene production that is auxin-induced, and that the cells are in reality of Type 2, or (2) that there is an ethylene enhancement of endogenous auxin levels, and that the cells are in reality Type 1.

The following evidence suggests that the shoot cells of these water plants are a unique class that will enlarge in response to either auxin or ethylene.

E. Evidence for a Type 3 Class of Cells

I. Absence of an Effect of Auxin on Ethylene Production

1. In *Regnellidium* there is no enhancement of ethylene production above the control level, by concentrations of IAA from 10^{-7}-10^{-3} M. The increased cell growth in auxin is therefore not mediated by auxin-induced ethylene. (J. Walters and C. Cookson, unpublished).

2. In *Ranunculus sceleratus*, there is modest (20-30%) auxin-enhanced ethylene production (only above 10^{-4} M IAA), but the addition of an analogue of rhizobitoxin, 1-2-amino-4-(2'-aminoethoxy)-trans-3-butenoic acid, completely inhibits the auxin-induced ethylene production of the tissue, without reducing the extent of auxin-enhanced growth (C. Cookson, unpublished).

II. Ethylene-enhanced growth is unlikely to be mediated solely by an ethylene-enhanced auxin level, for in *Regnellidium*,

1. The rate of metabolism of applied IAA-1-^{14}C or IAA-2-^{14}C is not altered by a prior 24-h exposure of the rachis to 10 µl/l of ethylene (J. Walters, unpublished).

2. Extraction of the endogenous IAA in the rachis and analysis of the indolo-α-pyrrone by spectrophotofluorimetry (12, 26) indicate that no increase in extractable IAA follows a 24-h exposure of the shoot to ethylene (J. Walters, unpublished). Using similar methods, no increase in auxin levels has been observed in *Ranunculus* in ethylene. It must be concluded from such evidence that both auxin and ethylene can enhance cell growth in shoots of water plants. Furthermore, the cells represent a recognisable and distinct target cell type (Type 3).

F. Conclusion

Although the IAA and ethylene regulation of cell expansion has been defined for the three classes of cells, the mechanism by which cell expansion is achieved by the two hormones in any of the target cell types remains unclear. In all, evidence suggests that cell wall relaxation is a primary event.

Cell wall-associated 1-4-glucanases increase in Type 1 cells with IAA, but the rise occurs later than the IAA-enhanced cell expansion and wall relaxation, and would seem therefore to be related to the long-term response to auxin rather than to early events. Ethylene neither modifies cell size nor glucanase activity in Type 1.

In Type 3 cells both IAA and ethylene enhance elongation and wall extensibility, but only ethylene induces enhanced activity of glucanases or glucosidases. Therefore cell growth and wall relaxation events may either follow different pathways with IAA or ethylene or alternatively, these enzymes are not directly involved in cell growth responses (5, 15). However, because in vitro assays of enzymes are normally carried out at optimum conditions for maximum activity, the absence of an increase in vitro does not exclude the possibility that in vivo, the hormones could regulate enzyme activities associated with cell expansion by locally shifting endogenous conditions nearer the enzyme optimum (e.g., a pH shift).

Only in the ethylene-induced growth and separation of Type 2 cells does it appear that enhanced glucanase activity could play a direct role. The ethylene-stimulated induction (8, 13) and secretion of enzymes through the plasmalemma could be causatively related to the ensuing degradation of middle lamellae structure (22, 23) prior to cell separation. The ethylene-induced synthesis and secretion of 1:3 and 1:4-glucanases in Type 2 cells is therefore analogous to the gibberellin-induced synthesis and secretion of α-amylase in the gibberellin target cells of the barley aleurone (27).

The identification of three classes of target cells for auxin or ethylene-enhanced growth, and the evidence for a new and perhaps unique class of cell that will enlarge in response to both auxin and ethylene, now affords new opportunities for physiologists to study the biochemical and physical changes involved in cell growth. In terms of plant development, the concept of positional differentiation of special kinds of target cells in precise regions of plants, and the implication of different hormonal molecular receptor systems permitting different metabolic changes, should provide new leads for the study of the hormonal control of plant growth. The factors that control positional differentiation of the different cell types and the question whether the cell types are determined at cell division, or undergo a post-division differentiation permitting transition from one cell type to another will be a rewarding study for the future.

References

1. Beyer, E.M., Morgan, P.W.: Plant Physiol. 48, 208 (1971)
2. Burg, S.P., Burg, E.A.: Botan. Gaz. 126, 200 (1965)
3. Cleland, R.: Planta 74, 197 (1967)
4. Craker, L.E., Abeles, F.B.: Plant Physiol. 44, 1144 (1969)
5. Evans, M.: Plant Physiol. 54, 213 (1974)
6. Frenkel, C., Klein, I., Dilley, D.R.: Plant Physiol. 43, 1146 (1968)
7. Hallaway, H.M., Osborne, D.J.: Science 163, 1067 (1969)
8. Horton, R.F., Osborne, D.J.: Nature (London) 214, 1086 (1967)
9. Jackson, M.B., Hartley, C.B., Osborne, D.J.: New Phytologist 72, 1251 (1973)
10. Jackson, M.B., Osborne, D.J.: Nature (London) 225, 1019 (1970)
11. Kang, B.G., Newcomb, W., Burg, S.P.: Plant Physiol. 47, 504 (1971)
12. Knegt, E., Bruinsma, J.: Phytochemistry 12, 753 (1973)
13. Lewis, L.N., Varner, J.E.: Plant Physiol. 46, 194 (1970)
14. MacLachlan, G.A., Davies, E., Fan, D.F.: In: Biochemistry and Physiology of Plant Growth Substances. Wightman, F., Setterfield, G. (eds.). Ottawa: Runge Press 1968, p. 443
15. Masuda, Y., Yamamoto, R., Tanimoto, E.: Plant Growth Substances 1970, Carr, D.J. (ed.). Berlin-Heidelberg-New York: Springer-Verlag, 1972, p. 17
16. Murray, A.K., Bandurski, R.S.: Plant Physiol. 56, 143 (1975)
17. Musgrave, A., Jackson, M.B., Ling, E.: Nature (New Biol.) 238, 93 (1972)
18. Musgrave, A., Walters, J.: Planta 121, 51 (1974)
19. Neljubow, D.N.: Beitr. Botan. Zentr. 10, 128 (1901)
20. Nevins, D.J.: Plant Physiol. 46, 458 (1970)
21. Osborne, D.J.: Perspectives Exp. Biol., Vol. 2, Botany. Sunderland, N. (ed.). New York: Pergamon Press 1976, p. 89

22. Osborne, D.J., Sargent, J.A.: Planta 130, 203 (1976)
23. Osborne, D.J., Sargent, J.A.: Planta (in press)
24. Ridge, I., Osborne, D.J.: Nature (London) 223, 318 (1969)
25. Sargent, J.A., Atack, A.V., Osborne, D.J.: Planta 109, 185 (1973)
26. Stoessl, A., Venis, M.A.: Analyt. Biochem. 34, 344 (1970)
27. Varner, J.E.: Plant Physiol. 39, 413 (1964)
28. Went, F.W.: Rec. Trav. Botan. Neerl. 25, 1 (1928)
29. Wright, M., Osborne, D.J.: Planta 120, 163 (1974)

Paper Demonstration References (see Section IX):

4 - 11 - 13 - 18 - 34 - 39 - 41 - 52 - 53 - 56 - 59 - 74 - 76 - 91
116 - 117 - 124 - 159 - 179

19. On the Role of Ethylene in Aging

H. KENDE and A. D. HANSON

Abstract

Excised tissue of the morning glory flower (*Ipomoea tricolor*, Cav.) was used to determine the sequence of events leading to ethylene production and senescence. Results are presented in support of a working hypothesis according to which changes in membrane permeability lead to the terminal surge in ethylene synthesis. Ethylene generation during senescence is not regarded as the triggering mechanism for aging but as a mechanism that accelerates and integrates the aging process of a whole organ.

A. Introduction

Climacteric fruits and certain flowers undergo strikingly similar metabolic changes during aging. In both instances, the senescent phase is characterized by a rise in respiration, by a surge in ethylene production, and by increased permeability of cellular membranes. Also, aging of such fruits and flowers can be induced prematurely by treatment of the respective organ with low levels of exogenous ethylene. From the very onset of research on fruit ripening, there was a controversy concerning the causal relationship between increased membrane permeability, or loss of "organization resistance", as Blackman (3) has termed it, and the aging process. In other words, is ripening a result of the degradation of membranes or vice versa? A similar question has been raised concerning the role of ethylene. Is ethylene the trigger of senescence, i.e., the ripening hormone proper, or is its production a consequence of earlier processes of aging? The significance of increased membrane permeability and of ethylene production in relation to aging has been summarized and brought into the proper historic perspective by Pratt and Goeschl (17). In the past years, the general tendency has been to regard permeability changes as a consequence of aging (1, 4) and to assign ethylene the role of a trigger that induces senescence (16, 17, 18), even though Sacher does cite instances where changes in membrane permeability precede other signs of aging (18), and even though the significance of ethylene as the ripening hormone has been challenged recently in at least two cases (10, 14).

The following summary of our work on the fading of the morning glory flower addresses itself mainly to the questions mentioned above: what are the causal relationships between aging, ethylene evolution, and increased membrane permeability? At present, we do not know whether it is possible to apply conclusions gained from our experiments to climacteric fruits or even to other flowers. However, the many basic similarities in fruit and flower aging indicate that the biochemical processes underlying aging are of general occurrence.

B. Materials and Methods

I. Plant Material

Plants of *Ipomoea tricolor* Cav. were grown in a growth chamber with 23°C day and 18°C night temperatures and a 16-h photoperiod. In work with isolated tissue, 1-cm segments of the corolla ribs were excised either from open flowers or from flower buds. The whole procedure was performed as rapidly and gently as possible to minimize injury to the tissue. In the text, we have used the following abbreviations to describe the age of flower tissues: day 0, day of flower opening and fading; day -1, one day before flower opening; day -2, two days before flower opening. The excised segments were floated on 2-5 ml of incubation medium, with additives and for time periods as specified for each experiment. More details on these procedures were given earlier (5, 9).

II. Measurement of Rolling Up of Rib Segments

To assess the progress of aging, the sides of rib segments in the flasks or syringes were aligned alongside a protractor and the degree (α) of rolling was measured. When segments were flat, α was scored as 180°; as rolling up occurred, α increased to > 360°, with the segment ends overlapping (5).

III. Ethylene Determination

Rib segments floating on solution produce little ethylene. For this reason, segments were transferred to small Erlenmeyer flasks (25 or 50 ml) containing 5-10 ml agar, or, in some instances, to plastic syringes with a lining of wet filter paper. The containers were stoppered with serum vial caps, flushed with ethylene-free air, and gas samples were withdrawn using 1-ml disposable syringes. The ethylene concentration in the samples was determined by gas chromatography. In calculating the amount of ethylene evolved, a correction was applied for the repeated removal of samples and their replacement with ethylene-free air (9).

IV. Analytical Methods

Compartmental analysis, determinations of methionine and S-methylmethionine (SMM) metabolism and of the specific radioactivities of ethylene and methionine were performed as described earlier (5, 6).

C. Results and Discussion

I. The Experimental System

Intact flowers of the morning glory *Ipomoea tricolor* open at 5:00 to 6:00 a.m. and begin to fade at about 1:00 p.m. of the same day. During the fading process, the corolla rolls inward driven by curling up of its five ribs, and a sharp increase in the rate of ethylene evolution occurs; the rolling up and ethylene evolution can be induced prematurely by the application of exogenous ethylene (8). Rolling up of the corolla is accompanied by increases in the activities of several hydrolases, by a rise in respiration, and by a decrease in vacuolar pH that causes a color change from blue to purple (19). Since the preceding changes all take place within less than 10 h, morning glory flowers are suitable for investigating the precise timing of the reactions that are thought to be involved in the senescence process.

Experiments have either been conducted with intact flowers or buds (8) or, more recently, with 1-cm segments excised from the ribs of the corolla (5, 6, 9). Isolated rib segments exhibit the same symptoms of aging as the respective intact tissue. They roll up at the same time of the day when the flowers begin to fade, they evolve ethylene during senescence, and their aging can be hastened by applied ethylene. We turned to the use of rib segments because they readily take up labelled percursors, ions, and inhibitors from the incubation medium, while it is very difficult to introduce any compounds into the intact flower. Excised tissue also permits measuring the efflux of cellular contents into the bathing solution.

II. The Working Hypothesis

In rib segments of the morning glory corolla, the following developmental stages leading to senescence can be distinguished: two days before flower opening (day -2), the segments are juvenile and do not roll up in response to exogenous ethylene. On the day before flower opening (day -1), rib segments roll up but do not evolve ethylene as a result of ethylene treatment. On the day of flower opening (day 0), rolling up and ethylene production can both be induced prematurely by exogenous ethylene. Hence, the sensitivity to ethylene and the capacity to produce ethylene "autocatalytically" develop sequentially over a two-day period.

We suggest that metabolic processes leading to aging precede the rise in ethylene production and the rolling up of rib segments. We propose that compositional changes of membranes, most likely of the tonoplast and perhaps also of the plasmalemma, are among the earliest events in flower aging. We further propose that in the juvenile tissue, components of the ethylene-generating system are spatially separated. Such a compartmentation may entail sequestering of ethylene-synthesizing enzyme(s), substrate(s), and cofactor(s), and the most likely compartments involved appear to be the cytoplasm and the vacuole. The proposed changes in membrane composition would, according to our hypothesis, lead to increased membrane permeability, which, in turn, would result in mixing of formerly separated components of the ethylene-generating system and therefore in the observed rise in ethylene production. Endogenously evolved and exogenously applied ethylene are thought to hasten the loss of cellular compartmentation, thereby accounting for the "autocatalytic" stimulation of ethylene synthesis.

In the following, we shall summarize evidence supporting this model.

III. Cellular Compartmentation During Aging

The change in cellular compartmentation was investigated by prelabeling rib segments with radioactive ions or organic molecules and observing the efflux of these substances during natural and ethylene-induced aging (5). Using $^{36}Cl^-$ as marker and analyzing the efflux kinetics according to MacRobbie and Dainty (13) and Pitman (15) by plotting the logarithm of the percent $^{36}Cl^-$ remaining in the tissue versus the time elapsed, at least two phases of ion loss could be discerned. After an initial rapid loss of $^{36}Cl^-$, efflux reached a constant rate at about 20 min that was then maintained for 7 h. The slow component had a half-life of about 6 h, and extrapolation of the slope of the slow phase to zero time showed that it accounted for some 80% of the $^{36}Cl^-$ absorbed (5). The $^{36}Cl^-$ released during the first 20 min is thought to originate from the free space and the cytoplasm, and the slowly exchanging $^{36}Cl^-$,

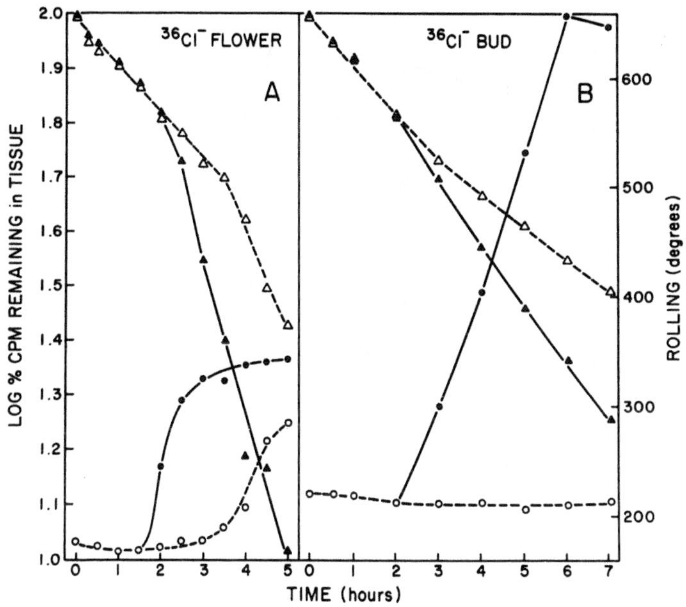

Fig. 1 A,B. Effects of ethylene on rolling up and loss of $^{36}Cl^-$ from rib segments of flowers (A) and buds (B). Segments were incubated under gas phase of air only (control) or of 10 µl/l ethylene added at zero time. o---o, rolling up of control segments; ●——●, rolling up of ethylene-treated segments; Δ---Δ, $^{36}Cl^-$ efflux from control segments; ▲——▲, $^{36}Cl^-$ efflux from ethylene-treated segments. $^{36}Cl^-$ present in tissue at start of efflux period = 1.1 × 10^4 cpm (*flowers*), 3.5 × 10^4 cpm (*buds*) per 10 segments. From Hanson and Kende (5)

from the vacuole. In subsequent experiments (Fig.1), the rib segments were washed twice for 10 min before ion efflux was measured. Hence, only the slow phase of ion release was investigated. Because of the 90 min lag in the action of ethylene, the effect of the gas on the rapid efflux phase could not be assessed.

Figure 1 shows the slow $^{36}Cl^-$ efflux component observed in both flower and bud rib segments as they rolled up. In flower rib segments (Fig.1A) $^{36}Cl^-$ efflux increased sharply after 2 h of ethylene treatment as premature rolling up was developing. In untreated segments, an increase in efflux rate occurred later, as spontaneous rolling up took place.

In bud rib segments (Fig.1B), which do not roll up spontaneously, efflux from untreated segments was maintained at a fairly constant rate. After 2 h of ethylene treatment, however, the efflux rate increased as the rolling up proceeded. Similar results were obtained when $^{86}Rb^+$ was used instead of $^{36}Cl^-$, or when the efflux of radioactive sucrose or organic acids was observed (5).

The compartmental analysis of $^{36}Cl^-$ and $^{86}Rb^+$ efflux shows very clearly an abrupt change in cellular compartmentation during natural and ethylene-induced aging. The data also indicate that the permeability of the tonoplast may be affected first, although similar permeability changes of the plasmalemma cannot be excluded at this point. The results with the organic assimilates further support the proposition that compartmentation between the vacuole and the cytoplasm is lost since both su-

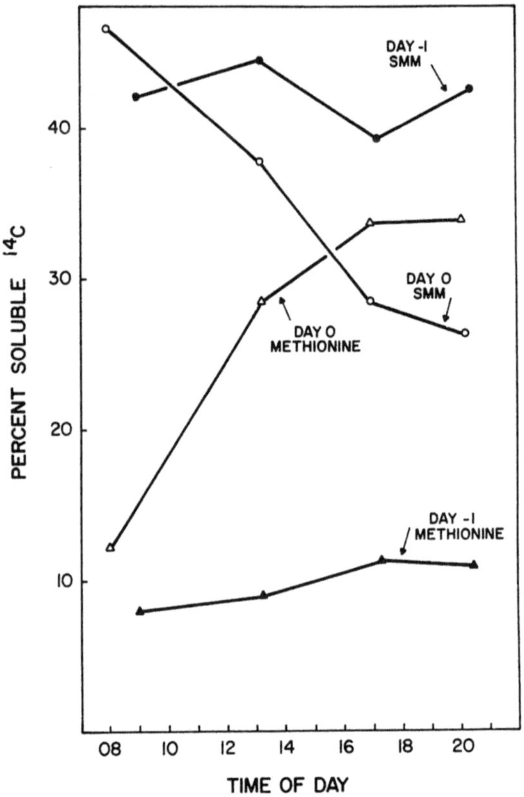

Fig. 2. Distribution of ^{14}C in ethanol-soluble fractions of segments cut on day -2 and day -1, incubated overnight on ^{14}C-methionine and extracted at various times during the following day. Concentration of ^{14}C-methionine was 6 μM (day -1) and 12 μM (day -2); radioactivity distribution was determined by scintillation counting of zones cut from TLC plates. From Hanson and Kende (6)

crose and organic acids have been reported to be stored in the vacuole (7, 12). Loss of the semipermeability of the tonoplast is also indicated by ethylene-induced release of anthocyanine from *Tradescantia* petals (Suttle and Kende, unpublished results) and by the ultrastructural investigations of Winkenbach using aging corolla tissue of the morning glory flower (19). Last, the rolling up of the rib segments itself is an indication of permeability changes of membranes, because this curling movement was shown to be due to an asymmetric loss of turgor that is controlled by either the tonoplast or the plasmalemma or both (5).

IV. The Pathway of Ethylene Biogenesis During Aging

We have begun to study the pathway of ethylene synthesis in morning glory tissue in the hope that this will help us to understand how changes in compartmentation may regulate ethylene formation. This part of our work is far from completed, but it indicates new approaches to the foregoing problem. Our results can be summarized as follows:

1. All ethylene evolved from rib segments during aging can be accounted for as originating from carbon atoms 3 and 4 of methionine. This was

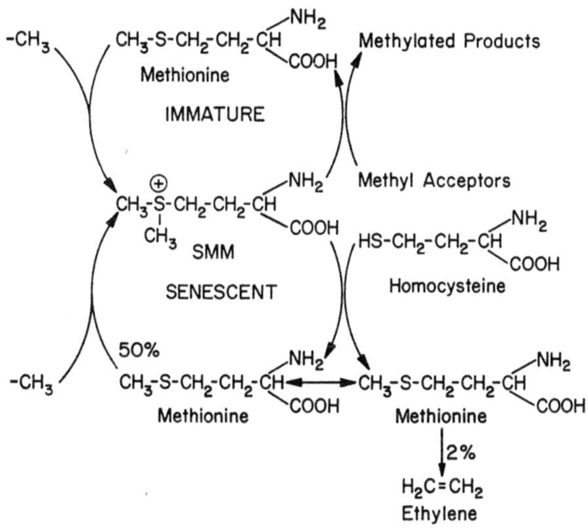

Fig. 3. Scheme of methionine metabolism in immature and senescent flower tissue of morning glory. From Hanson and Kende (6)

shown by prelabeling rib segments with methionine-U-^{14}C and by measuring the specific radioactivities of the ethylene evolved as well as that of carbon atoms 3 and 4 of methionine extracted from the tissue (6).

2. During the senescent phase there is a tenfold rise in the level of free methionine in rib segments. This increase is much greater than that of other amino acids during the same period (6).

3. Rib segments incubated on ^{14}C-methionine during the night between day -2 and day -1 incorporated about 60% of the total ^{14}C absorbed into ethanol-insoluble products. The major labeled ethanol-soluble product was S-methylmethionine (SMM); only a small portion of the ^{14}C remained in methionine (6). During the subsequent incubation of these immature segments, the distribution of ^{14}C within the ethanol-soluble fraction did not change significantly (Fig.2). Rib segments incubated on ^{14}C-methionine during the night between day -1 and day 0 showed a pattern of distribution of ^{14}C similar to that found in immature segments; about 45% of the ^{14}C was in ethanol-insoluble form, and in the ethanol-soluble fraction SMM contained considerably more radioactivity than methionine. However, during incubation on day 0, ^{14}C was lost from SMM and appeared in free methionine (Fig.2).

Results from dual-label experiments using methionine-U-^{14}C and methionine-methyl-^{3}H showed that SMM was metabolically active in immature tissue in that it was serving continuously as a methyl donor. While the identity of the methyl acceptors was not identified, the results indicated that the transmethylation reactions did not lead to a net increase in the level of methionine (6). It was concluded that the metabolic pathway summarized in the upper part of figure 3 was operative in immature rib segments. In senescing ribs, homocysteine became the predominant acceptor of methyl groups stemming from SMM (6). This was demonstrated by assessing the fate of the radioactive methyl groups of SMM and by tracing the products of ^{35}S-homocysteine metabolism. This transmethylation reaction summarized in the lower part of Figure 3 yields for each molecule of SMM two molecules of methionine, and the operation of this cycle, in which half of the methionine is re-methylated to SMM, results in a net increase of the methionine concentration. The SMM-homocysteine transmethylation reaction can be induced prematurely by treatment of the tissue with ethylene or with homocysteine (6). Both of these treatments lead to premature aging of the rib segments and on day 0 to a premature rise in ethylene synthesis. The effect of homocysteine is particularly noteworthy since applied methionine did not advance the aging of the tissue or stimulate ethylene production.

To explain the observed changes in the metabolism of methionine and their role in the synthesis of ethylene, we suggest the following working hypothesis. Ethylene production is dependent on methionine derived from SMM and homocysteine, possibly because the ethylene-generating and methionine-synthesizing systems are located in the same cellular compartment. During aging, the supply of homocysteine to this system increases, either through direct stimulation of homocysteine production or because of a breakdown in the compartmentation between stored homocysteine and the methionine-synthesizing system. Ethylene treatment may provoke a premature increase in homocysteine availability and thus lead to earlier ethylene production.

V. Is Ethylene the Trigger of Aging?

We propose that the aging process in morning glory flowers precedes the increase in ethylene production. According to this view, ethylene does not trigger aging but accelerates it. The following observations support this hypothesis:

1. An increase in the rate of ethylene evolution was never detected before the onset of rolling up. The two processes usually began together and, in experiments where aging was induced by ethylene or propylene application, ethylene evolution lagged behind rolling up (9). Rolling up is a turgor-driven movement resulting from a change in the ability of the cells on the inner rib side to retain solutes (5). It is therefore a manifestation of irreversible cell degeneration that must have been preceded by more subtle changes; our data indicate that increased ethylene production was not one of these changes.

2. The conversion of SMM to methionine appears to be a crucial reaction in the senescence process of morning glory flowers. It was found to commence on the morning of day 0 before any signs of senescence were visible and before the increase in ethylene generation was detectable (Ref.6, Fig.2).

3. The level of membrane phospholipids declined steadily during the night between day -1 and day 0. This may be indicative of changes in membrane composition that ultimately lead to the observed changes in membrane permeability. At the time when rib segments roll up and when they start to produce copious amounts of ethylene, the rate of phospholipid loss was accelerated about sixfold (2).

4. Rhizobitoxine and its ethoxy analog at concentrations of 10^{-5} M to 10^{-4} M are potent inhibitors of ethylene production in apple tissue (11). The capacity of rib segments to produce ethylene was also severely reduced by the ethoxy analog of rhizobitoxine; however, senescence, as scored by rolling up of the segments, was only delayed but not eliminated. In order to assess accurately to what extent ethylene synthesis was inhibited, 100 rib segments, either pretreated with the rhizobitoxine analog or not, were each enclosed in a syringe leaving only the minimum volume of free gas space from which samples for ethylene

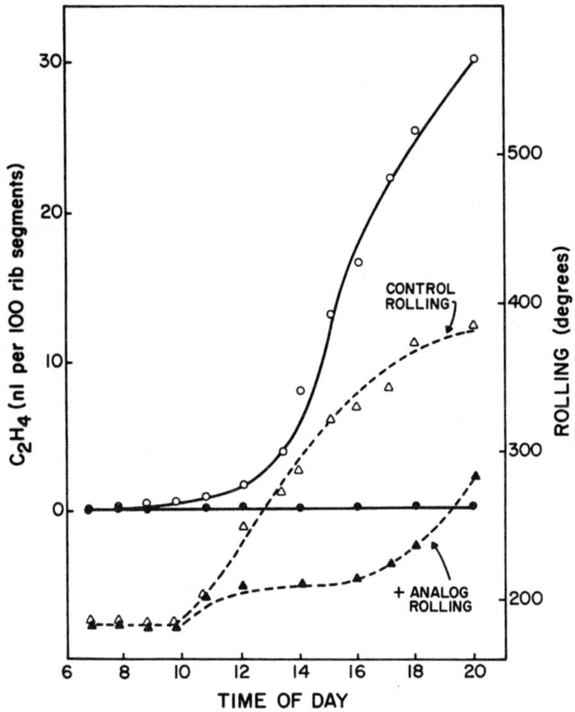

Fig. 4. Inhibition of spontaneous ethylene production and rolling up by rhizobitoxine analog. Batches of 100 segments were cut on day -1 and incubated overnight (5:00 p.m. to 5:30 a.m.) on 5 mM KCl (control) or 5 mM KCl containing 3×10^{-5} M rhizobitoxine analog. Each batch was enclosed in a 10-ml syringe for measurement of ethylene production. Δ---Δ, rolling up of control segments; ▲---▲, rolling up of inhibitor-treated segments; o——o, ethylene production by control segments; ●——●, ethylene production by inhibitor-treated segments. Total amount of ethylene evolved by inhibitor-treated segments was 0.22 nl. From Kende and Hanson (9)

analysis were withdrawn. As is evident from Figure 4, the ethoxy analog of rhizobitoxine inhibited ethylene formation by more than 99%, yet rolling up was again only delayed and slowed but not eliminated. In experiments where treatment with the rhizobitoxine analog was combined with ethylene trapping to remove trace amounts of residual ethylene, or with exposure of segments to CO_2 to counteract the effects of residual ethylene, rolling up still occurred at rates comparable to those of the control, albeit with some delay (9). It thus appears likely that this rolling up was independent of ethylene action.

5. When benzyladenine was applied to rib segments during the night between day -1 and day 0, senescence was retarded (9). This points to the possibility that benzyladenine delayed some early events in senescence, thus postponing the later sequence of symptoms.

D. Conclusions

We propose that the following sequence of events leads to senescence of the morning glory flower. Metabolic processes related to aging precede visible signs of senescence and the terminal increase in ethylene synthesis. One of these processes appears to lead to changes in membrane composition and, ultimately, to changes in membrane permeability. The resulting loss in cellular compartmentation may be the basis for the loss of turgor that causes rolling up of the segments and the surge in ethylene production. Ethylene itself, through some biological mechanism, acts to increase membrane permeability, thus promoting its own synthesis and accelerating aging. Accordingly, we do not regard ethylene as the substance that triggers aging on the cellular level. However, the significance of ethylene in initiating senescence is more complex when viewed at the level of the whole organ, which is comprised of many types of cells that probably do not age synchronously. Because of its unique capacity to induce both senescence and its own synthesis, ethylene originating from a few cells that age precociously could trigger the remaining tissues in a fruit or flower to enter the senescent phase, thereby integrating the aging of the whole organ. Thus the view that ethylene production is a consequence of cellular aging can readily be reconciled with the concept that ethylene acts as the ripening or senescence factor in climacteric fruits and in many flowers.

Acknowledgement

This research was supported by the U.S. Energy Research and Development Administration under Contract E(11-1)-1338.

References

1. Abeles, F.B.: Ann. Rev. Plant Physiol. 23, 259 (1972)
2. Beutelmann, P., Kende, H.: Plant Physiol. (Suppl.) 57, 97 (1976)
3. Blackman, F.F., Parija, P.: Proc. Royal Soc. (London), Ser. B. 103, 412 (1928)
4. Burg, S.P.: Plant Physiol. 43, 1503 (1968)
5. Hanson, A.D., Kende, H.: Plant Physiol. 55, 663 (1975)
6. Hanson, A.D., Kende, H.: Plant Physiol. 57, 528 (1976)
7. Hardy, P.J., Norton, G.: New Phytologist 67, 139 (1968)
8. Kende, H., Baumgartner, B.: Planta 116, 279 (1974)
9. Kende, H., Hanson, A.D.: Plant Physiol. 57, 523 (1976)
10. Kosiyachinda, S., Young, R.E.: Plant Cell Physiol. (Tokyo) 16, 595 (1975)
11. Lieberman, M., Kunishi, A.T., Owens, L.D.: In: Facteurs et Régulation de la Maturation des Fruits. Ulrich, R. (ed.). Paris: Editions du C.N.R.S., 1974, p. 161
12. Lips, S.H., Beevers, H.: Plant Physiol. 41, 709 (1966)
13. MacRobbie, E.A.C., Dainty, J.: Physiol. Plantarum 11, 782 (1958)
14. McGlasson, W.B., Dostal, H.C., Tigchelaar, E.C.: Plant Physiol. 55, 218 (1975)
15. Pitman, M.G.: Australian J. Biol. Sci. 16, 647 (1963)
16. Pratt, H.K.: In: Facteurs et Régulation de la Maturation des Fruits. Ulrich, R. (ed.). Paris: Editions du C.N.R.S., 1974, p. 153
17. Pratt, H.K., Goeschl, J.D.: Ann. Rev. Plant Physiol. 20, 541 (1969)
18. Sacher, J.A.: Ann. Rev. Plant Physiol. 24, 197 (1973)
19. Winkenbach, F.: Ber. Schweiz. Botan. Ges. 80, 374 (1970)

Paper Demonstration References (see Section IX):

14 - 40 - 49 - 85 - 152

20. Phenolic Substances and Their Possible Role in Plant Growth Regulation

V. I. KEFELI and M. KUTACEK

Abstract

In relation to their effect on growth processes, plant phenols may be divided into three groups: promotive, inhibitory, and inactive. Promotion of plant growth by phenols may proceed through the modulation of either IAA biosynthesis or its destruction. Inhibition of plant growth by phenols may be effected in one of the following ways:

1. Depression of IAA biosynthesis and/or activation of IAA degradation

2. Lowering of the growth-stimulating activity of IAA, gibberellins, or cytokinins

3. Uncoupling of respiration and oxidative phosphorylation

4. Interaction of the quinone forms of phenols with proteins and hence inhibition of metabolic pathways.

Phenolic substances do not possess antihormonal specificity; rather, they modify the activity of any of the known plant hormones.

A. Introduction

The physiologic role of phenolic substances is controversial. Some investigators argue against any role of phenols in plant growth regulation, because these compounds are localized in closed compartments of the cell, such as the vacuole, and hence remote from control of physiologic processes. Other scientists feel that endogenous phenols possess only stimulatory properties and thus can promote growth; they have in mind the auxin-like properties of caffeic, ferulic, and chlorogenic acid; lastly, a third group of investigators feels that at least some phenols, such as chalcones, flavanones, isoflavones, and derivatives of p-coumaric acid, may interfere with plant growth and thus act as endogenous growth regulators.

The authors of this report belong to the last of these three groups and attempt in the following to present evidence for their viewpoint.

The experimental work referred to was carried out in collaboration with numerous colleagues from various countries including G. Sembdner, V. Datte, and W. Bleiss (GDR), J.S. Knypl (Poland), K. Vackova, J. Eder, and M. Kralova (CSSR), Le-Tchi-Muoi (Vietnam) and N. Amrhein (FRG).

B. Materials and Methods

Experimental details are indicated under Results and Discussion where necessary or will be found in the references.

The following is a brief survey of the diverse techniques used to investigate the role of phenolic substances in plant growth regulation:

1. Determination of quantity and quality of phenols in growing and resting tissues and organs (6, 10)

2. Isolation of phenols and their application to growing plant material (3, 10)

3. Investigation of the localization of phenols in cell organelles (11)

4. Analysis of phenols in the bleeding sap, phloem, and xylem (9)

5. Reciprocal grafts between plants containing different amounts of phenols (1)

6. Studies of the effects of applied phenols on: (a) the biosynthesis of auxin from radioactive precursors (7); (b) the growth response of plant materials evoked by auxins, gibberellins, or cytokinins (4)

7. Application of phenols to enzyme preparations catalyzing auxin destruction (19)

8. Study of the metabolism of phenols by growing and resting tissues (6)

C. Results and Discussion

I. Relation Between Phenolic Content and Growth

Usually, when the growth of a plant stops, the concentrations of phenols in the tissues increase, in particular, of those that serve as precursors for lignin biosynthesis and have not been utilized in cell wall constitution during the growth period. There are, however, cases in which the phenolic concentration increases while the plant is growing.

When pea plants were exposed to high-intensity light (xenon arc lamp), growth was depressed, but concurrently, the quantities of quercetin derivatives, of which quercetin-3-glucosyl-p-coumarate (QGC) was present in the greatest amount (75%), increased (Table 1). A similar inverse correlation between growth and QGC content was observed for various dwarf pea mutants (Table 2). Analysis of the biological activity of the individual components of the complex QGC molecule in growth tests with wheat coleoptile and pea epicotyl sections shows that only one of the components, namely p-coumaric acid, possesses inhibitory properties.

Thus these data may exemplify the correlation between the high level of phenols and growth depression. Another example of this inverse relationship is provided by the buds of the willow (*Salix purpurea* L.): when these buds enter dormancy in the fall, the concentration of the naringenin derivative, isosalipurposide, increases three-fold (6). At concentrations of 0.25 mg/ml and above, this compound inhibits growth of wheat coleoptile sections (Table 3).

The general type of experiment considered so far has at least one debatable aspect in that a situation is produced by the experimentator that does not exist in the intact plant. Nevertheless, it can be hypothesized that at least some phenols participate in growth regulation, although the mechanism of this participation remains obscure.

Table 1. Effect of xenon-arc light on growth and QGC content of 20-day-old pea (cv. Early Green) plants

Response	Light intensity ($erg/cm^2/sec \times 10^3$) 0	25	200	300	420
Plant height (cm)	42	25	13	8	5
QGC content (in % of control)	100	200	263	364	390

Table 2. Relation between QGC content and growth of pea plants (cv. Torsdag) and mutants derived from this cultivar

Cultivar	Plant height (cm)	QGC content (mg/g dry wt)
Torsdag	41.4	1.8
Mutants:		
Compact	28.2	3.7
Branched	17.0	3.9
Dwarf	5.4	4.9

Table 3. Effect of isosalipurposide (ISP) on growth of wheat coleoptile sections

	ISP concentration (mg/ml) 0	0.25	0.50	1.0	2.0	4.0
Growth (in % of control)	100	72	40	8	0	0

II. Compartmentation of Phenolics

It is well established that the major portion of a cell's phenolic constituents is localized in the vacuole and the cell wall. On the other hand, the presence of phenols in willow chloroplasts has been demonstrated, albeit at concentrations some 50 times less (on a dry weight basis) than those found in the cytosol (11). Results of other investigators have supported this finding (14, 20). Photophosphorylative reactions might thus be influenced by chloroplast-localized phenols. There is no evidence for long-distance transport of phenols in plants, and phenols seem to be absent from the bleeding sap of birch trees, even though phloem and xylem tissues contain large amounts of phenolic products (9). Also, when apical segments of normal pea plants (low level of QGC) were grafted onto basal segments of dwarf pea mutants (high level of QGC), no evidence was obtained for the transfer of a growth inhibitory factor from the dwarf stock; and the same was true for reciprocal grafts (1). If phenols are not translocated in plants, the question remains open as to how they exert their effect on the growth of the total plant.

III. Phenolics and Auxin Biosynthesis and Destruction

Various steps of the biosynthetic pathway leading to the formation of indoles may be subject to interference from phenolic compounds. Both classes of compounds have in fact a common precursor, namely, chorismic acid. The branching of the shikimate pathway and the steps leading to the formation of indole and phenol derivatives are depicted in the following scheme:

Shikimic acid
↓
Chorismic acid
↙ ↘

Anthranilic acid	Prephenic acid
↓	↓
Indole glycerophosphate	L-phenylalanine
↓	↓
Indole	Cinnamic acid
↓	↓
L-tryptophane	p-coumaric acid
↓	
Indole-3-acetic acid (IAA)	
↓	
(IAA-glucose, IAA-asp, etc.)	

One of the most probable targets of the action of phenols in the cell is the interference with the pathway of IAA biosynthesis. It has been shown that ^{14}C-labeled anthranilic acid is converted uniformly to labeled tryptophan in plant tissues. Furthermore, labeled IAA was formed from 3-^{14}C-tryptophan. Both metabolic sequences are subject to interference by indole metabolites (intermediates of IAA biosynthesis) and by phenols (12). p-coumaric acid - one of the phenolic substances that remarkably inhibit cell elongation - depresses formation of tryptophan from ^{14}C-anthranilic acid, whereas caffeic acid, which enhances elongation growth, stimulates the incorporation of anthranilic acid into tryptophan (Table 4).

Thus one control point in the action of phenols on plant growth may be the depression of L-tryptophan biosynthesis. Another control point may be in the metabolic sequence leading from L-tryptophan to IAA. Natural phenolic growth inhibitors were checked for their influence on indole biosynthesis in the plants from which they were isolated, and were found to change the incorporation pattern of the radioactive precursor (Table 5). Although growth promotion and inhibition by phenols could thus partially be explained by the action of phenols on IAA biosynthesis, an effect on IAA destruction also seems feasible. It is known that phenols can modify IAA oxidase activity and consequently the effect of IAA on growth, e.g., the rooting of bean stem cuttings (19). IAA promotes the rooting of these cuttings; this effect can be modified by the simultaneous addition of phenols: ferulic, chlorogenic, and caffeic acids enhance the effect of IAA, while p-coumaric acid diminishes the effect of IAA. In the absence of IAA, the phenolic substances have hardly any effect (Table 6). The promotion of the IAA effect by ferulic, caffeic, and chlorogenic acids is correlated with their protective effect on IAA destruction by IAA oxidase (i.e., inhibition of IAA oxidase), while p-coumaric acid enhances IAA oxidase activity, which is correlated with their inhibitory action on IAA-induced rhizogenesis.

Some of those phenols that are able to modify IAA effects on growth are present in plants not only in free but also in bound forms, as is the case with QGC in peas (see above). A similar situation exists in *Brassica oleraceae* var. gongyloides, from the cotyledons of which we have isolated a bound form of sinapic acid (as sinapine derivative). This substance is highly unstable in alkaline solution and releases, among

Table 4. Effect of p-coumaric and caffeic acids on elongation of hypocotyl sections of *Brassica oleracea* var. gongyloides and on L-tryptophan synthesis from ^{14}C-anthranilic acid (data indicate % of control)

Substance	Concentation (M)	Growth	Incorporation of label
p-coumaric	5×10^{6}	99	78
acid	5×10^{3}	48	46
caffeic	5×10^{6}	98	113
acid	5×10^{3}	186	230

Table 5. Effect of phenols on idole biosynthesis from 3-^{14}C-L-tryptophan (in % deviation from control)

Applied compound	Test plant	Radioactivity incorporated into metabolite			
		Indole-acetamid	IAA-glucose	L-trypto-phan	gluco-brassicin
p-coumaric acid 0.7 mg/ml	*Zea mays*	+32[a]	-40[b]	+34	-
QGC 8 mg/ml	*Pisum sativum*	-72	-88	+65	-
Unknown phenolic acid from *Brassica oleracea* 0.5 mg/ml	*Brassica oleracea*	-22	-49	-8	+19

[a] +: enhanced incorporation of precursor

[b] -: diminished incorporation

Table 6. Effect of IAA and phenolcarbonic acids on rhizogenesis of bean stem cuttings (*Phaseolus vulgaris*, cv. Zelenostruchnaya). Concentrations are indicated in mg/l

Treatment	Roots per cutting
Water control	4.6
IAA 70	34.1
IAA 100	60.3
Ferulic acid	5.8
Ferulic acid + IAA, 70	49.3
Caffeic acid	6.0
Caffeic acid + IAA, 70	40.9
Chlorogenic acid	5.9
Chlorogenic acid + IAA, 70	35.0
p-coumaric acid	6.1
p-coumaric acid + IAA, 70	24.8

p = ±15%

Table 7. Effect of p-coumaric acid (PCA) and coumarin (COU) on growth elicited by auxin (IAA), gibberellic acid (GA_3), and kinetin in standard biotest systems. Data are in % of control (respective hormone and inhibitor absent). Concentrations used were with wheat coleoptiles: IAA, 7mg/l; PCA, 8 mg/l; COU, 8 mg/l; with pea seedlings: GA, 0.2 mg/l; PCA, 150 mg/l; COU, 20 mg/l; with the soybean callus: K, 1 mg/l; PCA, 100 mg/l; COU, 100 mg/l

Test system and phytohormone applied	Relative growth in the presence of				
	Phyto-hormone	PCA	COU	Phyto-hormone + PCA	Phyto-hormone + COU
Wheat coleoptile sections (cv. Ulyanovskaya) IAA	303	100	107	216	218
Pea seedlings (cv. Torsdag) mutant K-202 GA_3	357	110	100	214	271
Soybean callus (cv. Komsomolka) Kinetin	1015	160	198	602	242

the degradation products, sinapic acid which possesses weak growth stimulatory activity. This activity may again be due to a protective effect on IAA oxidation.

IV. Other Possible Modes of Action of Phenols on Plant Growth

We cannot exclude the possibility that phenolic substances act on plant growth in other ways, that is, by interference with processes not related to auxin. The results of the experiments presented in Table 7 show that phenols interfere not only with the action of auxin, but also with the actions of gibberellins and cytokinins on plant growth, i.e., they act nonspecifically. Such nonspecific action is illustrated in the following scheme:

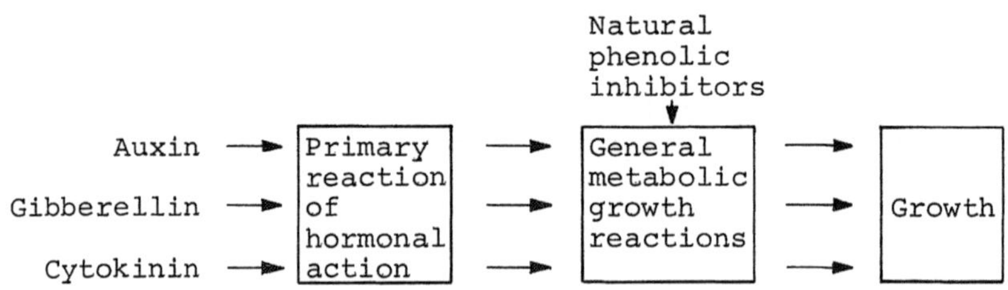

The multiple effects of some phenols might thus be explained by their action on general metabolic processes involved in the expression of growth, such as oxidative or photosynthetic phosphorylation, biosynthesis of nucleic acids and protein, etc. An example of the action of a natural phenolic inhibitor, phloridzin, on oxidative phosphorylation is presented in Table 8. Similar effects of phloridzin and other phenolics were observed by Stenlid (16). In oxidized (quinone) form, phenols show a remarkably enhanced activity, as is evident from Table 9, in which the action of quinones and the corresponding hydroquinones on

Table 8. Effect of phlorizin on oxidation and phosphorylation by pea mitochondria. Data are in % of control (17)

Phlorizin (M)	ΔO	ΔP
0	100	100
2.7×10^{4}	105	108
5.3×10^{4}	108	70
8.0×10^{4}	40	1
1.0×10^{3}	30	0.5

Table 9. Concentration of phenolic substances (mg/l) required to suppress growth of wheat coleoptile sections and rooting of bean cuttings by 50% (C_{50})

Substance	Coleoptile growth	Rooting
p-benzoquinone	120	500
Hydroquinone	1825	1000
Resorcinol	1725	1000
Catechol	475	125

the growth of coleoptile sections and on rhizogenesis is compared. Clearly, in the case of both phenols, the oxidized (quinone) form is more inhibitory to growth processes than the reduced (hydroquinone) form. Conversion of catechol to the O-benzoquinone (resorcinol) during contact with coleoptile sections was established by the stabilization of the quinone as O-dihydroxydiphenylsulfone (13).

D. Conclusions

Let us consider the evidence that supports, or at least does not contradict, the idea of participation of phenols in plant growth:

1. Phenols (QGS and isosalipurposide) accumulate in plant tissue under both natural and experimental conditions (dormancy in willow buds, reduction of pea stem growth)

2. Exogenously applied phenols depress the growth of plant tissues

3. While the bulk of the phenolic constituents of a cell is localized in the vacuole, phenols are also found in chloroplasts and cell walls, where they may act upon phosphorylation or cell elongation, respectively

4. Biosynthesis and destruction of IAA can be modified by phenols. Exogenously applied phenols can be shown to modify the growth response to gibberellins and cytokinins

5. Conversion of phenols to their respective quinone forms enhances their growth-depressing effect

Summing up these arguments, we feel justified in our view that phenolic substances participate in plant growth regulation. In the final elucidation of their mechanism of action there are, however, a number of uncertainties and unresolved congradictions. The localization of phenols mainly in the vacuole and cell walls provokes the question, will their concentration in the cytoplasm reach levels that might affect auxin metabolism and/or phosphorylative reactions in vivo? Exogenous phenols usually interfere with growth only at high concentrations, and it cannot be excluded that under these conditions reactions occur in the cell that are absent under in vivo conditions.

These questions remain unresolved and must be kept in mind and taken into account in further investigations.

References

1. Chailakhyan, M.Kh., Loznikova, V.N., Khlopenkova, L.P., Kefeli, V.I., Sidorova, K.K.: Fiziol. Rast. (Russ.), in press
2. Furuya, M., Galston, A.W., Stowe, B.B.: Nature (London), 193, 450 (1962)
3. Kefeli, V.I.: Dokl. Akad. Nauk SSSR 162, 462 (1965)
4. Kefeli, V.I.: Wiss. Z. Univ. Rostock 17, 383 (1968)
5. Kefeli, V.I., Kadyrov, C.S.: Ann. Rev. Plant Physiol. 22, 185 (1971)
6. Kefeli, V.I., Kof, E.M., Knypl, J.S., Bukhanova, L.V., Jarviste, K.L.: Biokhimiya 34, 891 (1969)
7. Kefeli, V.I., Kutacek, M., Vackova, K.: Biol. Plant. 12, 181 (1970)
8. Kefeli, V.I., Loznikova, V.N., Khlopenkova, L.O., Kof, E.M., Sidorova, K.K., Khvostova, V.V., Turetskaya, R.Kh., Chailakhyan, M.Kh.: Izv. Akad. Nauk SSSR Ser. Biol. 5, 681 (1973)
9. Kefeli, V.I., Sembdner, G., Vlasov, P.V., Datte, V.: Dokl. Akad. Nauk SSSR, in press
10. Kefeli, V.I., Turetskaya, R.Kh.: Fiziol. Rast. (Russ.) 12, 638 (1965)
11. Kefeli, V.I., Turetskaya, R.Kh.: Dokl. Akad. Nauk SSSR 170, 472 (1966)
12. Kefeli, V.I., Turetskaya, R.Kh., Kutacek, M.: In: Hormonal Regulation of Plant Life. Irkutsk: 1974, p. 35
13. Le-Tchi-Muoi, Stom, D.I., Kefeli, V.I., Turetskaya, R.Kh., Timofeeva, S.S., Vlasov, P.V.: Fiziol. Rast. (Russ.) 21, 164 (1974)
14. Monties, B.: Bull. Soc. Franc. Physiol. Vég. 15, 29 (1969)
15. Nitsch, J.P., Nitsch, C.: Ann. Physiol. Vég. 4, 211 (1962)
16. Stenlid, G.: Physiol. Plantarum 21, 882 (1968)
17. Sarapuu, L.P., Kefeli, V.I.: In: Phenolic Substances and Their Biological Functions. Nauka: 1968, p. 129
18. Tomaszewski, M.: In: Régulateurs Naturels de la Végétale, Paris: CNRS, 1964, p. 335
19. Turetskaya, R.K., Gusskov, A.V., Bleiss, V., Kof, E.M., Kefeli, V.I., Kutacek, M.: Fiziol. Rast. (Russ.) in press
20. Zaprometov, M.N., Kolonkova, S.V.: In: Phenolic Substances and Their Biological Functions. Nauka: 1968, p. 175

Paper Demonstration References (see Section IX):

10 - 54 - 78 - 94 - 99 - 100 - 138 - 150 - 157 - 190

21. Some Effects of Dikegulac on the Physiology of Whole Plants and Tissues; Interactions with Plant Hormones

P. F. Bocion and W. H. de Silva

Abstract

Various plant physiologic properties of sodium 2,3:4,6-di-O-isopropylidene-2-keto-L-gulonate (ISO accepted common name: dikegulac) were investigated on a number of bioassay systems. Dikegulac retarded growth of wheat, pea, and *Avena fatua*. This effect could be overcome by GA_3. The growth of tomato callus was retarded at lower concentrations and completely inhibited by higher concentrations of dikegulac. This effect was counteracted by GA_3 and, at certain rates, dikegulac even promoted the growth stimulation activity of GA_3. Parthenocarpic fruit set of pears was stimulated by dikegulac. Apical dominance was suppressed in *Poa pratensis*, *Lolium perenne*, and azalea, interactions with indole-3-acetic acid (IAA) and kinetin were seen in azalea. Translocation was studied on chrysanthemum. Dikegulac is the active ingredient of a new growth regulator (trade name: Atrinal-trademark of Dr. R. Maag Ltd., CH-8157 Dielsdorf, Switzerland) that is used as a pinching agent for ornamentals and as a growth retardant and branching agent for hedge plants.

A. Introduction

Although the importance of monosaccharides to the metabolism of plants has long been recognized, and several plant growth regulators have been synthesized, few of these are carbohydrates and none is a monosaccharide. We were therefore surprised to find that sodium 2,3:4,6-di-O-isopropylidene-2-keto-L-gulonate (Fig. 1, common name: dikegulac) acts as a plant growth regulator.

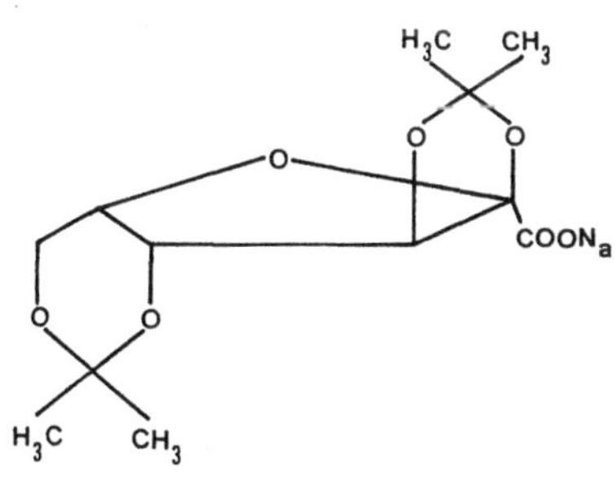

Fig. 1. Dikegulac

It was demonstrated (4) that dikegulac has a wide range of growth regulating effects. Histoautoradiographic studies showed that DNA synthesis was inhibited in the apical meristem, and normal cytohistologic zonation was no longer apparent (3). No such effects were seen in the axillary meristem. The aim of the present investigations was to gain some knowledge of the physiologic processes involved in the activity of dikegulac especially in respect to interactions with plant hormones.

B. Materials and Methods

If not otherwise stated the plants were grown in a soil mixture (80% peat, 20% clay) in 500-ml plastic pots. Dikegulac was dissolved in water, GA_3 was dissolved in methanol (purum), indole-3-acetic acid (IAA) in acetone (purum), and 6-furfurylamino purine (kinetin) in acetic acid (purum) and then made to volume with water. For foliar applications α(-p-nonylphenyl)-ω-hydroxy poly(oxyethylene) was added to the spray solution to increase wetting.

I. Translocation

Translocation of dikegulac was studied on *Chrysanthemum morifolium* cv. Princess Anne, grown in a growth chamber (temperature 22°C light, 19°C dark; relative humidity (R.H.) 70%; photoperiod 16 h). The first fully developed leaf at the top, a leaf in the middle, or the lowest leaf of the shoot was treated with either 3 or 6% dikegulac solution. Ten plants were used for each treatment.

II. Tillering of Grasses

P. pratensis and *L. perenne* were grown in the greenhouse (temperature 20-24°C light, 18°C dark; photoperiod 16 h) in 500-ml plastic pots and cut back to a height of 4 cm before treatment. Three days after the last cut, solutions containing 750-6000 mg/l dikegulac were applied using a spray volume of 1000 l/ha. Each treatment was repeated six times. Fourteen days after application, tiller number was recorded.

III. Parthenocarpy of Pears

The effect of dikegulac on parthenocarpic fruit formation was investigated on *Pyrus communis* cv. Williams under field conditions. Shortly before bloom the flower clusters were enclosed in insect-proof cotton bags to prevent pollination. The compound was sprayed to run-off either at full bloom only or in some cases 20 days later as well. GA_3 was used as a standard. After harvest the number of parthenocarpic fruits was recorded.

IV. Interactions with Plant Hormones

1. Callus of *Lycopersicon esculentum* cv. San Marzano was cultured on medium B_5 (6) at 27°C in the dark. The parent stock was subcultured every two weeks. Dikegulac and GA_3 were incorporated into the medium just after sterilization before the medium was poured into petri dishes. When the medium was cold, 100-mg callus were transferred to each petri dish, and 14 days later the weight of the callus was determined.

2. Leaf elongation of *Avena fatua*: The influence of dikegulac and GA_3 was tested on *A. fatua* seedlings, which were germinated in wet sand in an incubator (22°C; R.H. 80%). Two days after sowing, when the coleoptiles were 1.5 cm long, they were mounted on "Styropor" (porous polystyrene) strips and floated on water or aqueous solutions of the compounds. The length of the first leaf was recorded two and four days after treatment.

3. Growth of wheat and pea: The influence on growth of dikegulac, GA_3, kinetin, and IAA was studied on *Triticum aestivum* cv. Svenno and *Pisum*

Table 1. Translocation of dikegulac on *Chrysanthemum morifolium* cv. Princess Anne

% Leaf surface chlorotic			
Axillary shoots subtended by	Position of treated leaf		
	Top	Middle	Bottom
1[a]- 4 leaf	60	48	0
5 - 8 leaf	30	20	0
9 -12 leaf	20	5	5

[a] First visible leaf

On each plant one leaf was painted with 3% dikegulac solution. Chlorosis of axillary shoot indicated translocation of compound. Chlorosis was estimated visually 29 days after treatment (mean value of 10 plants). Position of assessed axillary shoots is identified by leaf subtending it

sativum cv. Petit Provençale. The seeds were germinated in wet sand in a growth chamber (for conditions see Sect.B.I), and the seedlings were transferred to half-strength Knop's solution at the 2-(wheat) or 4-leaved stage (pea). One week after transfer to nutrient solution, the compounds were added. The length of wheat tillers was recorded 21 days and the length of pea shoots 10 days after treatment.

4. Axillary shoot formation of azalea: To demonstrate the influence of dikegulac alone and in mixture with kinetin, IAA, and GA_3, *Rhododendron simsii* cv. Reinhard Ambrosius was used. The plants were grown in the greenhouse (18-26°C; R.H. 50-80%; photoperiod 16 h) in plastic pots containing peat. The plants were sprayed when the axillary shoots were 4-6 cm long. The number of axillary shoots was recorded six weeks after treatment.

5. Ethylene biosynthesis: Four pea seeds, presoaked in water for 24 h, were placed in 50-ml Erlenmeyer flasks. Two ml of aqueous solution of the test compound and a vial of 1*N* KOH were added to each flask, which was then sealed and kept at room temperature under artificial light. After 24 h, a 2-ml sample of the atmosphere of the flask was removed and analyzed for ethylene content using a gas chromatograph with an activated alumina column.

C. Results and Discussion

I. Translocation

One important aspect of the efficacy of a plant growth regulator is the degree to which the absorbed compound is translocated to its site of action. Chlorosis on axillary shoots (Table 1) indicated the translocation of the compound, which apparently occurred in the phloem. Translocation was also demonstrated by severe stunting of the main shoot (Figs. 2, 3) and by the induction of axillary shoot formation. Maximal effect on the apex occurred following treatment of the uppermost leaf, followed by the middle and lower leaves in that order. The reduced translocation from the bottom leaf may also be due to an age-related lowering of its photosynthetic capacity. When [6-^{14}C] dikegulac was applied to leaves beneath the apex, it was detected in the apex within a day (3).

II. Tillering of Grasses

Growth-retardant activity of dikegulac was demonstrated on *P. pratensis* and on *L. perenne* (4). The influence on tillering was also investigated on these two grasses. Tillering of both species was stimulated at all concentrations tested (Fig.4). Since the inflorescence number of gras-

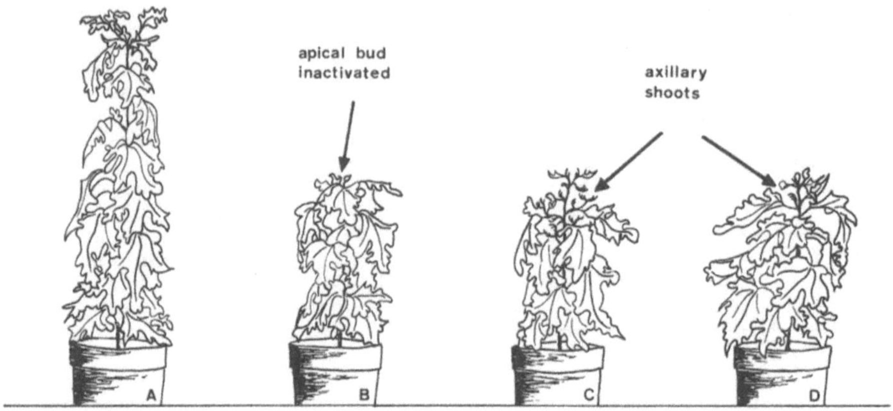

Fig. 2. Influence of dikegulac on apical dominance and axillary shoot development of *Chrysanthemum morifolium* cv. Princess Anne. One fully expanded leaf near the apex, in the middle, or at the lower end of the shoot was painted with 6% solution of dikegulac. Plants were sketched 3 weeks after treatment. *A*: untreated (not pinched); *B*: top leaf (12)[a] treated; *C*: middle (6)[a] leaf treated; *D*: bottom (1)[a] leaf treated.

[a] Numbers indicate position of treated leaf on stem

Fig. 3. Influence of dikegulac on apical dominance and axillary shoot development of *Chrysanthemum morifolium* cv. Princess Anne. One fully expanded leaf near apex, in middle, or at lower end of shoot was painted with 6% solution of dikegulac. Plants were sketched three weeks after treatment. *A*: untreated (pinched 1 day before application); *B*: top leaf (12)[a] treated; *C*: middle (6)[a] leaf treated; *D*: bottom (1)[a] leaf treated.

[a] Numbers indicate position of treated leaf on stem

ses is reduced by dikegulac treatment (Bocion and de Silva, unpublished data), it is possible that the meristem of the main shoot is inactivated (see below) and consequently that apical dominance is removed.

III. Parthenocarpy of Pears

Parthenocarpic fruit formation on pears cv. Williams was induced by a single application of dikegulac at concentrations ranging from 75-300 mg/l. Two applications at an interval of 20 days resulted in more fruits

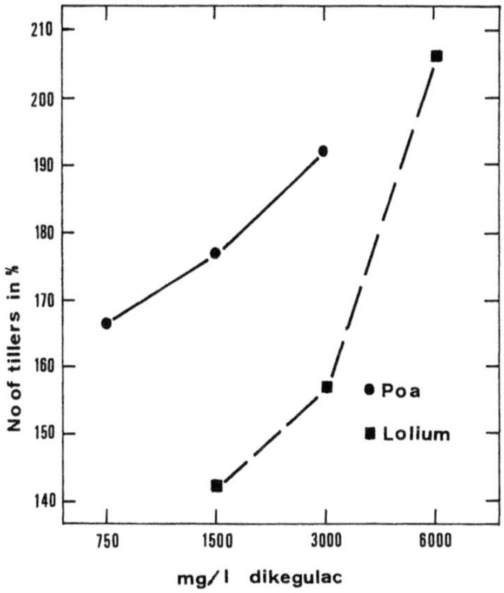

Fig. 4. Stimulation of tillering with dikegulac on *Poa pratensis* and *Lolium perenne*. Number of tillers was recorded 14 days after treatment and are given as % relative to untreated

Table 2. Effect of dikegulac and GA_3 on parthenocarpic fruit set of pears cv. Williams

Compounds	Concentration mg/l	Number of fruits per 100 clusters
GA_3	0	9
	45	42
Dikegulac	75	17
	150	11
	300	15
Dikegulac	75 + 75	20
	150 + 150	23
	300 + 300	28

Compounds were applied at full bloom and in some treatments again 20 days later. Number of parthenocarpic fruits per 100 flower clusters was recorded at harvest

than one application (Table 2). The best effect was obtained at 300 mg/l. GA_3 treatment produced more parthenocarpic fruits but, in contrast to dikegulac treatments, the fruits were badly deformed.

IV. Interaction with Plant Growth Regulators

Interactions of synthetic plant growth regulators, e.g., daminozide (succinic acid-2,2-dimethylhydrazide), chlormequat (2-chloroethyl trimethylammonium chloride), ancymidol [2-cyclopropyl-2-(4-methoxyphenyl)

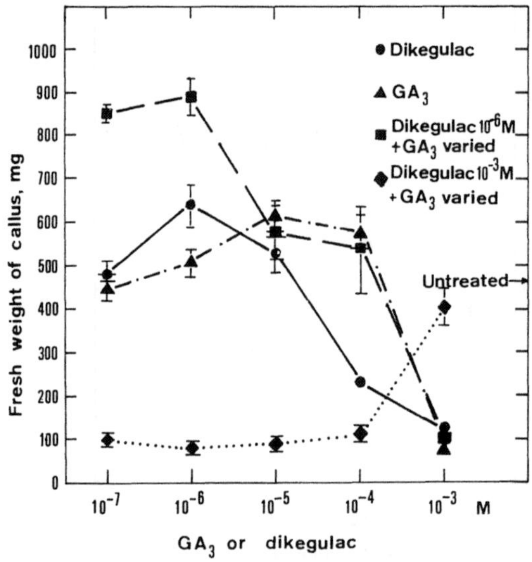

Fig. 5. Interaction between dikegulac and GA_3 on *Lycopersicon esculentum* cv. San Marzano on white callus. Weight of callus was assessed two weeks after treatment. Standard errors are indicated. Initial weight of callus 100 mg

-5-pyrimidine methanol], and TIBA (2,3,5-triiodobenzoic acid) with plant hormones like GA_3, IAA, kinins, ethylene, and abscisic acid (ABA) have been demonstrated in various bioassay systems. Daminozide and chlormequat caused a substantial increase in ethylene production in squash seedlings (1), which may in part account for the effect of these compounds in inhibiting stem elongation.

The elongation of lettuce hypocotyls caused by 10 μM GA_3 could partly be overcome by chlormequat and ancymidol, but daminozide was ineffective (10). It was also found that ancymidol strongly inhibits gibberellic acid-induced elongation of *Avena* segments at a concentration of 1 mM (11). Several types of growth retardants, including members of the choline, piperidine, hydrazide, and phosphonium classes, are known to supress GA biosynthesis (9, 18). Growth stimulation of oat coleoptile sections induced by 10 μM IAA was counteracted by ancymidol but not by chlormequat and daminozide (10). Conversely, TIBA inhibits polar transport of auxins (13).

To gain insight into interactions with plant hormones, the following experiments were initiated:

1. Tomato callus growth: The first effects were seen 4-6 days after treatment, when the color of the tissue turned from white to yellow and later brownish at the higher concentrations. The effects on growth were recorded visually 7-8 days after application. Dikegulac promoted callus growth at 10^{-6} M (Fig.5). At higher concentrations a strong inhibitory effect was seen resulting in cell death at 10^{-3} M. GA_3 at 10^{-3} M also killed the cells, although at lower concentrations growth was promoted. The effect of higher concentrations of dikegulac, e.g., 10^{-3} M, could only be counteracted by similar concentrations of GA_3. However, 10^{-6} M of dikegulac in mixture with 10^{-7} and 10^{-6} M GA_3 stimulated growth more than GA_3 alone. This would suggest that at lower concentrations the effect of mixtures of dikegulac and GA_3 was additive while at higher concentrations the effect was antagonistic.

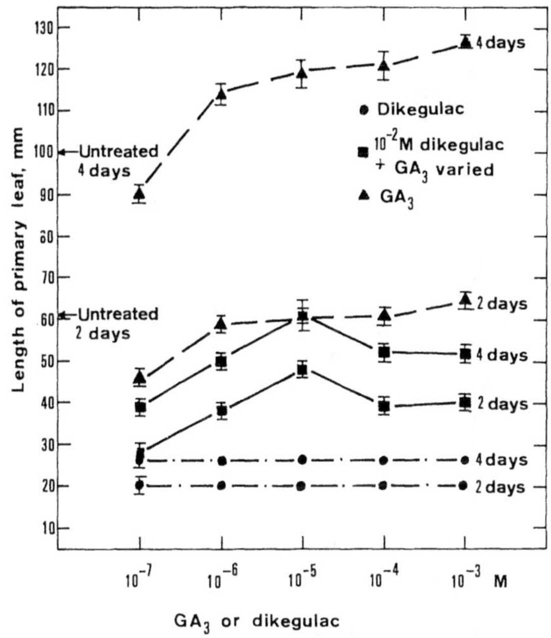

Fig. 6. Influence of dikegulac and GA_3 alone and in mixture applied through roots on elongation of primary leaf of *Avena fatua*. Length of leaf was recorded two or four days after treatment. Standard errors are indicated

2. Leaf elongation of *Avena fatua*: Dikegulac retarded leaf elongation 2 and 4 days after application. This could be overcome by GA_3 application at rates from 10^{-6} to 10^{-3} M (Fig.6). GA_3 alone did not promote leaf elongation two days after treatment, but a strong growth-stimulating effect was noticed two days later at concentrations ranging from 10^{-6} to 10^{-3} M. No distortion and chlorosis of the leaves were seen, as was the case with *Zinnia* and *Helianthus* (3) and various other species tested in our laboratory. The reported behavior of dikegulac could be the result of an effect on the mode of action of GA_3.

3. Growth of wheat and pea: The effects of dikegulac on the growth of wheat and dwarf peas were counteracted by GA_3 but not by IAA at the rates tested (Table 3 and Fig.7, respectively). Neither chlorosis nor deformation was seen on either species. No stimulatory effects were found as was the case on tomato callus. This latter phenomenon may be confined to tissue lacking chloroplasts.

4. Axillary shoot formation of azalea can be triggered by dikegulac, as we reported previously (4, 15). It is known that application of cytokinins to axillary buds of apple overcomes apical dominance (17), whereas IAA may or may not enhance dominance (8, 16). Therefore the influence of IAA and kinetin on the activity of dikegulac was investigated on azalea. Two weeks after application, the apex and surrounding leaves turned slightly chlorotic, and leaf deformation was noticeable. These two effects were less pronounced if IAA was applied in combination with dikegulac. Three to four weeks after treatment the axillary shoots began to develop, and six weeks after treatment chlorosis and deformation gradually disappeared and the leaves turned dark green again. Because IAA can maintain apical dominance (16) and dikegulac re-

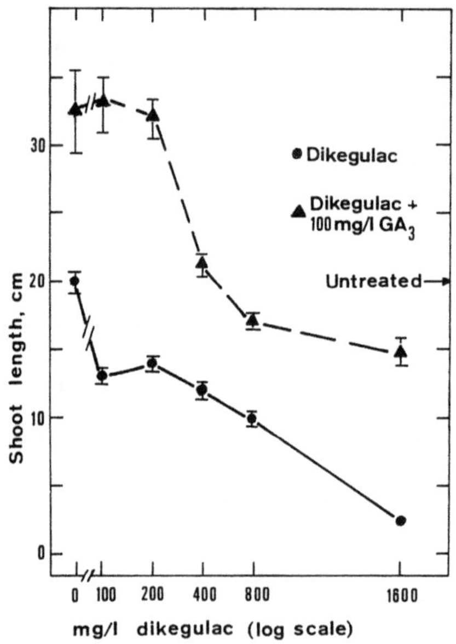

Fig. 7. Influence of dikegulac and GA_3 alone and in mixture applied on roots on growth of *Pisum sativum* cv. Petit Provencale. Length of shoot in cm recorded 10 days after treatment. Standard errors are indicated

Table 3. Influence of mixture of dikegulac with GA_3 and IAA on growth of *Triticum aestivum* cv. Svenno

Compound	Dosage	Dikegulac				
		0	10^{-5} M	10^{-4} M	10^{-3} M	10^{-2} M
GA_3	0	9.4	9.3	7.5	7.1	5.6
	10^{-6} M	10.6	12.0	10.4	8.0	6.6
	10^{-5} M	15.3	14.8	12.6	8.6	6.2
	10^{-4} M	16.4	16.4	15.2	9.8	6.5
IAA	10^{-6} M	11.4	9.4	8.4	6.3	6.4
	10^{-5} M	9.0	8.3	8.0	6.3	6.4
	10^{-4} M	8.9	8.0	7.8	5.8	6.3

LSD (P = 0.05) 0.92

leases it, an interaction between these two regulators was not unexpected (Table 4), since exogenous IAA would substitute for the auxin previously supplied by apices inactivated by dikegulac. The pinching effect of dikegulac at 4000 mg/l was enhanced when 2000 mg/l kinetin was applied simultaneously (Table 4).

5. Ethylene biosynthesis: Despite IAA stimulation of ethylene biosynthesis in some test systems (2), it has also been shown to impede abscission (7, 12). In our experiments 10^{-3} M dikegulac increased ethylene biosynthesis 6-fold. However, in the presence of 10^{-4} M IAA, ethylene production of treated seeds dropped 70% (Table 5).

Table 4. Influence of mixtures of dikegulac with IAA and kinetin on axillary shoot number of *Rhododendron simsii* cv. Reinhard Ambrosius six weeks after treatment

No. of axillary shoots per plant								
Dikegulac mg/l	IAA							
	0	s.e. ±	400 mg/l	s.e. ±	2000 mg/l	s.e. ±	10,000 mg/l	s.e. ±
0	0		0		0		0	
4000	30	6.3	11	3.8	12	3.1	0	
8000	51	8.5	24	4.7	15	1.9	3	1.5
	Kinetin							
0	0		0		0		0	
4000	30	6.3	39	4.5	54	5.6	24	2.6
8000	51	8.5	51	7.9	51	6.6	48	2.6

Table 5. Effects of mixture of dikegulac and IAA on ethylene production of pea seedlings

Compound	Dosage	Dikegulac							
		0	s.e. ±	10^{-2} M	s.e. ±	10^{-3} M	s.e. ±	10^{-4} M	s.e. ±
Untreated	0	0.45	0.09	0.34	0.08	2.87	0.41	0.60	0.10
IAA	10^{-4} M	0.44	0.01	0.23	0.05	0.86	0.14	0.19	0.02
IAA	10^{-6} M	0.42	0.06	0.35	0.01	1.12	0.06	0.33	0.01

µl Ethylene after 24-h incubation

D. Conclusions

Dikegulac, a new plant growth regulator, was tested in the laboratory and greenhouse to gain an insight into its physiologic effects. In the present investigations, growth-retarding activity was demonstrated on tillers of wheat, shoots of pea, leaves of *A. fatua*, and callus of tomato. Tillering of *P. pratensis* and *L. perenne* was enhanced. Apical dominance was released in azalea. Parthenocarpic fruit formation was induced on pears cv. Williams. Translocation from treated chrysanthemum leaves, at various positions on the stem, to the apical region was demonstrated. Interactions of dikegulac with the plant growth regulator IAA as well as kinetin were recorded.

GA_3 counteracted the retardation effect of dikegulac on wheat, *A. fatua*, and pea, which leads to the conclusion that either the biosynthesis of GA_3 or its mode of action is antagonized by dikegulac in a manner similar to that of chlormequat and other quaternary ammonium compounds (5, 14). Conversely, relatively low concentrations of dikegulac stimulated promotion of growth by GA_3 on tomato callus and produced parthenocarpic fruits in pears. The enhancement of axillary branching and ethylene biosynthesis by dikegulac were counteracted to a certain extent by IAA.

Kinetin at 2000 mg/l enhanced the apical dominance suppression of 4000 mg/l of dikegulac. The mode of action of dikegulac appears to be related mainly to its interaction with gibberellins and/or nucleic acids.

Acknowledgements

We gratefully acknowledge the constructive criticisms of many of our colleagues at Dr. R. Maag Ltd., as well as those of Dr. J. Gressel and Mr. S. Zilkah who also kindly supplied us with the tomato callus. We are also thankful for the expert assistance of many technicians.

References

1. Abdel-Samad, H.A., Ketellapper, H.J., Yang, S.F.: Plant Physiol. 47, (Suppl.): 16 (1970)
2. Abeles, F.B.: Ann. Rev. Plant Physiol. 23, 259 (1972)
3. Arzee, T., Langenauer, H., Gressel, J.: Submitted to Botan. Gaz.
4. Bocion, P.F., Hüppi, G.A., de Silva, W.H., Szkrybalo, W.: Nature (London) 258, 142 (1975)
5. Dennis, D.T., Upper, C.D., West, C.A.: Plant Physiol. 40, 948 (1975)
6. Gamborg, O.L., Eveleigh, D.E.: Can. J. Biochem. 46, 417 (1968)
7. Jackson, M.B., Hartley, C.B., Osborne, D.J.: New Phytologist 72, 1251 (1973)
8. Jacobs, W.P., Danielson, J., Hurst, V., Adams, P.: In: Developmental Biology. New York: Academic Press, 1959, p. 534
9. Lang, A.: Ann. Rev. Plant Physiol. 21, 537 (1970)
10. Leopold, A.C.: Plant Physiol. 48, 537 (1971)
11. Montague, M.J.: Plant Physiol. 56, 167 (1975)
12. Morgan, P.W.: Planta 129, 275 (1976)
13. Niedergang-Kamien, E., Skoog, F.: Physiol. Plantarum 9, 60 (1975)
14. Ninnemann, H., Zeevart, J.A.D., Kende, H., Lang, A.: Planta 61, 229 (1964)
15. de Silva, W.H., Bocion, P.F., Walther, H.R.: HortScience (in press)
16. Wickson, M., Thimann, K.V.: Physiol. Plantarum II: 62 (1958)
17. Williams, M.W., Stahly, E.A.: Hort. Sci. 3, 68 (1968)
18. Wylie, A.W., Ryugo, K., Sachs, R.M.: J. Am. Soc. Hortic Sci. 95, 627 (1970)

Paper Demonstration References (see Section IX):

17 - 19 - 28 - 52 - 68 - 78 - 112

VII. Gravity and Light Effects

22. Geotropic Response Mechanisms in Roots and Shoots

M. B. WILKINS

Abstract

A critical assessment is made of the present state of knowledge of the geotropic response mechanism in dicotyledonous shoots and roots. Studies of the involvement of auxin, gibberellins and inhibitors in the response of shoots are reviewed, and those aimed at identifying the inhibitor involved in root geotropism are evaluated.

In a recent review of geotropism (26), attention was confined largely to the responses in coleoptiles and primary roots of *Zea mays* seedlings. I imposed this limitation because of the many detailed studies that have been made with this species in a number of different laboratories, and because of the importance I attach to argument and discussion of mechanisms being based on studies of the same organ in the same species of plant. It is becoming increasingly evident that differences in mechanism do occur between different organs, and between similar organs in different plant species. Plant physiologists seem too eager to seek an all-embracing general mechanism to explain the response of all organs of a particular type regardless of the species of plant involved. To assume that all roots, or indeed all shoots, have the same geotropic mechanism could be as wrong as assuming that all plants fix carbon dioxide by one and the same biochemical mechanism. Study of one species of plant is as justified as that of another, and I hold no particular brief for the *Z. mays* seedling other than the fact that it is a very convenient laboratory object, easily obtainable and one of the world's most important agricultural crops. But the progress achieved in the elucidation of the geotropic response mechanisms of coleoptiles or roots of *Z. mays* does not necessarily assist greatly in establishing the geotropic response mechanism of, let us say, the shoot of a young dicotyledonous seedling or that of the leaf sheath base ("node") of a mature graminaceous stem (7-9). It is to emphasise this point that I have tried to adopt a broader, comparative approach in this report rather than deal in detail with essentially one species as in my earlier review (26).

The geotropic response of a *Zea* coleoptile involves at least two changes in the transport of apically synthesised indole-3-acetic acid (IAA). The first is the induction, by the stimulated gravity-sensing system, of a downwards lateral transport of IAA into the lower half of the organ. The second is an increase in the basipetal movement of IAA along the lower half of the organ. These two changes can lead to an increase in the amount of IAA in the lower half of the growing zone of the horizontal coleoptile and a decrease in the amount in the upper half. This difference in IAA distribution leads to differential growth and upwards curvature (26).

To what extent are these findings generally applicable to geotropic responses of shoots, particularly those of dicotyledonous plants? The answer to this question is just beginning to emerge and, as more work with dicotyledonous shoots is published, it appears increasingly likely that the mechanism involved differs in a number of respects from that in coleoptiles, and may indeed be much more complex, involving more than one type of growth regulating hormone.

Several years ago, my colleague I.G. Bridges and I briefly studied the movement of IAA in geotropically stimulated hypocotyl segments of *Phaseolus vulgaris* and *Helianthus annuus* to ascertain whether or not lateral transport occurred. We found a small but significant downwards lateral movement of radioactivity in horizontal hypocotyl segments supplied asymmetrically with either ^{14}C-IAA or ^{3}H-IAA. The lateral movement was similar in the case of each isotope, but was very much less than that found in *Zea* coleoptiles. Subsequently, IAA has been shown to be present in the apical bud of *P. vulgaris* seedlings (23) and it is transported basipetally. While these facts pointed to the involvement of IAA in the regulation of growth, Bridges and I felt that the magnitude of the lateral transport of IAA which we observed in the hypocotyl segments was scarcely large enough in itself to establish an adequate asymmetry in distribution to give rise to the differential growth underlying the geotropic response.

Recently Phillips and Hartung (19) made a detailed study of the possible occurrence of lateral transport of IAA and GA in green, first-internode segments of *H. annuus* seedlings. The hormones were supplied either symmetrically or asymmetrically in agar blocks to the apical end of the segments. When symmetrical donors containing ^{14}C-GA_1 were applied, no significant difference in the distribution of radioactivity, which was shown to be confined exclusively to the GA_1 molecule, was observed in the horizontal and vertical segments. After asymmetric donation, no significant effect of orientation in the gravitational field could be detected on the amount of GA_1 reaching the non-donated half of the internode segments. When the segments were supplied either symmetrically or asymmetrically with agar blocks containing ^{14}C-IAA, virtually identical results were obtained. These very careful studies thus provide no evidence whatsoever that lateral transport of IAA or GA_1 occurs in geotropically stimulated green internode segments of *Helianthus*, a pattern of behaviour, at least for IAA, which was different from that found in the hypocotyl segments by Bridges and myself.

In an earlier paper Phillips (18) had studied the application of IAA or gibberellin to decapitated shoots of *H. annuus* and *P. multiflorus*. Decapitation strongly suppressed internode elongation compared with the intact control. Gibberellin (GA_3) applied in lanoline at concentrations between 100 and 1000 ppm to the cut surface of the decapitated plants caused elongation to occur at a rate at least as great as in the intact control. In contrast, IAA applied at similar concentrations did not increase the growth rate significantly above the very low value in the decapitated plants. These effects are not peculiar to a particular stage in internode development. Of considerable interest is the observation that the application of IAA in addition to GA_3 virtually abolished the growth induced by the latter hormone. Since Jones and Phillips (15, 16) and Phillips (17) have shown that the young leaves of the apical bud are the sites of synthesis of gibberellins which appear to be concerned with the regulation of cell elongation that underlies internodal growth, it is clear that the regulation of differential growth in the geotropic response of a green dicotyledonous shoot may in fact be attributable to an asymmetry in the distribution of gibberellins being established on geotropic stimulation. It is not known whether an asymmetrical distribution of naturally occurring IAA develops in horizontal internodes of intact shoots; certainly no lateral transport of externally applied IAA occurs in internode segments. However, if an asymmetry in IAA distribution were to occur, it would be unlikely to give rise to differential growth since this hormone does not appear to have the capacity to promote cell extension in the green *Helianthus* internode. What is even more interesting in the light of the findings of Phillips (18) is that if an asymmetric distribution of IAA in favour of the lower half of a horizontal shoot of *Phaseolus* or *Helianthus* were to occur, it might abolish the effectiveness of an asymmetry in the distribution of gibberellins in inducing differential growth and hence upwards curvature.

The experiments of Phillips (18) have shown that in excised apical bud preparations of *H. annuus* seedlings an asymmetrical distribution of gibberellin-like activity does develop following gravitational stimulation. Using the lettuce-hypocotyl bioassay procedure to assess the amounts of hormone-like activity in two agar blocks in contact with the basal cut surface of the apical bud preparations, Phillips (18) found there to be much more gibberellin-like activity in the lowermost of the two blocks, when the preparation was horizontally orientated. The ratio of growth-hormone activity between the lower and upper blocks was, on average, virtually 10:1. Despite substantial variation in the yield of gibberellin-like activity per bud between different experiments, the ratio of distribution between upper and lower receiver blocks was remarkably constant. The foregoing results were obtained when the horizontal bud preparation was arranged so that the larger pair of apical leaves were vertical. When the bud preparations were arranged so that these leaves were in the horizontal plane, the difference in the distribution of gibberellin-like activity between the upper and lower agar receiver blocks was even greater. In fact, in several experiments no detectable gibberellin activity could be found in the upper block while large amounts were present in the lower one. A further finding of critical importance was the comparison of the total yield of gibberellin-like activity from apical bud preparations that were maintained either horizontally or vertically. On a per bud basis, the yield of GA_3-equivalents was approximately three times as great from the horizontal preparation as from the vertical one. No significant asymmetry of gibberellin-like activity was detected between the two receiver blocks in contact with the basal ends of the vertical apical bud preparations.

There can be no doubt, therefore, that an asymmetric distribution of net gibberellin-like activity occurs in the basal receiver blocks in contact with detached geotropically stimulated apical buds of *H. annuus*. There are, however, a number of questions that need to be considered before these findings can be used to draw inferences on the involvement of gibberellins in the mechanism of the geotropic response in the green shoot itself. First, the data of Phillips (18) were based on the *accumulation* of gibberellin-like activity in the agar receiver blocks after an incubation period of 20 h. Geotropic curvature of *Helianthus* shoots begins to develop within an hour or so of their being placed in the horizontal position, and any hormone redistribution that is causally related to this curvature must precede the onset of bending. It would be of substantial importance to know whether the enhanced release of net gibberellin-like activity into the receiver blocks, and the development of the asymmetric distribution between the upper and lower receiver blocks in horizontal, apical bud preparations occurred in the time period before the onset of geotropic curvature.

Second, it is necessary to interpret the apparent hormone content of agar receiver blocks with extreme caution since they do not necessarily reflect the amounts of hormone in the adjacent tissues with which they are in contact. Differences in the rates of longitudinal transport and secretion of gibberellins and of other substances that may antagonise the activity of the gibberellin molecules, by the upper and lower halves of the horizontal shoot, will in time lead to differential *accumulation* of activity in the two blocks without there being any real difference in the amount of hormone per cell in the two halves of the tissue. Although Phillips (18) attempted to separate chromatographically the gibberellin activity in the agar receiver blocks, activity was found to be spread over the whole of the chromatogram in some cases, and on 7 or 8 of the Rf zones in virtually all of the published results. Clearly the total gibberellin-like content of the agar block could be appreciably affected by the presence on the starting line of any acidic-inhibiting substance that might have come through the partitioning procedures, wherever it might run on the chromatogram. The presence of components of the β-inhibitor complex might well interfere with the overall gibberellin-like activity levels in the bioassay system. If more β-inhibitor complex substances were present in the upper receiver blocks than in the lower ones, then the smaller amounts of gibberellin-like activity in the upper blocks, and especially its virtual absence from Rf 0.6-1.0, evident quite often in Phillips' (18) data, might give rise to differences in total gibberellin-like activity between the two receiver blocks. It is obvious, however, that such an occurrence would give rise only to a difference in gibberellin-like activity in a bioassay system and not a real difference in gibberellin content. This is a critical point, the importance of which needs urgent investigation with physicochemical analytical techniques now available for the gibberellins.

Third, if it is accepted that Phillips' (18) receiver block data reflect a real redistribution of gibberellins within the tissues, by what mechanism is this redistribution achieved? Obviously the situation is a complicated one, since horizontal placement of the apical bud leads to a threefold increase in the amount of gibberellin-like activity in the receiver blocks. A gravity-stimulated increase in synthesis or release of gibberellins may be occurring, or an increase in the longitudinal basipetal transport and release of gibberellins might be taking place. It is quite possible that any one or all of these changes might be occurring to a greater extent in the lower half than in the upper half of the horizontal bud preparations and might give rise to the asymmetrical distribution of gibberellin-like activity in the receiver blocks. Phillips (18) himself draws attention to his data being insufficient to establish how the observed asymmetry in gibberellin-like activity is developed. Another possibility is that there is an overall decrease and an upward asymmetric redistribution of an acidic complex of inhibiting substances, with little or no change in the level or distribution of gibberellins.

It is possible, of course, that a downwards lateral transport of gibberellin into the lower half of the horizontal shoot might be taking place, but Phillips and Hartung (19) could find no evidence of this in green internodes of *H. annuus* when using labelled gibberellin GA_1. However, the lateral transport of gibberellin in *Helianthus* shoots may, in fact, take place in the extreme apex of the bud and might not, therefore, be found in excised internode segments. On the other hand, it is by no means certain that GA_1 occurs naturally in *Helianthus* hypocotyls. Even if it does, however, the absence of lateral movement of this one gibberellin does not preclude the possibility that lateral movement of other gibberellins takes place on geotropic stimulation. Phillips' (18) data showing gibberellin activity to be broadly spread over the chromatogram from the starting line to the solvent front must indicate that a number of different gibberellins are present, and the fact that one cannot be shown to undergo lateral transport does not eliminate the possibility that one or several of the others present do.

Obviously the findings of Phillips (18) and Phillips and Hartung (19) have opened up a most interesting field of enquiry and raise many most important questions. Progress is, however, going to be limited until more sophisticated analytical techniques are employed on the problem and, in particular, until the individual gibberellins present in the tissues of the apical buds and subapical parts of the shoot are identified by rigorous physicochemical methods. This is technically rather difficult since relatively large amounts of tissue will be required initially to extract sufficient gibberellins for mass spectrometric analysis. Rather less tissue could be used subsequently to study the development of asymmetry in the distribution of different gibberellins within the tissues, using high-performance, liquid chromatography, which might be adequate to resolve the distribution of different gibberellins within the upper and lower halves of geotropically stimulated dicotyledonous shoots. Clearly this field is on the threshold of substantial development and the exploitation of Phillips' important primary work will undoubtedly lead to substantial advance in our understanding of the mechanisms of the geotropic response of dicotyledonous shoots.

Although Phillips and Hartung (19) could find no evidence of lateral transport of IAA in green fourth internode segments of *H. annuus*, it is not yet clear whether a gradient in naturally occurring IAA develops between the upper and lower halves of the intact horizontal shoot. It is important that this point be resolved by analytical procedures, since the absence of lateral transport does not rule out the development of an asymmetrical distribution of IAA on geotropic stimulation.

The lateral transport of IAA in horizontal *Helianthus* hypocotyl segments observed by Bridges and myself is in agreement with an earlier report that after symmetrical application of ^{14}C-IAA to the apical ends of horizontal *H. annuus* hypocotyl segments, a gradient of ^{14}C was established between the upper and lower halves of the organ (10). More recently Abrol and Audus (1) reported that ^{14}C-2,4-D applied asymmetrically to the apical end of horizontal, 6-mm segments of light-grown *Helianthus* hypocotyls underwent lateral displacement towards the lower side of the organ. The movement of 2,4-D was, however, very slow; no radioactivity emerged into the receivers until at least 5 h had elapsed since the application of the donor blocks. Incorporation of N-1-naphthylphthalamic acid into the donor and receiver blocks markedly reduced this gravity-induced lateral transport of 2,4-D (2). Evidence of a lateral movement of radioactivity from ^{14}C-IAA, ^{3}H-IAA, and from ^{14}C-2,4-D in *hypocotyl* segments of *Helianthus* has thus been reported on several occasions, and stands in contrast to the lack of detectable lateral transport in segments of internodes above the insertion of the cotyledons (19). These findings are perhaps suggestive of the involvement of auxin in the geotropic response of the hypocotyl but not in that of the internodes above the cotyledons, in which, on present evidence, the gibberellins may play a significant role. By way of comparison, there is certainly no evidence of the involvement of lateral transport of IAA or any other hormone from the upper to the lower half of the organ in the geotropic response of the leaf sheath base of mature green stems of wheat *Triticum vulgare* (7, 8, 9).

Although most interest has been centred on the asymmetric distribution of growth-regulating molecules in geotropically stimulated shoots, there have been several studies of the establishment of asymmetry in other cellular components. In particular, the distribution of the cations K^+ and Ca^{2+} have been investigated in geotropically stimulated hypocotyls of *H. annuus*. Bode (5, 6) found a higher proportion of K^+ in the ash of the lower half of the hypocotyl than in that of the upper half, and there was an indication that Ca^{2+} showed the reverse trend. The distribution of Ca^{2+} was also investigated by Arslan-Cerin (3). The seedlings were supplied with $^{45}CaCl_2$ and then placed horizontally for 1, 2, or 3 h. More $^{45}Ca^{2+}$ was found in the upper half of the hypocotyl than in the lower half, especially after 3 h. No such asymmetry could be detected when geotropic stimulation took place under anoxic conditions, suggesting that metabolic energy was involved in the establishment of the asymmetry. Confirmation of these findings has recently been published by Goswami and Audus (11), who have investigated the redistribution kinetics. The maximum difference in Ca^{2+} concentration of about 20% in favour of the upper half of the hypocotyl was attained after about 1 h in the horizontal position, by which time geotropic curvature was just beginning. An asymmetry in K^+ was also observed in favour of the lower half of the horizontal hypocotyl; it increased linearly with time, having reached significance after 1 h. An asymmetry in ^{32}P appeared to develop in favour of the lower half and was similar in its kinetics to that of K^+ ions. Further study of the $^{45}Ca^{2+}$ asymmetry showed that it did not develop at 4°C or when the hypocotyls were treated with N-1-naphthylphthalamic acid, which also abolishes the geotropic response.

The role of these ion asymmetries and the mechanism by which they arise are not yet understood. It appears that they could be a consequence of the asymmetry in IAA that arises in geotropically stimulated hypocotyls. Goswami and Audus (11) established artificial IAA gradients in vertical *Helianthus* hypocotyls using the unilateral application procedure of Wilkins and Woodcock (27) and found more $^{45}Ca^{2+}$ in the non-donated concave side, and more $^{32}K^+$ and ^{32}P in the IAA-rich donated side of the hypocotyl, as would have been anticipated on the basis of a downwards lateral transport of IAA in horizontal hypocotyls. These ionic asymmetries do not appear to arise as a consequence of the differential expansion of cells on the two sides of the geotropically responding organ, but rather to be caused by the redistribution of growth hormones, particularly IAA. Their function is unknown, and the phenomenon of ion redistribution may fall into a similar category to the geoelectric effect, which is undoubtedly a consequence of the changes in IAA levels in the two sides of the organs giving rise to differences in ion (proton) transport across the plasma membrane (27-31). The asymmetry in Ca^{2+}, K^+, and ^{32}P distribution may involve both lateral redistribution and differential longitudinal movement along the seedling axis.

Several recent studies have focussed attention on the importance of the epidermis in regulating the rate of extension of the two halves of a geotropically stimulated shoot, and the control of epidermal cell extension by endogenous hormones. Yamamoto et al. (33) and Tanimoto and Masuda (22) found that auxin-induced extension of pea stem sections is controlled by regulation of the growth rate of the epidermal cells. Auxin was shown to modify the extensibility of the epidermal cell walls. Iwami and Masuda (13) reported that a strip of filter paper saturated with a buffer solution at pH = 3.0 placed on the upper side of a horizontal, light-grown hypocotyl of *Cucumis sativus* eliminated the negative geotropic response by promoting the growth of the cells with which it was in contact. Similar unilateral treatment of a vertical hypocotyl caused curvature and induced cell wall loosening of the epidermis.

Iwami and Masuda (14) studied in further detail the geotropic response in *Cucumis* hypocotyls especially in relation to epidermal cell extension and its control by auxin. Application of auxin to the apical end of intact seedlings promoted elongation of the hypocotyl especially in the uppermost 5-mm portion (Zone I) but also in the next two 5-mm portions (Zones II and III) beneath. In horizontal seedlings geotropic curvature began within 30 min, commencing in Zone I and moving basipetally. Removal of the apex and the cotyledons did not appreciably affect the geotropic response, but removal of Zone I of the hypocotyl substantially reduced the response, which could be restored by the application of IAA. The extensibility of the walls of the lower epidermis cells in horizontal hypocotyls was increased significantly after 15-30 min of geostimulation, that is, before the onset of geotropic curvatures. No change in the extensibility of the upper epidermal cells was observed. Iwami and Masuda (14) conclude that Zone I or the region immediately above Zone I is the source of IAA which is necessary for the geotropic response, that it undergoes downwards lateral transport on geotropic stimulation, and that the increased amount of IAA in the lower epidermal cells increases proton secretion, which then gives rise to an increase in cell wall extensibility and hence growth.

The geotropic response of primary roots has been investigated with particular reference to those of *Z. mays* seedlings, and the findings up to the beginning of 1976 have been incorporated in my earlier reviews (24, 26). In general, there is now a substantial body of evidence from several laboratories that indicates that downwards curvature of the primary root of *Z. mays* is due to the downwards lateral transport of an inhibitory substance which is synthesised in the root cap and which moves basipetally through the root apex and into the growing zone of the root. There is strong evidence that this inhibitor is abscisic acid, although the involvement of other inhibitors cannot be excluded at the present time, and that lateral transport takes place largely in the cap, though it might also occur in the root apex. The root apex, as distinct from the cap, is not normally the site of production of any growth regulators, and no growth-regulating compounds involved in the geotropic response move acropetally into the growing zone of the root from its more basal parts.

Comparative studies made on primary roots of *Z. mays* and *P. sativum* by Shaw and Wilkins (20) revealed no substantial difference between the behaviour of these two species. Pea roots thus seemed to have in principle the same kind of mechanism as *Zea* roots, with the important exception that they bend downwards in the dark whereas roots of most varieties of *Zea* do not. Detailed analytical studies of the chemical nature of the inhibitors present in pea root caps have not been made. Since Wilkins and Wain (32) found two inhibitors in addition to abscisic acid in *Zea* root caps, it is possible that in pea roots the principal inhibitor responsible for the geotropic response may not be abscisic acid.

In the case of *Zea* seedlings, as pointed out in my recent review (26), the one piece of evidence required to establish the validity of the mechanism outlined above was a *direct* demonstration of the occurrence of lateral transport of abscisic acid in *Z. mays* root caps. A recent paper by Hartung (12) using the microapplication technique developed by Shaw and Wilkins (21) reports just such studies but on the primary roots of seedlings of *P. coccineus*. Unilateral microapplication of ^{14}C-ABA to the upper or lower side of the root cap revealed that after 1 h and 3 h there was a significantly greater *upward* lateral movement of ^{14}C in the tissue. No significant difference was detectable after 6 h. The observation of an upwards lateral transport of ^{14}C is contrary to the result expected on the basis of the many previous investigations on the lateral transport of naturally occurring growth inhibitors in the root cap of *Zea*. There are several points which need to be kept in mind, however, in evaluating Hartung's (12) results. Firstly, they were carried out on *P. coccineus* and not on *Z. mays* roots. Secondly, ^{14}C-ABA is available only at a very low specific activity, 45 µCi mg^{-1}, and as a result the compound had to be prepared at an extremely high concentration for use with the micropipette method. Hartung used a concentration of 1 mg/ml (1000 ppm) of ABA and the pipette was left in the root for 20 s. Obviously this is such a high concentration of a physiologically active molecule that the amount in the tissue after 20 s might well attain toxic or damaging levels. Thirdly, it is necessary to establish whether or not the ^{14}C in the two halves of the root is still confined to the ABA molecule when supplied to the tissues at this concentration, and this analysis has not been carried out. Fourthly, the point of application was stated to be 1-1.5 mm behind the root tip. Since the root cap does not extend more than about 1.5 mm behind the apical extremity of the root, the application of ABA was apparently made to the most basal regions of the cap. It might have been better to apply the ABA to the most apical region of the cap so that it would have passed through the cap tissues, in which the gravity-sensing mechanism is located. Obviously Hartung (12) has attempted a critical test of the current hypothesis of root geotropism, but the technical difficulties arising from the low specific activity of the ABA available makes the firm interpretation of his findings somewhat difficult at the present time.

An excellent recent review on root geotropism has been published by Audus (4) in which recent studies on different species of plants are described, and the involvement of different hormones discussed. It must be emphasised that there are now very serious doubts indeed about the involvement of IAA in root geotropism, and these have been expressed elsewhere (24, 25, 26). Indeed, the role of the known growth hormones in regulating root growth and development is at best speculative at the present time and presents a potentially fruitful field of research, provided that rigorous physicochemical analytical methods are used, and hormone applications are critical and realistic. External application of growth hormones to the entire external surface of roots is fraught with danger and yields results of little interpretable value.

Studies of geotropism in shoots and roots have made significant progress in recent years, but as I have tried to show in this paper, much remains to be elucidated, even in orthogeotropic organs, before we can rise to the challenge presented by the mechanisms involved in plagiotropic and diageotropic organs.

References

1. Abrol, B.K., Audus, L.J.: J. Exp. Botany 24, 1209 (1973)
2. Abrol, B.K., Audus, L.J.: J. Exp. Botany 24, 1224 (1973)
3. Arslan-Cerin, N.: J. Exp. Botany 17, 236 (1966)
4. Audus, L.J.: In: The Development and Function of Roots. Torrey, J.G., Clarkson, D. (eds.). London: Academic Press, 1975, p. 327
5. Bode, H.R.: Planta 53, 212 (1959)
6. Bode, H.R.: Planta 54, 15 (1960)
7. Bridges, I.G., Wilkins, M.B.: Planta 112, 191 (1973)
8. Bridges, I.G., Wilkins, M.B.: Planta 114, 331 (1973)
9. Bridges, I.G., Wilkins, M.B.: Planta 117, 243 (1974)
10. Gillespie, B., Thimann, K.V.: Plant Physiol. 38, 214 (1963)
11. Goswami, K.K.A., Audus, L.J.: Ann. Botany (London) 40, 49 (1976)
12. Hartung, W.: Planta 128, 59 (1976)
13. Iwami, S., Masuda, Y.: Plant Cell Physiol. (Tokyo) 14, 757 (1973)

14. Iwami, S., Masuda, Y.: Plant Cell Physiol. (Tokyo) 15, 121 (1974)
15. Jones, R.L., Phillips. I.D.J.: Plant Physiol. 41, 1381 (1966)
16. Jones, R.L., Phillips. I.D.J.: Planta 72, 53 (1967)
17. Phillips. I.D.J.: J. Exp. Botany 22, 465 (1971)
18. Phillips, I.D.J.: Planta 105, 234 (1972)
19. Phillips, I.D.J., Hartung, W.: New Phytologist 76, 1 (1976)
20. Shaw, S., Wilkins, M.B.: Planta 109, 11 (1973)
21. Shaw, S., Wilkins, M.B.: J. Exp. Botany 25, 199 (1974)
22. Tanimoto, E., Masuda, Y.: Plant Cell Physiol. (Tokyo) 12, 663 (1971)
23. White, J.C., Medlow, G.C., Hillman, J.R., Wilkins, M.B.: J. Exp. Botany 26, 419 (1975)
24. Wilkins, M.B.: Current Adv. Plant Sci. No. 13, 317 March (1975)
25. Wilkins, M.B.: In: Perspectives in Experimental Biology. Sunderland, N. (ed.). Oxford-New York: Pergamon Press 2, 77 (1976)
26. Wilkins, M.B.: Sci. Prog. (Oxford) 63, 187 (1976)
27. Wilkins, M.B., Woodcock, A.E.R.: Nature (London) 208, 990 (1965)
28. Woodcock, A.E.R., Wilkins, M.B.: J. Exp. Botany 20, 156 (1969)
29. Woodcock, A.E.R., Wilkins, M.B.: J. Exp. Botany 20, 687 (1969)
30. Woodcock, A.E.R., Wilkins, M.B.: J. Exp. Botany 21, 985 (1970)
31. Woodcock, A.E.R., Wilkins, M.B.: J. Exp. Botany 22, 512 (1971)
32. Wilkins, H., Wain, R.L.: Planta 121, 1 (1974)
33. Yamamoto, R., Shinozaki, K., Masuda, Y.: Plant Cell Physiol. (Tokyo) 11, 947 (1970)

Paper Demonstration References (see Section IX):

2 - 11 - 31 - 41 - 72 - 82 - 124 - 133 - 148 - 166

23. Ultrastructural Aspects of Georeceptors in Roots

A. SIEVERS and D. VOLKMANN

Abstract

The statocytes of the root caps are characterized by sedimentable amyloplasts as well as by a striking endoplasmatic reticulum. Those of *Lepidium sativum* L. are strongly polarized whereby the rough endoplasmatic reticulum (ER complex) is arranged at the distal cell pole. In normal position of the root, the amyloplasts are sedimented on it. Functional relationships between both organelles indicate that they bring about graviperception. After experimental influence on the root by rotation on clinostats and forced inversion, the original polarity of the statocytes remains basically unchanged. However, in addition to the distal ER complex, the formation of further ER complexes is inducible, depending on the direction of gravity. Abscisic acid inhibits the development of cell polarity. Long-term rotation on the classical clinostat and an electrical field disturb the polar arrangement of the cell organelles in the statocytes.

A. Introduction

In recent years the question of the localization of graviperception in roots has been more and more defined. There is almost general agreement that the root cap is the organ that, by means of hormones, regulates both normal and gravitropic growth (1, 29). The perception of gravity depends on the presence of sedimentable amyloplasts in the central cap cells (7).

When considering the role of sedimented particles in a cell during graviperception, the extremely high sensitivity of the roots to mass accelleration has to be taken into account. The threshold value for gravireactions with respect to acceleration lies between 10^{-3} and 10^{-4} g (19, 24). At 1 g the presentation time for roots only amounts to some seconds (1, 12). Deviations from the perpendicular of 5° are already perceived (3). Thus, only a very short time (fractions of sec, 14) and space (fractions of µm ?) are available for the movement of a statolith that could transduce the gravity stimulus into a physiologic signal.

For this reason, the assumption that a translocation of amyloplasts in the statocytes causes graviperception is unnecessary and perhaps even misleading. On the contrary, one should pay more attention to the idea that interactions of sedimented amyloplasts with their close cytoplasmic surroundings can already be considered as a signal transformer. It is important that, by understanding the ultrastructure of statocytes and its alteration through variation of the stimulus, one obtains information about graviperception.

B. Materials and Methods

The root caps of the plants shown in Table 1 were examined with regard to the ultrastructure of the statocytes. The conventional fixatives such as glutaraldehyde-OsO_4, OsO_4, and $KMnO_4$ led to different results,

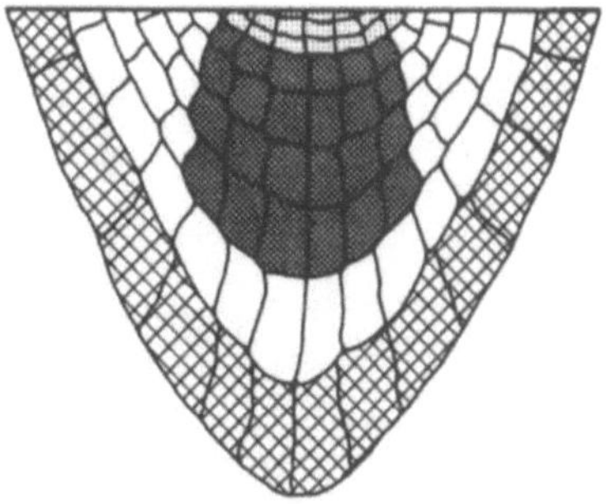

Fig. 1. Root cap of *Lepidium*. Statenchyma consists of 4 storeys

Table 1. Plants investigated with regard to the ultrastructure of their root statocytes

Family	Plants	References
Apiaceae (Umbelliferae)	*Daucus carota* L.	25
Brassicaceae (Cruciferae)	*Lepidium sativum* L.	8, 22
	Raphanus sativus L.	16
Caryophyllaceae	*Vaccaria pyramidata* Med.	16
Fabaceae (Papilionaceae)	*Genista tinctoria* L.	25
	Lens culinaris Med.	13, 25
	Pisum sativum L.	11
	Vicia faba L.	4
Liliaceae	*Allium cepa* L.	25
Poaceae (Gramineae)	*Zea mays* L.	10

so that none of them could provide an optimal picture of the statocytes. In particular, the endoplasmatic reticulum (ER) of the statocytes is frequently broken down into vesicles after fixation with glutaraldehyde-OsO_4 and OsO_4 (this may be a function of its physiologic status, see Ref.20) so that its exact shape and localization within the cell are badly preserved. It is a fact that fixation with $KMnO_4$ maintains the ER cisternae; however, it causes, for instance, a strong shrinkage of the small vacuoles. Possibly the high sensitivity of statocytes toward exogenous influences is already shown by the lability toward all chemical fixatives.

C. Results and Discussion

The root cap consists of three tissues: the basal meristem, the central statenchyma, and the peripheral secretion tissue (Fig.1). Only the statenchyma contains cells with sedimentable amyloplasts. A special characteristic of the statocytes is the presence of large complexes of rough ER, which are reminiscent of the classical ergastoplasm of animal cells. In its capacity and regularity, it is truly unique in the plant kingdom. The plants in Table 1 show this phenomenon nearly with-

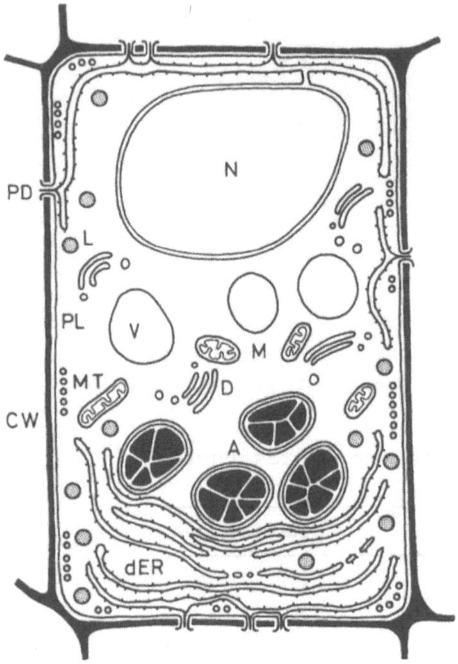

Fig. 2. Drawing of central statocyte. *A*: amyloplast; *CW*: cell wall; *D*: dictyosome; *dER*: distal endoplasmatic reticulum; *L*: lipid droplet; *M*: mitochondrion; *MT*: microtubule; *N*: nucleus; *PD*: plasmodesmos; *PL*: plasmalemma; *V*: vacuole

out exception. In particular, a standard pattern of the ER in the statocytes of *Lepidium sativum* L. was found, so the term distal ER complex was introduced (22, 28). The special regularity in the ultrastructure of cress statocytes led us to the conclusion that the polar arrangement of cell organelles is a condition for graviperception.

I. Polarity of Statocytes

Figure 2 represents a central statocyte of *Lepidium* (22, 25). One can see that the amyloplasts are sedimented on the distal ER complex. The nucleus is localized in the proximal pole of the cell. Mitochondria, dictyosomes, and some small vacuoles are distributed at random in the cytoplasm, but they do not occur in between the ER cisternae. Microtubules run perpendicularly to the root axis, mainly near the anticlines. Further structural components of the ground cytoplasm have so far not been shown. Lipid droplets are preferably located in the cell periphery. They are embryonic storage material and do not occur in lateral roots. There are many more plasmodesmata in the periclines compared with the anticlines (in *Zea mays* 10: 1; Ref.9).

In addition to the existence of movable amyloplasts, the distal ER complex is a second criterion for cress statocytes (Fig.3). It develops during germination of the radicle (27). At the same time starch is synthesized and the amyloplasts begin to sediment. Figure 4 indicates that the number of the cisternae in the ER complexes increases from storey I to storey IV of the statenchyma. Thirty-six hours (Fig.4b) after the beginning of soaking there are more cisternae in the ER complexes than is the case after 24 h (Fig.4a). During normal growth the individual cisternae in the ER complexes are closely stacked and arranged nearly parallel to the periclinal wall. The margin of each ER complex points upward and partly covers the anticlines.

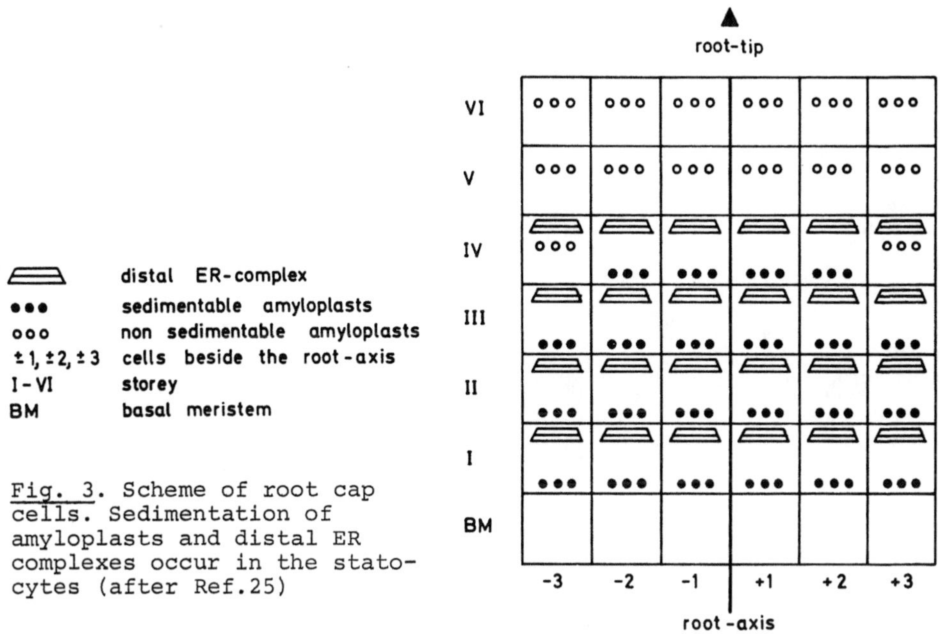

Fig. 3. Scheme of root cap cells. Sedimentation of amyloplasts and distal ER complexes occur in the statocytes (after Ref.25)

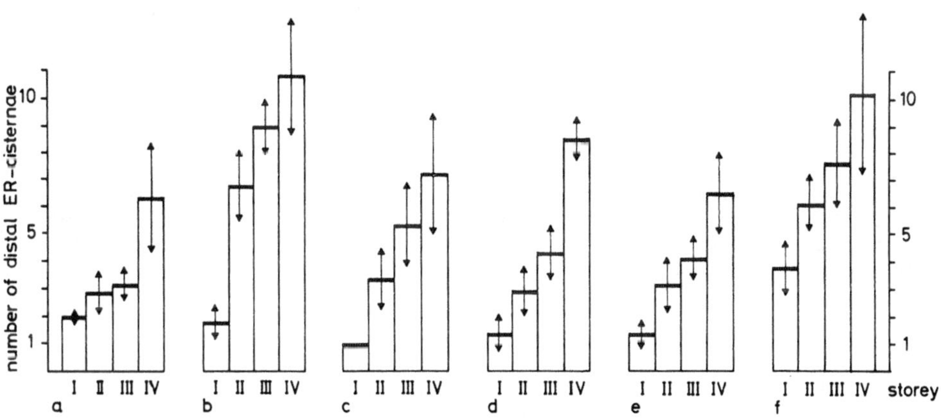

Fig. 4 a-f. Increase in number of distal ER cisternae from storey I to IV after different exposures of roots: (a), 24 h normal; (b), 36 h normal; (c), 24 h clinostated at 2 rpm (roots horizontally grown parallel to clinostat axis, Ref.6); (d), 24 h clinostated at 55 rpm (roots grown in clinostat axis, Ref.24); (e), 24 h clinostated at 120 rpm (roots grown in centrifugal direction, Ref.24); (f), 20 h clinostated at 120 rpm (roots grown in centrifugal direction, Ref.24) plus 4 h normal (roots grown downward in vertical direction, Ref.24)

The root cap of *Lepidium* is built up strictly geometrically. In each storey the periclinal cell walls have a parabolic shape (Fig.1). From this special geometry it is assumed that the contact between amyloplasts and ER complexes is the cause for graviperception (22). In case of normal, vertical growth (Fig.5a), there is a labile equilibrium (because of the radial symmetry) of the stress of all ER complexes by

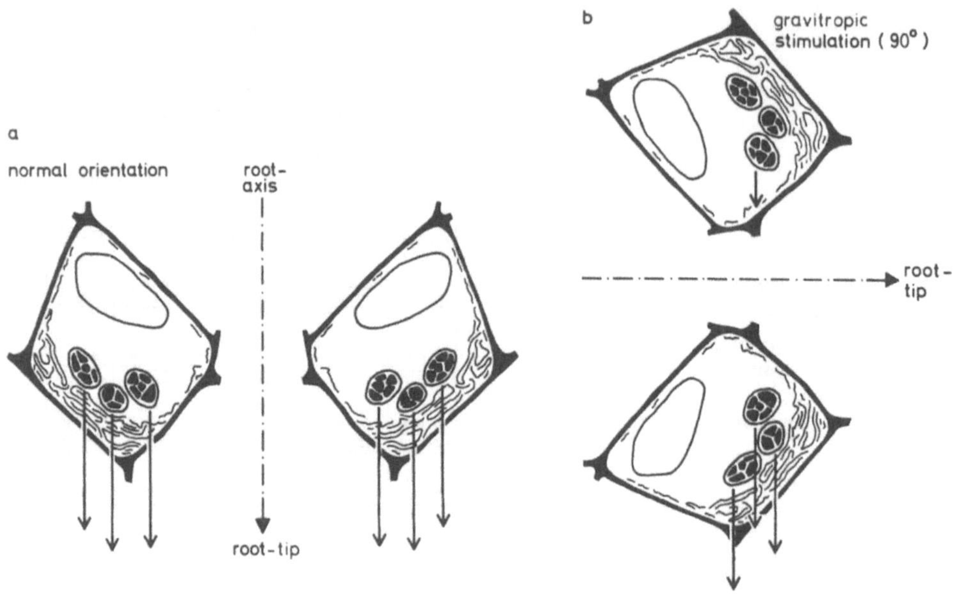

Fig. 5 a,b. During normal orientation of the root (a) the amyloplasts cause equal stress on ER complexes; in the case of gravitropic stimulation (b) they cause unequal stress (demonstrated in pair of peripheral statocytes, after Ref.22)

the sedimented amyloplasts. Each deviation from the perpendicular causes a drastic unequilibrium in symmetrical pairs of peripheral statocytes, especially in horizontal exposure (Fig.5b). On the physically lower side the ER complexes remain stressed by amyloplasts; on the physically upper side they are de-stressed at once. A differential signal could come about in this way. The question is whether or not the perception of the gravity stimulus depends on the interaction of amyloplasts and ER complexes.

A series of observations shows that there are close functional relations between amyloplasts and ER complexes in statocytes:

1. In normal position of the root, the amyloplasts, at least temporarily, press their own sculpture into the ER complex (deformation effect, 25). Applying glutaraldehyde-OsO_4 as a fixative, especially that part of the ER that is stressed by sedimented amyloplasts shows large cisternae while the ER frequently vesiculates where it is not stressed by amyloplasts (cf. Material and Methods).

2. In the case of deviations from the perpendicular (at 30°, even more at 60°), the amyloplasts not only settle downward in the direction of gravity but are also shifted upward (5). This can only be explained by the fact that the bed of the amyloplasts, i.e., the ER complex, is elastic up to a certain degree (Fig.6).

3. During the gravitropic latent and reaction time (approx. 2 h), the ER complexes remain located at the distal pole in all statocytes (25). The result is that the differential signal can be continuously reduced up to the reinstatement of the original labile equilibrium when the root curves.

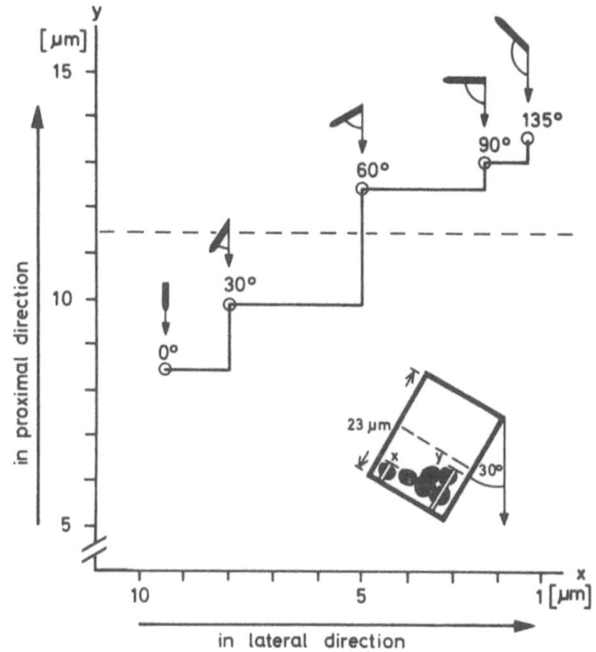

Fig. 6. Displacement of amyloplasts in central statocytes of cress. Roots were stimulated at different angles for 15 min. Inset demonstrates mode of measurement (after Ref.5)

4. If the ER complexes are freed from the weight of the amyloplasts, e.g., by rotation of a root on a clinostat (23) or by inversion (28), the cisternae are not strictly stacked any more and are not that parallel to the periclinal cell wall. This is a clear indication for the pressing effect of the amyloplasts on the ER complex.

5. Lateral roots reacting gravitropically possess sedimentable amyloplasts and an ER complex in their statocytes (21); ageotropic laterals contain sedimentable amyloplasts, but they do not contain an ER complex (15).

II. Experiments on the Polarity of Statocytes

1. Seeds of *Lepidium* are soaked for 30 min with a 10^{-4} M solution of abscisic acid. After 120 h the ultrastructure of the statocytes considerably deviates from that of the controls (Fig.7c). The elongation of the statocytes is inhibited, the electron transparency of its ground cytoplasm is less, the sedimentation of the amyloplasts is prevented, the formation of an ER complex at the distal pole is missing. Only short, unorientated ER cisternae are localized at the distal cell pole (26). This stage of development corresponds approximately to that of cells of very young control roots (10-14 h after soaking). Neither reacts gravitropically.

2. Cress roots were forced to grow in capillaries for 18 h in inverted position. During this time the statocytes produced a second ER complex at the proximal cell pole in addition to the distal ER complex (28).

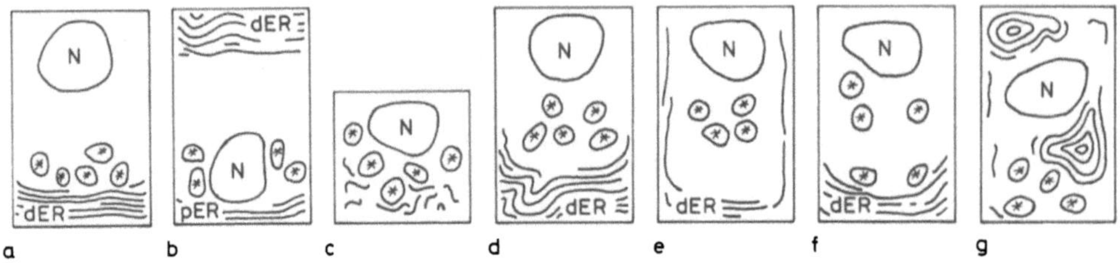

Fig. 7 a-g. Localization of ER (*dER*: distal ER complex; *pER*: proximal ER complex), amyloplasts (*), and nucleus (*N*) in central statocytes after different treatments of the roots: (a), normal; (b), 18 h normal plus 18 h inversion (after Ref.28); (c), 30 min abscisic acid (10^{-4} M) during soaking plus 120 h normal (after Ref.26); (d), 24 h clinostated at 2 rpm (roots grown horizontally, parallel to clinostat axis, Ref.6); (e), 24 h clinostated at 55 rpm (roots grown horizontally in clinostat axis, Ref.24); (f), 24 h clinostated at 120 rpm (roots grown in centrifugal direction, Ref.24); (g), 24 h normal plus 6 h electrical field (after Ref.2)

The new proximal ER complex originates underneath the already sedimented amyloplasts (cf. Fig.7a,b). Thus, it is possible to induce the localization of an ER complex in the cell by the direction of gravity. This supports the postulation that the contact between amyloplasts and an ER complex is a condition of orientation in space.

3. In order to exclude the gravity vector as the inductor for the distal ER complex, cress seeds were allowed to germinate on horizontal clinostats at 2 rpm and 55 rpm for 24 h. The straight growth of the roots shows that the polarity of the statocytes also develops if gravity has never had an effect on the root from the apical direction (23). The cisternae in the distal ER complex are not arranged strictly parallel to the periclinal cell wall, as is normally the case. The amyloplasts are not sedimented but are distributed between the nucleus and the ER complex (Fig.7d). The number of cisternae in the ER complexes increases from storey I to storey IV of the statenchyma (Fig.4c,d) - as is the case under normal conditions - and reaches the value of comparable controls (Fig.4a). This result proves that the formation of an ER complex at the distal cell pole is not induced by the direction of gravity. It must, therefore, be endogenously caused.

4. If 24-h-old vertically growing roots are rotated for 2 h on a horizontal clinostat at 2 rpm, the polarity of the statocytes corresponds in principle to that of those that germinated and grew on a clinostat (6). However, the margin of the cup-like distal ER complex reaches up to approximately half that of the anticlinal walls. The gravitropic reaction of the 2-h rotated roots starts earlier than that of the controls. This does not depend on different growth rates but on a stronger gravisensibility (Fig.8). In this case the graviperception is possibly caused by the fact that on the physically lower side the amyloplasts sediment on a more extensive ER complex than they do on the physically upper side. Due to the rotation on the clinostat the sum of the contact areas between amyloplasts and ER complexes is bigger in the lower side (Fig.9) than is the case during normal stimulation (Fig.5b). This could also explain the "pre-inversion effect" (8).

5. If 24-h-old roots are rotated for 16 h on the horizontal clinostat at 2 rpm the statocytes are extraordinarily changed (6). In comparison with the controls, the ultrastructural picture of the statocytes is characterized by the following criteria: the distal ER complex is de-

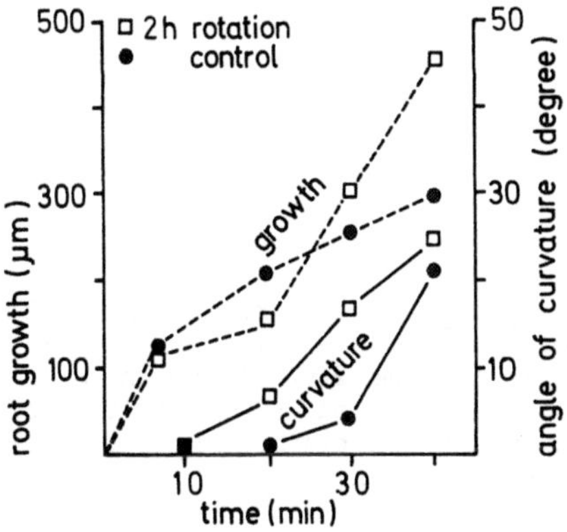

Fig. 8. After 2 h at 2 rpm, gravitropic reaction starts earlier compared with control. This is independent of growth rates (after Ref. 6)

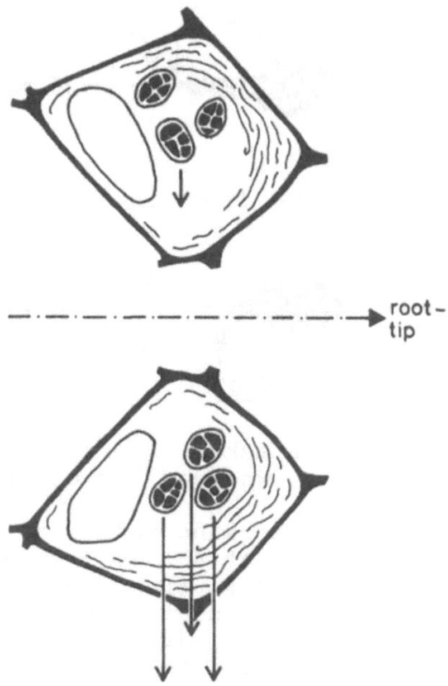

Fig. 9. Compare with Fig. 5b. Gravitropic stimulation after 2 h at 2 rpm (after Ref.6)

stroyed, the ER cisternae are distributed at random in the cell; the starch content of the amyloplasts is reduced; the number of lipid droplets is decreased, their size is strongly increased; numerous lysosomes appear; and anticlinal cell walls are perforated sometimes even up to a small rim. These observations prove that the constant gravitropic stimulation by longterm rotation on the classical clinostat destroys the statocytes of those roots that developed for 24 h under normal conditions. Forty-h-old, normally growing roots show an intact polarity.

6. Dry seeds of *Lepidium* were allowed to germinate on a fast rotating horizontal clinostat for 24 h at 55 rpm so that the clinostat axis ran through the longitudinal axis of the seeds (24). Approximately 20% of the roots did not react in the direction of the centrifugal force, but they grew in the clinostat axis (Fig.10a). Here they were subjected to a calculated centrifugal acceleration of masses of 4×10^{-4} g max, which is below the threshold for the plant's gravisensitivity (19). In the case of 120 rpm, all roots reacted gravitropically and grew in the centrifugal direction (Fig.10b). After 24 h their tips were subjected to a calculated centrifugal acceleration of masses of 4×10^{-2} g max. In order to exclude possible damage of the normal gravitropical activity, the roots were exposed horizontally for 2 h at 1 g after rotation. They showed a normal gravireaction.

These experiments indicate that, provided that a suitable rotation speed is given, it is possible to produce physiologic weightlessness in the axis of the fast rotating clinostat. However, immediately above the threshold the roots start to react gravitropically. The polarity of the statocytes corresponds in principle to that of the controls, as is the case after rotating on the classical clinostat. However, in uncurved roots (Fig.10a) the ER cisternae reach up to the proximal cell wall. The amyloplasts are not sedimented but are concentrated in the cell center (Fig.7e). In the statocytes of those roots that grow in centrifugal direction (Fig.10b), the ER cisternae in the distal cell

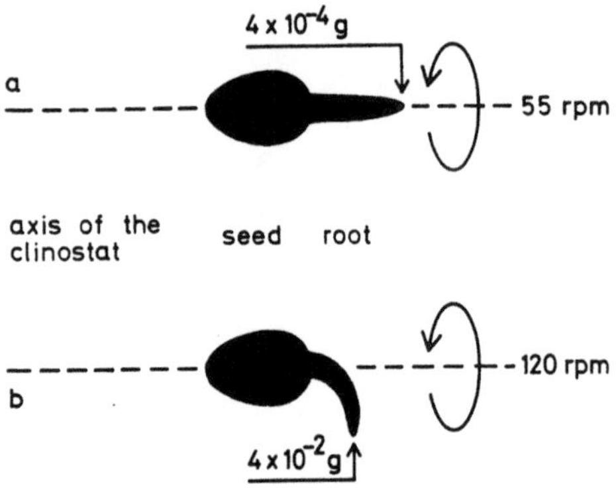

Fig. 10 a,b. Root growth on fast rotating clinostats. After 24 h (a), at 55 rpm 20% of roots do not curve; (b) at 120 rpm 100% of roots curve in centrifugal direction (after Ref.24)

pole are more stacked. Few amyloplasts are sedimented on it, especially in storey I, whereas the others are distributed at random (Fig.7f).

If instead of exposing roots for 24 h at 4×10^{-2} g max they are only exposed for 20 h at 4×10^{-2} g max and afterwards for 4 h at 1 g, the number of the distal ER cisternae increases significantly (Fig.4f). It is nearly doubled in storeys I-III in comparison with those roots that grew for 24 h at 1 g (Fig.4a) or at 4×10^{-2} g max (Fig.4e), respectively. This proves that the induction of the formation of ER cisternae is also realized at the distal cell pole. Furthermore, the induction depends not only on the direction of mass acceleration but also on its magnitude when a smaller amount is followed by a larger one.

7. Cress roots are for several hours subjected to an electric field (2270-3000 V/cm) that causes an electrotropic reaction (17). The distal ER complexes are removed (2). The ER can be found as lumps of ergastoplasm in the cell and is distributed at random (Fig.7g). This again proves the high sensitivity of statocytes to exogenous influences.

D. Conclusions

The study of the ultrastructure of cress statocytes shows an intimate correlation between amyloplasts and ER complexes. It can be assumed that sedimented amyloplasts do not stay motionless but that they are continuously in movement on the elastic ER complexes like a sportsman on a trampoline. Would it not be possible that alterations could be caused in or at the membranes involved? Might the permanent change of tensional stress and de-stress influence the permeability of membranes? Could it be possible that the whole statenchyma of a root cap might be able to integrate such amounts and, by deviation from the perpendicular, signal a difference? We do not yet know the answers. However, the search for answers to these questions seems to promise more success if we want to understand graviperception in roots rather than the study of the sedimentation of dictyosomes (18) or that of the translocation of ER cisternae (11). We could not observe either with *Lepidium*. Induction of the formation of ER complexes, which we found under defined experimental conditions, must not lead us to think that the endogenously developed ER complexes are movable. They are stably located in the cell.

Acknowledgements

The authors thank Harald Behrens, Hans Haberkorn, Wolfgang Hensel, Uta Schmitz, Volker Sobick, and Inge Stender for stimulating cooperation,

Inge Beging, Christa Grabert, and Eva Lieb for technical assistance. Special thanks are expressed to Renate Parkhouse for the translation of the manuscript. These studies were made possible by grants from the Deutsche Forschungsgemeinschaft.

References

1. Audus, L.J.: In: The Development and Function of Roots. Torrey, J.G., Clarkson, D.T. (eds.) London: Academic Press, 1975, p. 327
2. Behrens, H.M.: unpublished
3. Fitting, H.: Jahrb. Wiss. Botan. 41, 221 (1905)
4. Griffiths, H.J., Audus, L.J.: New Phytologist 63, 319 (1964)
5. Haberkorn, H.: Abstr. Intern. Assoc. Plant Physiology, Würzburg (1974)
6. Hensel, W., Sievers, A.: Abstract, 9th Intern. Conf. on Plant Growth Substances, Lausanne (1976)
7. Iversen, T.-H.: Physiol. Plantarum 22, 1251 (1969)
8. Iversen, T.-H., Larsen, P.: Physiol. Plantarum 28, 172 (1973)
9. Juniper, B.E., Barlow, P.W.: Planta 89, 352 (1969)
10. Juniper, B.E., French, A.: Planta 95, 314 (1970)
11. Juniper, B.E., French, A.: Planta 109, 211 (1973)
12. Larsen, P.: Physiol. Plantarum 10, 127 (1957)
13. Perbal, G.: C.R.Acad. Sci. Paris 273, 789 (1971)
14. Pickard, B.G.: Can. J. Botany 51, 1003 (1973)
15. Schmitz, U.: Abstr., Intern. Assoc. Plant Physiology, Würzburg (1974)
16. Schmitz, U.: unpublished
17. Schrank, A.R.: In: Encycl. Plant Physiol. Ruhland, W. (ed.). Berlin: Springer, 17 (1), 148 (1959)
18. Shen-Miller, J., Hinchman, R.R.: Bioscience 24, 643 (1974)
19. Shen-Miller, J., Hinchman, R.R., Gordon, S.A.: Plant Physiol. 43, 338 (1968)
20. Sievers, A.: Z. Zellforschung 64, 280 (1964)
21. Sievers, A., Schmitz, U.: Planta 114, 373 (1973)
22. Sievers, A., Volkmann, D.: Planta 102, 160 (1972)
23. Sievers, A., Volkmann, D., Hensel, W., Sobick, V., Briegleb, W.: Naturwissenschaften 63, 343 (1976)
24. Sobick, V., Sievers, A., Volkmann, D., Briegleb, W.: Abstr., 9th Int. Conf. Plant Growth Substances, Lausanne (1976)
25. Volkmann, D.: Protoplasma 79, 159 (1974)
26. Volkmann, D.: Planta 136, 89 (1976)
27. Volkmann, D.: Abstr., 9th Intern. Conf. Plant Growth Substances, Lausanne (1976)
28. Volkmann, D., Sievers, A.: Planta 127, 11 (1975)
29. Wilkins, M.B.: Sci. Prog. (Oxford) 63, 187 (1976)

Paper Demonstration References (see Section IX):

32 - 62 - 72 - 79 - 146 - 179

24. Hormonal Regulation of Phototropism in Dicotyledonous Seedlings

J. BRUINSMA

Abstract

The validity of the Cholodny-Went theory of phototropism is assessed, particularly its applicability to dicotyledonous seedlings. Spectrofluorometric analysis of extracts shows that during phototropic curvature of hypocotyls of light-grown sunflower seedlings no lateral redistribution of indole-3-acetic acid (IAA) occurs. No basipetal IAA translocation can be found either, but diffusates from hypocotyl bases contain unknown substances promoting and inhibiting cell elongation in the straight-growth test of *Avena* coleoptile segments.

Also because phototropic curvature of the hypocotyl requires the presence of illuminated cotyledons, and illuminated cotyledons inhibit cell extension in the hypocotyl, it is hypothesized that phototropic curvature in dicotyledonous seedlings is a matter of growth inhibition rather than caused by a growth-promoting substance.

A. Historical Background and Limitations of the Cholodny-Went Theory

The problem of phototropism stood at the cradle of hormone research. The very discovery of auxin by Went (52), half a century ago, was the eventual result of another fifty years of research on the phototropic curvature of grass coleoptiles, originating from the statement made by Darwin (8): "When seedlings are freely exposed to a lateral light, some influence is transmitted from the upper to the lower part, causing the latter to bend". Went demonstrated that the tip of the oat coleoptile released a growth-promoting substance, and also found about twice as much growth-stimulating activity occurring in the diffusate from the shaded side of an unilaterally illuminated coleoptile tip than in that from its illuminated side (53). This led to the extension to phototropism of the concept for geotropic curvature, developed by Cholodny in 1924. According to this concept, the curvature is caused by differential cell elongation because of asymmetrical distribution of a cell-elongating substance (6).

When the chemical nature of this substance was identified as indole-3-acetic acid (IAA), and labeling with radioactive markers became possible, the testing of the Cholodny-Went theory by the application of labeled auxins led to considerable controversy, partly because of overloading the tissue. With unilaterally illuminated etiolated coleoptiles, however, asymmetric distribution of exogenous auxin could be demonstrated (15, 40). Differential auxin distribution by light-induced changes in the synthesis or destruction of auxin could not be found (4), leaving a lateral movement mechanism the most plausible one.

Although most of the research in phototropism has been done with etiolated grass coleoptiles, the original statement by Darwin was based on experiments with light-grown plants as well, and particularly also with various dicotyledonous seedlings. Generalizations from the results on etiolated coleoptiles were partly derived from similar observations on sporangiophores of such fungi as *Phycomyces* and *Pilobolus* (42); however,

Table 1. Distribution of IAA in upright and in phototropically curving hypocotyls

	ng IAA per g fresh weight						
	Upright hypocotyls			Curving hypocotyls			
	Left half	Right half	No. of plants	Light half	Shadow half	°Curvature	No. of plants
Expt. 1	44	45	*55*	54	49	21	*44*
Expt. 2	64	68	*60*	63	62	23	*45*

From light-grown sunflower seedlings, aerial parts were sampled, halved longitudinally, cotyledons and apex removed, and halved hypocotyls extracted for spectrofluorometric IAA-determination according to (30). Curving hypocotyls were phototropically stimulated for 4 h prior to dissection. Data from four duplicate experiments

auxin does not affect fungal growth, and there is no evidence of asymmetric auxin distribution in these organs (18).

There are two more objections to a generalization of the Cholodny-Went theory. Firstly, the theory applies only to a limited part of the complex dose-response curve of etiolated coleoptiles. At higher dosages, the positive curvature diminishes, may even become negative and, subsequently, positive again. It is the illumination range of this second positive curvature that is ecologically relevant and to which also light-grown seedlings respond with a curvature (13). Secondly, the evidence in favor of the Cholodny-Went theory is completely indirect. It has been derived either from the fate of exogenous auxins or from tests in which endogenous substances were determined in a bioassay that is not specific for auxins but also responds to other materials, e.g., inhibitors. The direct measurement of a lateral displacement of endogenous IAA in a phototropically or geotropically stimulated organ had to wait for the development of a sensitive and accurate determination method.

In the mean time, two sources of information indicated that the phototropic system in light-grown seedlings, and especially in dicots, is likely to be far more complex than that in etiolated coleoptiles. On the one hand, data have recently become available about different pigments, possibly located in different organs, cells, or organelles, involved in the perception and guidance of the phototropic stimulus (12, 35). On the other hand, the anatomical structure of the leafy dicotelydonous stem with its epiderm, cortex, and stele, is so much more complex than that of a coleoptile and has such an array of possibilities for hormonal translocation, that an explanation on the basis of lateral transport from cell to cell is by no means compulsory (9, 39, 51). For these reasons, we set out to determine IAA levels in the shadow and illuminated sides of light-grown sunflower seedlings during their phototropic curvature.

B. Role of Growth Promotors in Phototropic Curvature

The cultivation and treatment of the *Helianthus annuus* L. seedlings and the spectrofluorometric determination of IAA are described elsewhere (5, 30). We compared the amounts of IAA from halves of upright and phototropically curving hypocotyls. Table 1 shows that the equal

Table 2. Auxin activity in hypocotyl diffusates

	ng IAA per plant	IAA-eq. bioassay, ng per plant		
		Average	Lowest	Highest value
Expt. 1	0.07 ± 0.03	0.85	0.35	1.58
Expt. 2	0.04 ± 0.04	1.93	1.07	3.15

Diffusates from 25 plants per sample, illuminated from above, were collected over a period of 6 h in 5 ml water (Expt. 1) or in buffer solution (Expt. 2), five replicates per experiment. Auxin was determined spectrofluorometrically, as IAA, or in *Avena* coleoptile straight-growth test according to (37)

distribution of IAA in straight hypocotyls is completely maintained after 4 h of phototropic stimulation, giving about 20° curvature, while after 6 h the curvature is about 30°. There is no accumulation of extractable IAA in the shadow half of the curving hypocotyl. Also at Wilkins laboratory, evidence has been obtained in experiments on geotropism that curvature can occur without participation of IAA (3, 17). Because part of the extractable IAA might not be diffusable and, thereby, not actively participating in cell elongation, the level of IAA diffusing from hypocotyl bases into distilled water or buffer solution was also determined (Table 2).

Diffusion of IAA from the hypocotyl bases was barely detectable, the amounts in the diffusate being of the order of the amount of radioactive IAA added as an internal standard in the spectrofluorometric determination. According to the IAA content of the hypocotyls and the known rate of transport of exogenous IAA in sunflower hypocotyl segments of 7 mm/h (34), the amounts detected are only a few per thousand of what might be expected. It is unlikely that decarboxylating enzymes, which are no longer present after removal of the seedlings from the diffusates (5), would so nearly completely have removed the diffused IAA from the solutions. This means that the IAA in the hypocotyls, being immobile in lateral and basipetal directions, is probably confined to such compartments as the vacuole and not actively participating in cell elongation processes. That IAA in sunflower hypocotyls is pooled in the vacuoles was suggested earlier by dela Fuente and Leopold (16). However, the often substantial auxin acitivity of the diffusate in the *Avena* bioassay (Table 2) indicates the presence of another cell-elongating substance. The chemical nature of this substance is still unknown, so that its distribution within phototropically stimulated hypocotyls cannot yet be determined. It may be another auxin, e.g., phenylacetic acid (43), or belong to one of the other groups of substances positively active in this test, e.g., a gibberellin (45). The activity of gibberellins in phototropic responses of sunflower was suggested by Phillips (38), and the participation of gibberellins in hypocotyl growth in general has been established by various research groups (1, 24, 27, 28, 29). The light-induced inhibition of hypocotyl growth can be removed by exogenous gibberellins in a variety of plant species, including sunflower, and lack of gibberellins is considered to limit the growth of these species in the light (44). The mode of action of gibberellins in the growth of illuminated hypocotyls is uncertain: removal of light-inhibition of IAA synthesis is mentioned as well as effects on the permeability of membranes (14, 23, 36).

In a recent report, however, Wakhloo (50) finds endogenous gibberellins ineffective in the growth of etiolated lettuce hypocotyls; he

Table 3. Phototropic curvature of hypocotyls observed after 6 h of unilateral illumination, with one or both cotyledons removed or covered; 25 plants per sample

Cotyledon(s)	°Curvature	
	Removed	Covered
Intact	30	25
Both cotyledons	7	7
Cotyledon toward light	16	16
Cotyledon away from light	8	10

therefore suggests the light-induced growth inhibition is caused not by a lowered level of gibberellins, but either by a decrease in some unknown growth promotor or, alternatively, by the appearance of a light-induced inhibitor that can be counteracted by exogenous gibberellin.

C. Role of Growth Inhibition in Phototropic Curvature

Tropistic movements of higher plants (but cf.26) are invariably growth responses caused by differences in cell elongation. Differential cell elongation, however, need not result from an uneven distribution of growth-promoting substances; it may also be due to the asymmetric distribution of an inhibitor, lowering the sensitivity to (25), or otherwise antagonizing equally distributed auxin to a different extent. It is interesting, therefore, to investigate the possibility of the participation of growth-inhibiting substances in phototropism, the more so because inhibitors apparently play an important role in the geotropic response of roots in which they are produced particularly upon illumination (41, 46, 49, 54, 55). We encountered growth-inhibiting substance(s) when studying the effects of the cotyledons on the phototropic curvature of the hypocotyl.

Darwin's statement, quoted previously, was also derived from experiments with dicotyledonous seedlings. Later Lam and Leopold (33) found that shading one or both cotyledons of light-grown sunflower seedlings promoted hypocotyl growth; when one cotyledon was shaded the hypocotyl bended in the direction of the illuminated one. Although Lam and Leopold could not detect a lateral distribution or dark-promoted release of radioactive IAA, applied to the cotyledons, they interpreted their results in terms of differential auxin supply and ascribed the effect of cotyledons in the phototropic response of the hypocotyl to an increased auxin release from darkend cotyledons. We confirmed the necessity of the presence of illuminated cotyledons for a proper phototropic response of the hypocotyl. When one or two cotyledons were removed or covered with aluminium foil, the response was considerably reduced (Table 3).

However, the fact that covering the cotyledons had the same influence as their removal led us to the reversal of the hypothesis of Lam and Leopold: not the shaded but the illuminated cotyledon is required for phototropic curvature and, consequently, not a dark-produced auxin but a light-produced inhibitor is released from the cotyledons into the hypocotyl (indeed, Lam and Leopold mentioned the presence of a strong inhibitor in their plant extracts). This view was further confirmed by measuring the extension growth of hypocotyls of seedlings in light and in darkness, intact, or with one or two cotyledons covered or removed (Fig.1).

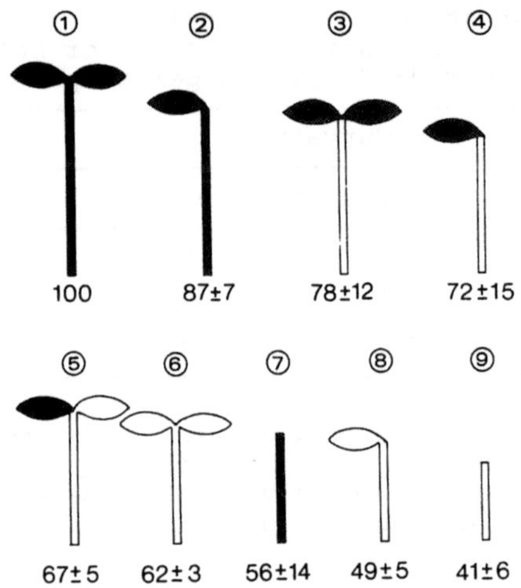

Fig. 1. Growth increase, in percentages, of light-grown seedlings, placed for 24 h in light or in darkness with their cotyledons intact, covered, or removed (closed parts darkened, open parts illuminated). Average values of four experiments with 10-20 seedlings per sample. Possible curvatures neglected

Table 4. Recovery in *Avena* coleoptile straight-growth test of IAA added to diffusates to final concentration of 3×10^{-6} M

Seed					
Seedlings:	Light-grown		Etiolated		No diffusate
Diffusion in:	Light	Dark	Light	Dark	(Control)
Elongation, in % of control	109	96	96	99	100
Same, + 3×10^{-6} M IAA	118c	135b	109d	128b	169a
Percentage IAA recovery	15	45	30	40	100

Diffusates from hypocotyl bases of light-grown and etiolated sunflower seedlings were collected after 6 h illumination from above or in darkness. Numbers followed by different indices differ significantly at P = 0.05. With permission of J.M. Franssen

By comparison of seedlings 1, 2 and 7, and of 6, 8 and 9 in Figure 1, it can be seen that cotyledons promote hypocotyl growth by the supply of nutrients and other positive factors (24, 48). The inhibition of hypocotyl growth by illumination (cf. 1 and 6, 2 and 8) is partly prevented by shading of cotyledons (cf. 6, 5 and 3, 8 and 4), which demonstrates the inhibition of hypocotyl growth by illuminated cotyledons. We interpret this finding as the production and release of an inhibiting substance during illumination. Comparison of 1 and 3, 2 and 4, and 7 and 9 shows that this inhibitor is also formed within the hypocotyl upon its illumination.

Further evidence for the existence of such an inhibiting substance released into, and translocated through the hypocotyl, was obtained by recovery experiments in which known amounts of IAA were added to hypocotyl diffusates without any IAA oxidase activity (Table 4). These preliminary experiments show a very incomplete recovery of IAA in the

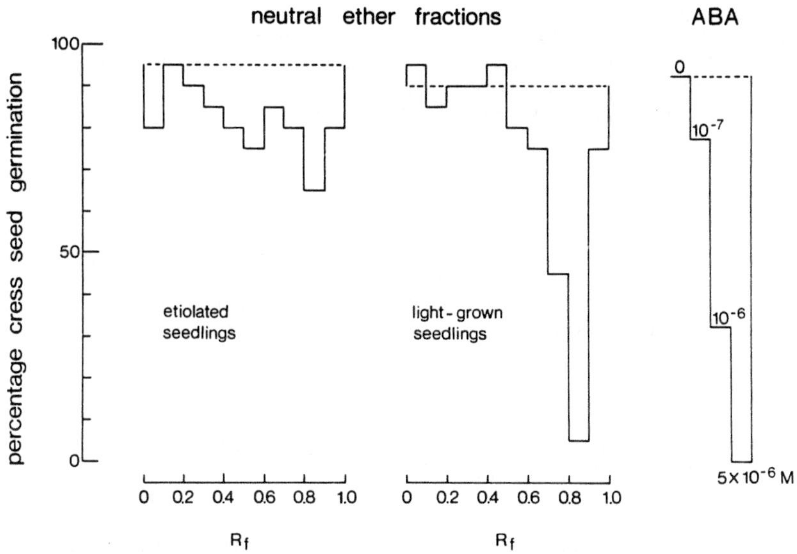

Fig. 2. Inhibition of germination of *Lepidium sativum* L. seed by paper-chromatographic fractions of neutral ether extracts from dark- and light-grown sunflower seedlings, and by abscisic acid solutions. With permission of A.G. Thompson

bioassay, particularly with diffusates obtained during illumination of the seedlings, demonstrating the presence of growth-inhibiting substance(s) in the diffusate. The presence of an auxin-antagonizing factor in extracts and diffusates of, e.g., sunflower tissues has been reported since 1956 (2, 7, 19, 47).

In the geotropic response of roots, abscisic acid (ABA) is likely to be involved (11, 31, 32, 54, 55). This growth inhibitor can be formed upon illumination (46), and its translocation can also be influenced by light (20). However, the light-induced inhibitor involved in phototropism of stems is derived from leaves rather than from roots and may well be of a different chemical nature. For instance, removal of cotyledons and apex, which reduces the phototropic response, fails to decrease the geotropic response of light-grown cucumber hypocotyls (22). The light-induced inhibition of stem growth of peas is independent of both endogenous and applied ABA (10). From peas, two inhibitors can be extracted, ABA and an unknown neutral inhibitor Y; this factor Y is absent in etiolated shoots, and dwarf plants contain three times as much of it as normal pea varieties (21). In accordance with these and other data from the literature (7, 19), we find a neutral, ether-soluble growth inhibitor in extracts from illuminated sunflower seedlings, which occurs in far lower amounts in etiolated material (Fig.2). This substance cochromatographes with xanthoxin. Its synthesis and distribution in the seedling are under study at present.

D. Conclusions

If the Cholodny-Went theory is applied to the phototropic curvature of sunflower seedlings, then the growth promotor involved must be different from the immobile IAA. However, the theory describing tropistic movements in terms of lateral transport of a growth-promoting substance is gradually becoming undermined. In geotropism as well as in the less studied field of phototropism, evidence is accumulating that permits a description of tropistic curvature in terms of differential growth inhibition rather than growth promotion. This view is in conflict with the Cholodny-Went theory, but still agrees with the original concept of Darwin.

Acknowledgements

The tables and figures of this paper are from results obtained by my past and present co-workers M. Benschop, J.B. van Dort, Miss J.M. Franssen, A. van Giffen, Dr. C.M. Karssen, Dr. E. Knegt, Mrs. Y.E. van Oosten-Legro, A.G. Thompson, Ph.D., and E. Vermeer.

References

1. Acton, G.J., Murray, B.B.: Planta 117, 219 (1974)
2. Bayer, M.: Planta 57, 258 (1961)
3. Bridges, I.G., Wilkins, M.B.: Planta 112, 191 (1973)
4. Briggs, W.R., Tocher, R.D., Wilson, J.F.: Science 126, 210 (1957)
5. Bruinsma, J., Karssen, C.M., Benschop, M., van Dort, J.B.: J. Exp. Botany 26, 411 (1975)
6. Cholodny, N.G.: Ber. Deut. Botan. Ges. 42, 356 (1924)
7. Cutler, H.G., Cole, R.J.: Plant Cell Physiol. (Tokyo) 15, 19 (1974)
8. Darwin, C.: The Power of Movements in Plants, London: Murray 1880
9. Digby, J., Firn, R.D.: Current Adv. Plant Sci. 8, 953 (1976)
10. Dörffling, K.: Z. Pflanzenphysiol. 70, 131 (1973)
11. Elliott, M.C.: Plant Physiol. 56, suppl. 39, (1975)
12. Elliott, N.B., Elliott, W.M.: Abstr. 12th Intern. Bot. Congress, Leningrad, 2, 287 (1975)
13. Everett, M.: Plant Physiol. 54, 222 (1974)
14. Field, R.J.: Roy. Soc. New Zealand Bull. 12, 673 (1974)
15. Fuente, R.K. dela, Leopold, A.C.: Plant Physiol. 43, 1031 (1968)
16. Fuente, R.K. dela, Leopold, A.C.: Plant Physiol. 50, 491 (1974)
17. Gardner, G., Shaw, S., Wilkins, M.B.: Planta 121, 237 (1974)
18. Gruen, H.: Ann. Rev. Plant Physiol. 10, 405 (1959)
19. Guttenberg, H.v., Zetsche, K.: Planta 48, 99 (1956)
20. Hartung, W., Behl, R.: Planta 122, 61 (1975)
21. Higgings, T.J.V., Bonner, B.A.: J. Exp. Botany 25, 705 (1974)
22. Iwami, S., Masuda, Y.: Plant Cell Physiol. (Tokyo) 15, 121 (1974)
23. Janistyn, B., Drumm, H.: Planta 125, 81 (1975)
24. Kamisaka, S.: Plant Cell Physiol. (Tokyo) 14, 747 (1973)
25. Kang, B.G., Burg, S.P.: Plant Physiol. 53, 445 (1974)
26. Kataoka, H.: Plant Cell Physiol. (Tokyo) 16, 439 (1975)
27. Kazama, H., Katsumi, M.: Plant Cell Physiol. (Tokyo) 14, 449 (1973)
28. Kazama, H., Katsumi, M.: Plant Cell Physiol. (Tokyo) 15, 307 (1974)
29. Kazama, H., Katsumi, M.: Plant Cell Physiol. (Tokyo) 15, 315 (1974)
30. Knegt, E., Bruinsma, J.: Phytochemistry 12, 753 (1973)
31. Kundu, K.K., Audus, L.J.: J. Exp. Botany 25, 479 (1974)
32. Kundu, K.K., Audus, L.J.: Planta 117, 183 (1974)
33. Lam, S.L., Leopold, A.C.: Plant Physiol. 41, 847 (1966)
34. Leopold, A.C., Lam, S.L.: In: Plant Growth Regulation, Klein, R.M. (ed.) Ames: Iowa State Univ. Press, 1961, p. 411
35. Mohr, H.: Abstr. 12th Intern. Botan. Congress, Leningrad, 2, 303 (1975)
36. Muir, R.M., Chang, K.C.: Plant Physiol. 54, 286 (1974)
37. Nitsch, J.P., Nitsch, C.: Plant Physiol. 31, 94 (1956)
38. Phillips, I.D.J.: Planta 106, 363 (1972)
39. Phillips. I.D.J., Hartung, W.: New Phytologist 76, 1 (1976)

40. Pickard-Gillespie, B., Thimann, K.V.: Plant Physiol. 39, 341 (1964)
41. Pilet, P.E.: Physiol. Plantarum, 55, 94 (1975)
42. Reinert, J.: Ann. Rev. Plant Physiol. 10, 441 (1959)
43. Schneider, E.A., Wightman, F.: Ann. Rev. Plant Physiol. 25, 487 (1974)
44. Silk, W.K., Jones, R.L.: Plant Physiol. 56, suppl. 43, (1975)
45. Sirois, J.C.: In: Biochemistry and Physiology of Plant Growth Substances, Wightman, F., Setterfield, G. (eds.) Ottawa: Runge Press, 1968, p. 1611
46. Tietz, A.: Biochem. Physiol. Pflanzen 165, 387 (1974)
47. Tillberg, E.: Physiol. Plantarum 31, 106 (1974)
48. Varga, M., Lenart, I.: Acta Agron. Acad. Sci. Hung. 23, 275 (1974)
49. Wain, R.L.: (this volume)
50. Wakhloo, J.L.: J. Exp. Botany 26, 841 (1975)
51. Wangermann, E.: New Phytologist 73, 623 (1974)
52. Went, F.W.: Rec. Trav. Botan. Néerl. 25, 1 (1928)
53. Went, F.W.: Rec. Trav. Botan. Néerl. 25, 483 (1928)
54. Wilkins, M.B.: (this volume)
55. Wilkins, H., Wain, R.L.: Planta 123, 217 (1975)

Paper Demonstration References (see Section IX):

30 - 31 - 41 - 106 - 110 - 115

25. Light, Rhythms, and the Problem of Florigen

A. W. GALSTON

Abstract

Photoperiodic induction in plants occurs when a change in state of phytochrome or other pigment is synchronized with circadian rhythmic events, leading to the export of a yet uncharacterized mobile stimulus (florigen) from leaf to bud. To analyze the mechanisms involved in this response, we have studied simple systems displaying phytochrome control, phytochrome-rhythm interaction, and the generation of mobile stimuli.

In the leguminous plants *Albizzia* and *Samanea*, rhythmic and photoinduced changes in leaflet orientation are caused by alternating net influx and efflux of K^+ and Cl^- ions into and out of pulvinal motor cells. Influx is characterized by a $Q_{10} > 1$ and efflux by a $Q_{10} < 1$ in the physiologic temperature range. Together with data from inhibitors and promotors, this implies the simultaneous operation within any one cell of two competing processes, whose alternating predominance leads to cycling. Each pulvinus acts as a unit, unaffected by its neighbors. Excised pulvini continue to oscillate if supplied with far-red phytochrome (P_{fr}) and sucrose, phytochrome conversion serving to set the clock.

In various plant systems, circadian time or the state of phytochrome can control electrical potential, the degree of electrical coupling, the transport of solutes, and photomorphogenesis. These facts, together with the growing evidence that phytochrome and hormones act on cellular membranes, lead to the postulate that the mobile stimulus connecting leaf and bud controls cellular transport phenomena.

A. Introduction

The basic facts about photoperiodism in plants have been known for almost sixty years (10), yet an understanding of the fundamental physiology of the process continues to elude us. The major advances in our understanding have come from (a) the discovery of phytochrome as a photoreceptor (5), (b) the demonstration that rhythms, as well as light signals perceived by phytochrome, are involved in timing (3), and (c) the evidence that a mobile stimulus (florigen) passes from leaf to bud following photoperiodic induction (6). Despite prodigious efforts by many investigators using sensitive bioassays (summarized in Ref.30), there are no generally accepted examples of the successful extraction or characterization of this hormone. This paradoxical result has led some workers to reject the concept of a mobile floral hormone, despite the convincing physiologic evidence for its existence. This paper is an attempt to apply newer knowledge of the physiology of phytochrome and rhythms to a re-analysis of the problem.

It is difficult to study cellular and metabolic events involved in the photoperiodic control of flowering, mainly because the process is so slow and complex. Accordingly, it has become the practice to use simpler, more rapid, and more convenient systems for the analysis of phytochrome-rhythm interaction in the hope that generalizations so derived can be applied to the more complex problems associated with floral induction.

For the past eight years, Dr. R.L. Satter and I, together with various colleagues, have used this approach in studying leaflet movements in the leguminous plants *Albizzia julibrissin* and *Samanea saman* (major conclusions reviewed in Ref.9). We chose to analyze this phenomenon because of the demonstrated correlation between vigorous circadian leaf movements and photoperiodic determinism (4) and because nyctinastic leaf movements are sometimes controlled by phytochrome (8).

B. Materials and Methods

Albizzia and *Samanea* plants were grown from seed in soil in the greenhouse and transferred to environmentally controlled chambers for several days prior to experimentation (25, 26). Measurements of leaf angles, either on or off the plant, were made by comparison with drawn standards. Explants of *Albizzia* pinnule pairs attached to a bit of rachilla continued oscillating when floated on appropriate solutions in a petri dish (26). Delaminated explants of *Samanea* pulvini continued movement when fed appropriately through rachillar stumps dipping into a solution contained in a small vial (25).

Etiolated Alaska pea seedlings were grown in a totally dark chamber for seven days at 27 ± 1°C, and the growth of internodes and terminal buds examined as described previously (11). *Avena* seedlings (19) were also grown in total darkness prior to experimentation.

K^+, Cl^-, and other ions were analyzed by electron microprobe in situ, in specific regions of pulvini that had been frozen, sectioned in a cryostat-microtome, lyophilized, and pressed flat between glass slides (26). Electrical measurements were made by connecting KCl-filled microelectrodes (tip diameter less than 0.5 µm, resistance 15-45 MΩ) with Ag-AgCl wire to an electrometer and comparing with grounded reference electrodes (20). Light sources were as previously described (26).

C. Results and Discussion

I. Pulvini and Leaflet Movement

In both *Albizzia* and *Samanea*, leaflets are generally open in light and folded together in dark (Fig.1) but continue to open and close with circadian rhythmicity when the entire plant or appropriate explant is maintained in continuous darkness (Fig.2). These movements are caused by changes in shape and size of ventral and dorsal motor cells (Fig.3). Nyctinastic closure occurs best when phytochrome is in the P_{fr} state at appropriate points in the circadian cycle (Fig.4). Failure to close in white light, despite the predominance of P_{fr}, is probably due to the opposing action of blue light (7, 13) possibly absorbed by flavoproteins (15, 18).

II. Ion Flux as a Basis for Changes in Motor Cells

In all cases studied, leaflet angle is correlated with the K^+ and Cl^- content of the motor cells on opposite sides of the pulvinus (Fig.5). No significant correlations have yet been established for Ca^{2+}, Mg^+, or Na^+. Changes in leaflet orientation, whether caused by light or

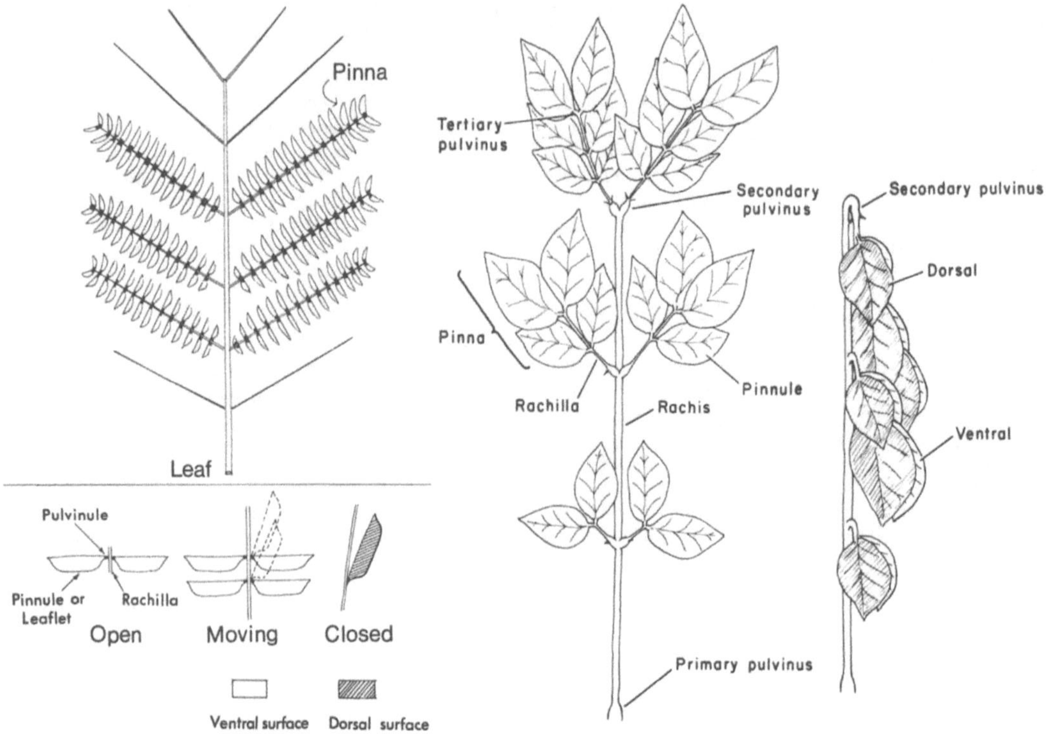

Fig. 1. Nyctinastic leaflet movements in *Albizzia julibrissin* (*left*) and *Samanea saman* (*right*). In both plants, excised pulvini attached to rachilla continue to move under appropriate conditions

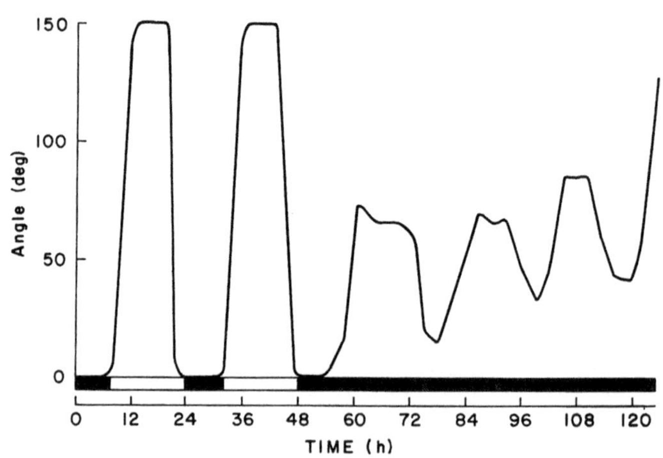

Fig. 2. In both *Albizzia* and *Samanea*, circadian leaf movements initiated by light-dark cycles can be maintained during subsequent long dark periods. Abscissa shows dark ▬ and light ▭ treatments with duration in hours

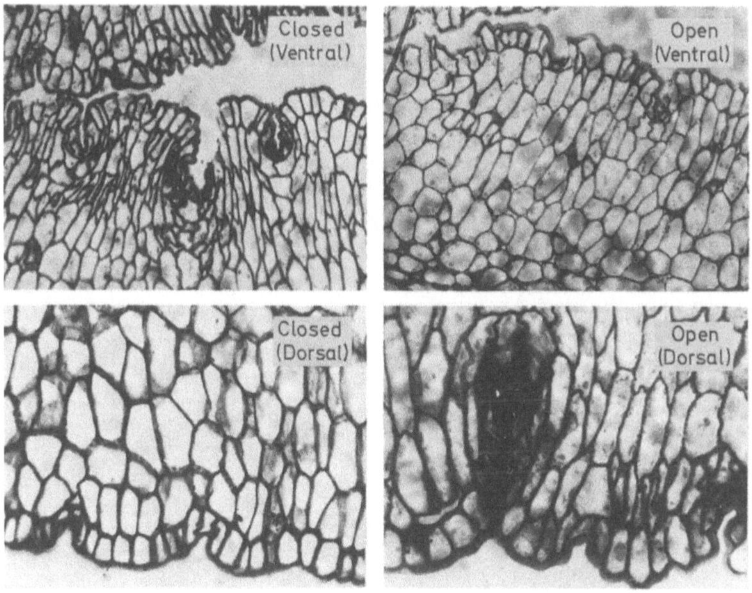

Fig. 3. Leaflet movements in *Albizzia* are due to changes in size and shape of ventral and dorsal motor cells in pulvinus

rhythm, are parallel with and apparently effected through control of K^+ (and Cl^-) movement (Fig.6).

In short-term experiments, rhythmic movement consists of two sharply differing phases. During opening, the effect of raising temperature from 24° to 37°C is to increase both the rate and final extent of the reaction; during closure a similar rise in temperature retards the rate of the reaction (Fig.7). Sensitivity to applied respiratory inhibitors, ions, and hormones also varies in the two phases (27). These data are consistent with the simultaneous operation within any one motor cell of two competing processes: (1) a highly temperature-sensitive ("active"?) process controlling influx, and (2) a less temperature-sensitive process controlling efflux of ions. In one phase of the cycle, influx predominates, in the other, efflux. This alternation could be due to a circadian oscillation in the magnitude of either component, or both. In that phytochrome mediated differences (P_r vs. P_{fr}) are greater at higher than at lower temperatures, phytochrome appears to affect the temperature-sensitive component. It is not known whether ion flux occurs via pumps and pores in membranes or via plasmodesmata of the symplast, but whatever the pathway, it is clear that the ions move between peripheral motor cells and inner cortical cells (Fig.8).

III. Electrical Correlates of Rhythmic and Photoinduced Changes

The electrical potential of *Samanea* flexor motor cells (those that contract during leaflet closure) correlates well with rhythmic (dark) leaflet movement in *Samanea*. Hyperpolarization from about -40 to -100 mV occurs during opening, and the reverse changes during closure (Fig. 9). The potential of extensor cells (those that expand during opening) also oscillates rhythmically, about 8 h out of phase with the flexor cells. Hyperpolarization from about -50 to -90 mV occurs just before

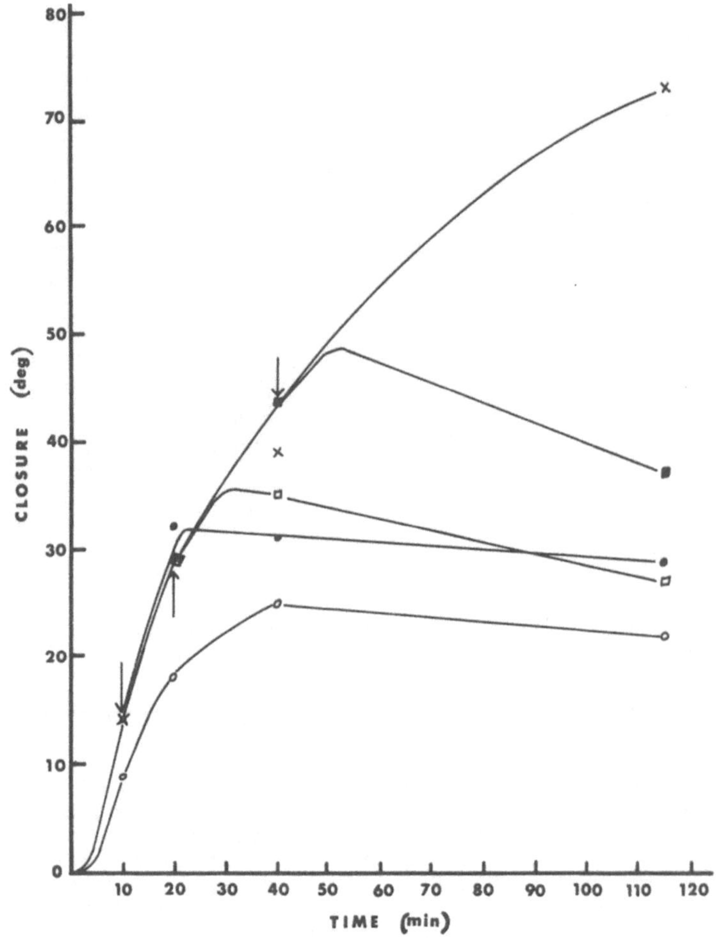

Fig. 4. *Albizzia* leaflets close rapidly when phytochrome is in P_{fr} state (×——×). Administration of far-red light at any time during reaction, thus converting phytochrome to P_r form, causes marked diminution in rate of closure (●——●; □——□; ■——■). Leaflets with P_r from the start close slowly and to least extent (o——o)

discernible opening (20). Bright white light depolarizes the flexor cells, low intensity red light followed by darkness leads to hyperpolarization of flexor cells, and this effect is reversed by far-red (Fig.10). Clearly, in electrical manifestations, as in ion flux and leaf movements, there is joint rhythmic and phytochrome control.

IV. The Role of Sucrose in Electrogenic Events

The nature of the electrogenic event is probably related to the observation that changes in external pH are without significant effect on internal potential unless sucrose is included in the external medium (Fig.11). This may indicate obligate inward cotransport of H^+ and sucrose; a similar phenomenon has been described in *Chlorella* (14). The potential generated by this cotransport might be involved in establishing the proper electrochemical gradient for subsequent ion movements.

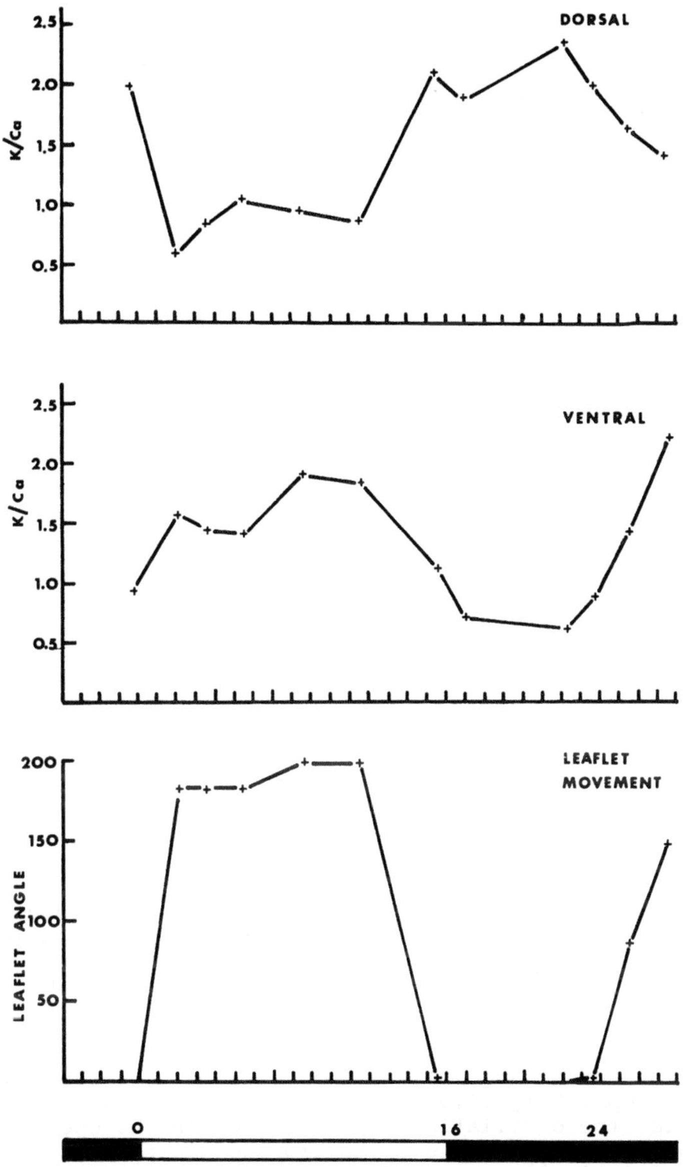

Fig. 5. Leaflet movement in *Albizzia* is directly correlated with $[K^+]$ content of ventral motor cells and inversely with the $[K^+]$ content of dorsal motor cells. Expression K/Ca corrects values to a per cell basis, as Ca^{2+} is virtually immobile. Abscissa shows dark ■ and light ▭ duration in hours

V. The Role of Sucrose and P_{fr} in Rhythmic Maintenance

In both *Albizzia* and *Samanea*, explants exposed to light-dark regimes and then transferred to extended darkness continue to oscillate without damping for at least five cycles provided (a) that they are supplied

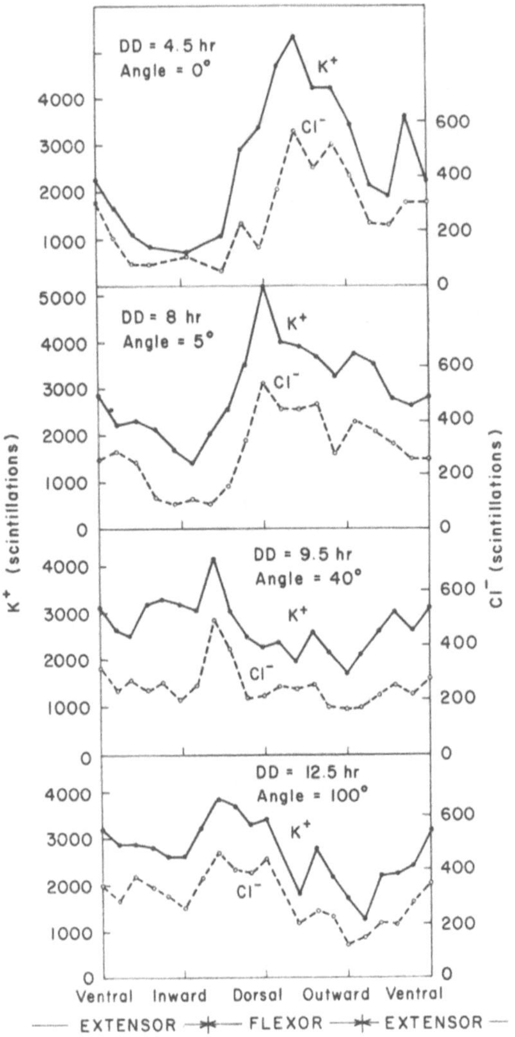

Fig. 6. In *Samanea*, leaflet movement is accompanied by changes not only in $[K^+]$, but in $[Cl^-]$. Ordinates show $[K^+]$ and $[Cl^-]$ scintillations, and abscissa, various regions around circumference of pulvinus

with about 50 mM sucrose and (b) that phytochrome is converted to the P_{fr} state by appropriately timed red light flashes (22, 29). In the absence of sucrose, the oscillation tends to damp in the open condition; and in the absence of P_{fr}, it tends toward the closed position (Fig.12). To keep the oscillation going in a given phase, P_{fr} must be properly timed; thus if red light flashes are given at times other than immediately preceding opening, they cause a phase shift (29).

VI. Phytochrome and Electrical Coupling Between Cells

In the etiolated *Avena* coleoptile, where cells are nicely aligned, the state of phytochrome controls the degree of electrical coupling between cells in the same vertical column (19). In darkness, with phytochrome in the P_r state, the coupling intensity between adjacent cells is of the order of 0.20, i.e., 20% of the direct current injected at one cell is detectable in the next cell. After red irradiation (P_{fr} phytochrome),

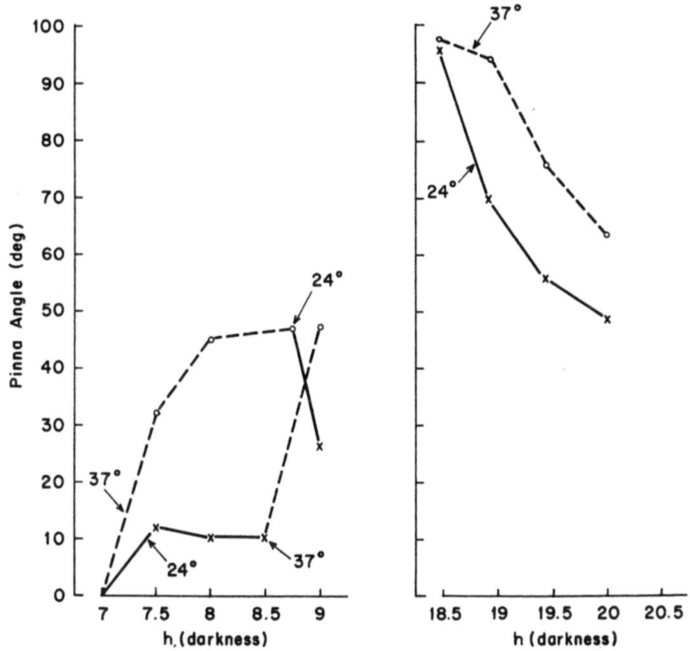

Fig. 7. In *Samanea*, both rate of rhythmic opening reaction and final angle achieved are increased by raising temperature from 24° to 37°C (*left*), while similar rise in temperature decreases rate of closure (*right*)

coupling intensity rises to about 0.50. No such correlations could be observed with transverse coupling. These results are consistent with a phytochrome-mediated change in the electrical resistance between cells and suggest effects on long-distance transport. This, in fact, had been noted in some earlier experiments with peas.

VII. Sucrose Transport as a Basis for P_{fr}-mediated Growth Reactions in Etiolated Peas

In etiolated peas, the red light-induced conversion of P_r to P_{fr} is associated with inhibition of growth of the epicotyl and promotion of growth of the terminal bud (17). If the epicotyl is excised from the seed, then neither promotion of terminal bud growth (11) nor inhibition of stem growth (2) by P_{fr} is seen if the growth medium lacks sucrose. With excised subapical epicotyl sections, it is the sucrose-induced growth that is inhibited by P_{fr}; by contrast, growth induced by either auxin or gibberellin is not photosensitive, and indeed, if either of these hormones is added together with sucrose, then photosensitivity of even the sucrose-induced growth reaction is annulled (2). Similarly, the P_{fr}-sucrose-induced promotion of bud growth, inhibition of stem growth, P_{fr}-induced altered auxin optimum for the growth of excised epicotyl sections, and the P_{fr}-mediated increase in flavonoid synthesis are all annulled by the application of GA_3 (12, 21).

In the excised etiolated pea system, there is a dual interdependence; not only is the growth manifestation of the $P_r \rightarrow P_{fr}$ conversion dependent on the presence of sucrose, but sucrose uptake, transport, and/or

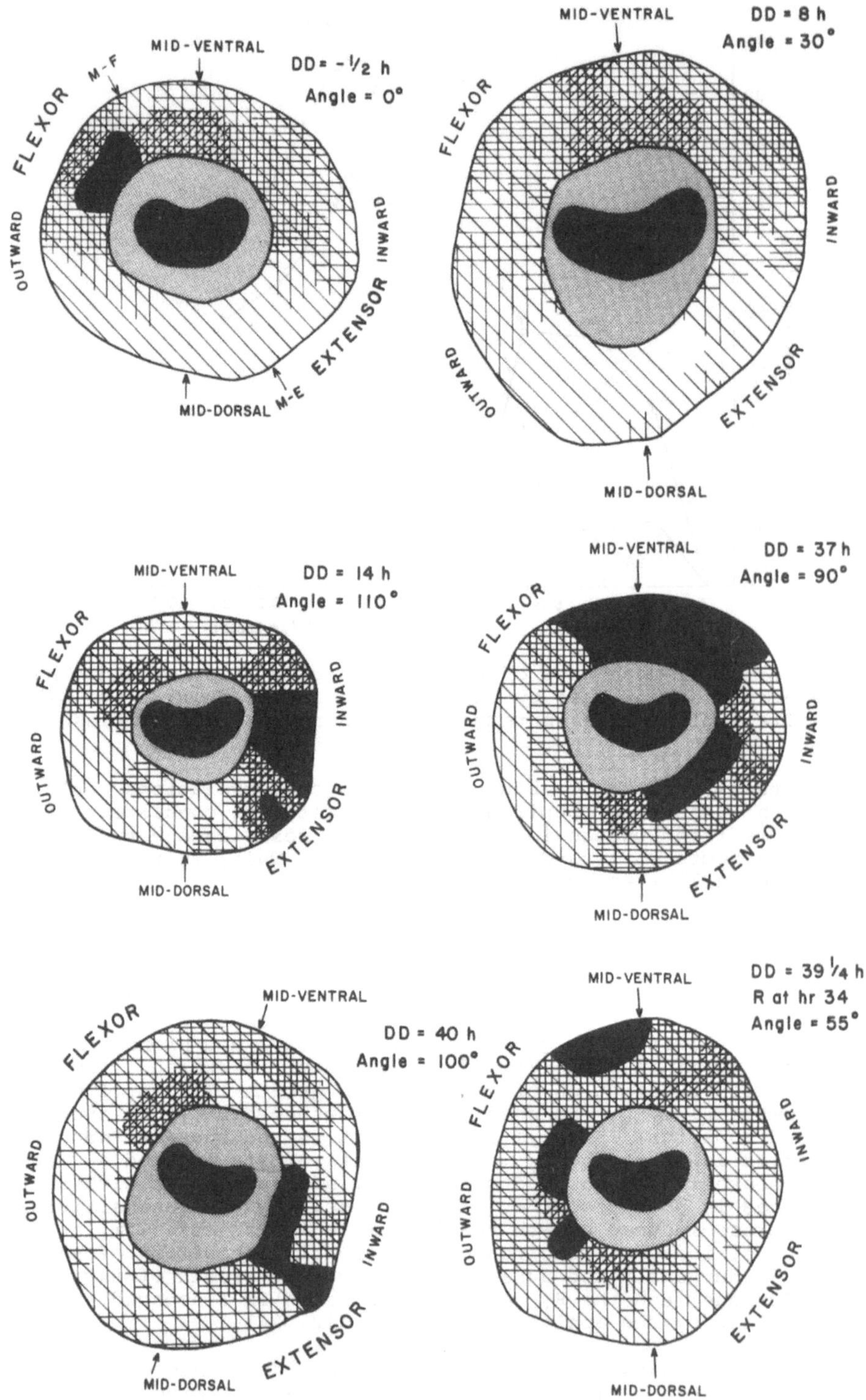
MID-VENTRAL
M-F
DD= -1/2 h
Angle = 0°
FLEXOR
OUTWARD
INWARD
EXTENSOR
M-E
MID-DORSAL
MID-VENTRAL
DD = 8 h
Angle = 30°
FLEXOR
INWARD
OUTWARD
EXTENSOR
MID-DORSAL
MID-VENTRAL
DD = 14 h
Angle = 110°
FLEXOR
INWARD
OUTWARD
EXTENSOR
MID-DORSAL
MID-VENTRAL
DD = 37 h
Angle = 90°
FLEXOR
OUTWARD
INWARD
EXTENSOR
MID-DORSAL
MID-VENTRAL
FLEXOR
DD = 40 h
Angle = 100°
OUTWARD
INWARD
EXTENSOR
MID-DORSAL
MID-VENTRAL
DD = 39 1/4 h
R at hr 34
Angle = 55°
FLEXOR
INWARD
OUTWARD
EXTENSOR
MID-DORSAL

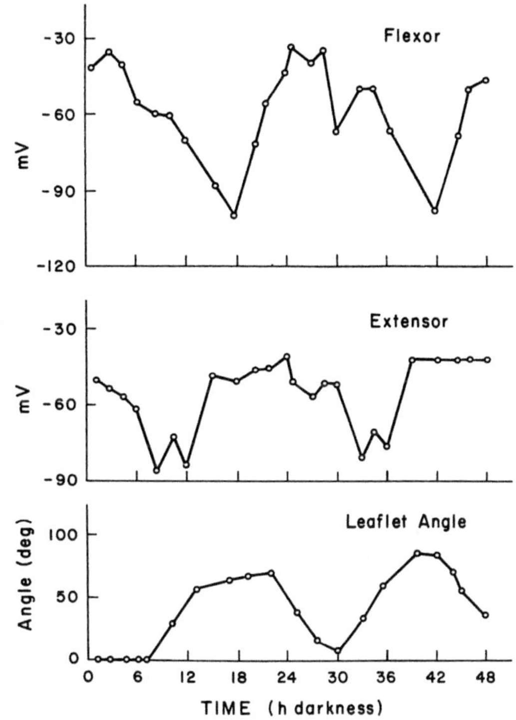

Fig. 9. Electrical potential changes in pulvinal motor cells of *Samanea* during rhythmic opening and closing in prolonged darkness

incorporation into upper epicotyl and terminal buds is dependent on P_{fr} (11). Thus if epicotyls are severed at the level of the seed and immersed in ^{14}C-sucrose in the presence of 0.05-0.10 M carrier sucrose, both the growth of the bud and its uptake of ^{14}C-sucrose are promoted by P_{fr}, provided that the sucrose passes through considerable epicotyl tissue on its way to the bud (Fig.13). In epicotyl tissue, where elongation is inhibited by P_{fr}, ^{14}C-sucrose content is depressed. In both stem and leaf tissues, therefore, a correlation exists between phytochrome effects on transport and phytochrome effects on growth. Furthermore when hormones interfere with or reverse the morphogenetic of light, they simultaneously modify light-induced sucrose transport.

Fig. 8. Topographic maps of K^+ scintillations in various cross-sectional regions of *Samanea* pulvinus as function of rhythmic movements during long dark period. K^+ in central vascular core was not measured.

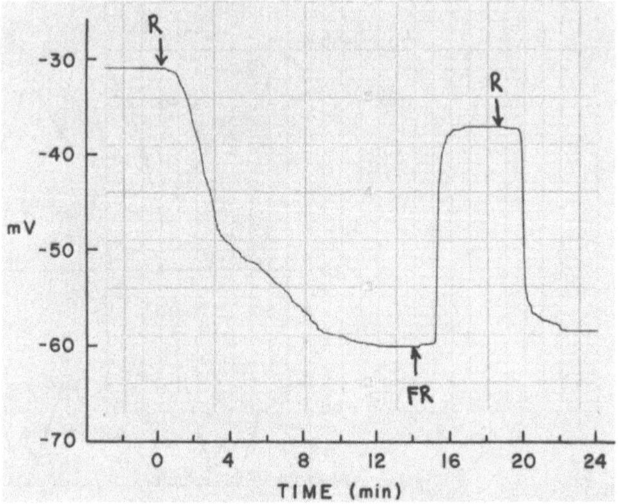

Fig. 10. Effect of phytochrome transformation by red and far-red light on electrical potential of flexor motor cells in *Samanea*

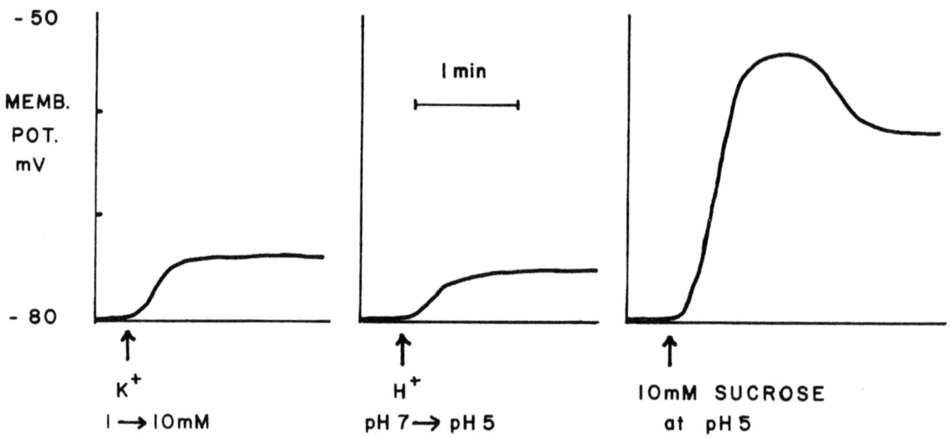

Fig. 11. Small depolarization of *Samanea* flexor motor cells caused by pH transition from 7 to 5 is greatly increased by 10 mM sucrose, itself without effect at pH 7. This is taken as evidence for inward cotransport of H^+ and sucrose. There is no similar interaction of K^+ and sucrose

D. Conclusions

It has long been recognized that florigen transport is closely correlated with sucrose movement in many but not in all plants (evidence reviewed in 30). Contrariwise, it has also been pointed out that under some conditions, florigen seems to be exported from darkened, juvenile, or mutilated leaves that are probably net importers of photosynthate. Is, then, the frequent correlation between sucrose and florigen trans-

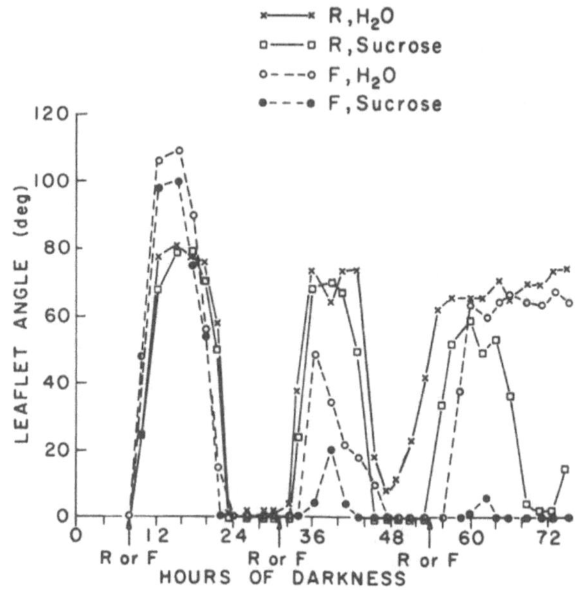

Fig. 12. Perpetuation of rhythmic oscillation of excised *Albizzia* leaflets requires P_{fr} and sucrose (□——□). With P_r and sucrose, oscillation gradually damps in closed condition (●——●). Without sucrose, whether phytochrome is in P_{fr} (×——×) or P_r (o——o) condition, oscillation tends to damp in open position

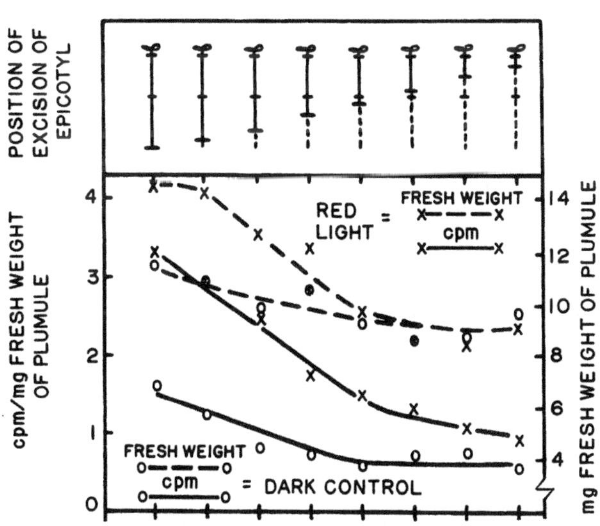

Fig. 13. Effect of residual epicotyl length (*solid line, above*) on (*below*) incorporation of basally fed ^{14}C-sucrose into terminal buds (*solid lines*) and fresh weight of buds (*dotted lines*). ×: red light; *o*: dark control

port trivial? The evidence presented in the present paper suggests that it is probably not, and that in fact sucrose transport may be a sine qua non for florigen movement, when properly coupled to other events regulated by the circadian oscillator.

The evidence from *Albizzia* and *Samanea* pulvini, as well as from etiolated *Pisum* epicotyls, leaves no room for doubt on the question of whether the state of phytochrome can control solute transport. It can, and does, and the substances transported under phytochrome control include such widely diverse materials as ions (K^+, Cl^-, and H^+) on the one hand and sucrose and related sugars on the other hand. Together with the evidence that phytochrome can control the intensity of electrical coupling between aligned cells, this can be taken to mean that whatever else phytochrome does, it opens up the long-distance transport lines of communication between source and sink cells. This is precisely what we have to deal with in the movement of florigen from leaf to bud. Furthermore, the sucrose requirement for rhythmic maintenance satisfies another known requirement for photoperiodic induction.

But solute movement alone apparently cannot produce the expected phytochrome-induced morphogenetic transformations. In the case of the promotion of terminal bud growth in etiolated peas, it appears that P_{fr} is ineffective unless P_{fr}-promoted sucrose transport traverses at least a required minimal length of epicotyl tissue en route from source to sink. This could mean that en route from source to sink, sucrose mobilizes other materials from the stem tissue, and that these materials, arriving at the bud, produce the altered growth patterns. Taken in this context, florigen becomes definable as a promoter of long-distance transport, elaborated in the leaf under photoinductive conditions. A corollary of this definition would be that florigen could not be assayed by direct application to the receptor bud, but only by judicious application to an uninduced leaf, debladed petiole, or an appropriate site on the stem.

The noted effects of known plant hormones in promoting or negating floral induction (discussed above) can be rationalized with their observed effects on sucrose-mediated, phytochrome-induced effects on growth. Thus, auxin and gibberellin each negate P_{fr}-reversal of sucrose-induced growth in pea sections (2), gibberellin prevents P_{fr}-mediated sucrose transport and the accompanying biochemical and morphogenetic changes in pea epicotyls (12, 21), and auxin has some effect on potassium flux in *Albizzia* pulvini in the dark (27), although it does not interact with control by phytochrome.

It may be argued that floral induction involves a differentiation event, not just an effect on growth rate. However, even the transformation of a bud from vegetative to reproductive could result from differential effects on growth of determinative tissue regions such as *tunica* and *corpus*. This, in turn, could result from a phytochrome-initiated cascade of nutrient transport from leaf to bud, in which the effective agent is florigen.

Acknowledgements

The author was aided by a succession of grants from the National Science Foundation. I am indebted to my colleague Ruth L. Satter for constant collaboration and many rewarding discussions. I thank Richard Racusen, Esther Simon, Martin Schrempf, Philip Applewhite, and Javade Chaudhri for the use of unpublished data.

References

1. Anand, R., Galston, A.W.: Am. J. Botany 59, 327 (1972)
2. Bertsch, W.F., Hillman, W.S.: Am. J. Botany 48, 504 (1961)
3. Bünning, E.: Ann. Rev. Plant Physiol. 7, 71 (1956)
4. Bünning, E.: Z. Naturforsch. 36, 457 (1948)
5. Butler, W.L., Norris, K.H., Siegelman, H.W., Hendricks, S.B.: Proc. Natl. Acad. Sci. U.S. 45, 1703 (1959)
6. Chailakhian, M.Kh.: C. R. (Doklady) Acad. Sci. USSR 12, 443 (1936)
7. Evans, L.T., Allaway, W.G.: Australian J. Biol. Sci. 25, 885 (1972)
8. Fondéville, J.C., Borthwick, H.A., Hendricks, S.B.: Planta 69, 357 (1966)
9. Galston, A.W., Satter, R.L.: In: Light and Plant Development, Smith, H. (ed.). London: Butterworth's 1976, pp. 159-184.
10. Garner, W.W., Allard, H.A.: J. Agr. Res. 18, 553 (1920)
11. Goren, R., Galston, A.W.: Plant Physiol. 41, 1055 (1966)
12. Goren, R., Galston, A.W.: Plant Physiol. 42, 1087 (1967)
13. Jaffe, M.J., Galston, A.W.: Planta 77, 135 (1967)
14. Komor, E., Tanner, W.: Europ. J. Biochem. 4, 219 (1974)
15. Muñoz, V., Butler, W.L.: Plant Physiol. 55, 421 (1975)
16. Njus, D., Sulzman, M., Hastings, J.W.: Nature (London) 248, 116 (1974)
17. Parker, M.W., Hendricks, S.B., Borthwick, H.A., Went, F.W.: Am. J. Botany 36, 194 (1949)
18. Poff, K.L., Butler, W.L.: Nature (London) 248, 799 (1974)
19. Racusen, R.H.: Planta 132, 25 (1976)
20. Racusen, R.H., Satter, R.L.: Nature (London) 255, 408 (1975)
21. Russell, D.W., Galston, A.W.: Plant Physiol. 44, 1211 (1969)
22. Satter, R.L., Applewhite, P.B., Chaudhri, J., Galston, A.W.: Photochem. Photobiol. 23, 107 (1976)
23. Satter, R.L., Galston, A.W.: Plant Physiol. 48, 740 (1971)
24. Satter, R.L., Galston, A.W.: In: Chemistry and Biochemistry of Plant Pigments, Goodwin, T. (ed.). New York: Academic Press 1976
25. Satter, R.L., Geballe, G.T., Applewhite, P.G., Galston, A.W.: J. Gen. Physiol. 64, 413 (1974)
26. Satter, R.L., Marinoff, P., Galston, A.W.: Am. J. Botany 57, 916 (1970)
27. Satter, R.L., Marinoff, P., Galston, A.W.: Plant Physiol. 50, 235 (1972)
28. Schrempf, M., Satter, R.L., Galston, A.W.: Plant Physiol. 58, 190 (1976)
29. Simon, E., Satter, R.L., Galston, A.W.: Plant Physiol. 58, 421 (1976)
30. Vince-Prue, D.: Photoperiodism in Plants. London: McGraw Hill 1975

Paper Demonstration References (see Section IX):

30 - 121 - 136 - 147 - 153 - 170 - 189

VIII. Hormonal Control

26. The Hormonal Regulation of the Cell Division Cycle

C. PÉAUD-LENOËL

Abstract

This report deals with the central question of whether plant growth regulators or hormones control the cell division cycle at the molecular level via major changes in the metabolism of nucleic acids or proteins of plant tissues or whether a more subtle effect of a specific hormone triggers one or several discrete steps of cell division, leading in turn to mass changes of the physiologic status of the tissues.

Concepts and plausible models of the division cycle will first be discussed to clarify the possible experimental approaches to the previous question. These models are essentially distinguished by the central postulates: (*1*) the mitotic events are a linear sequence of events arranged by their direct causal connection, or (*2*) several lines of independent causal sequences contribute to the structure and functions of the mitotic apparatus.

The effects of auxin on cell division, studied with simplified biological systems, suggest that auxin may have an indirect impact on the cell cycle. Directly or indirectly, auxin specifically affects the S phase of DNA synthesis.

In various model biological systems and especially in suspended plant cell cultures, cytokinins do not affect the S phase. The machinery of protein synthesis seems to be functionally independent of the cytokinin effect. The biosynthesis of specific proteins involved in the functions of the mitotic apparatus may be triggered by the cytokinins with the consequence that in the absence of the hormone, cells are belated or blocked at the metaphase step.

A. Introduction

About forty years ago, permanent plant tissue cultures were first established by Gautheret and Nobecourt. They could be maintained only in the presence of the hormone factor indole-3-acetic acid. At the present time, it is safe to suggest that most, if not all, plant tissues require at least an auxin-like factor to multiply their cells, whether or not this factor is exogenously supplied.

Despite an auxin supply, some plant tissues would not grow until kinetin was isolated from a DNA acid hydrolysate and shown to be an active cell division factor in many tissues, the best known being tobacco pith tissue (9). Substituted adenines with cytokinin activity were later proved to be natural substances (31); it is now very likely that the

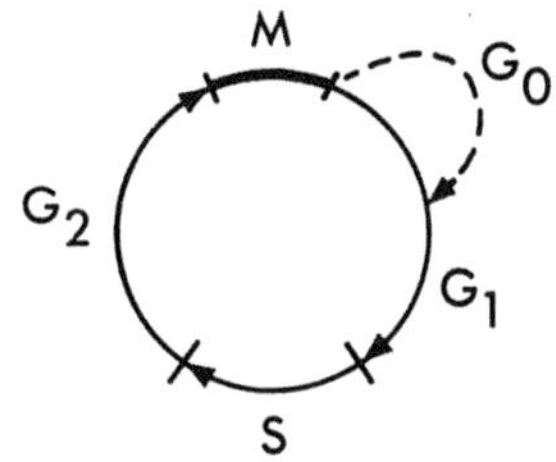

Fig. 1. The classical scheme

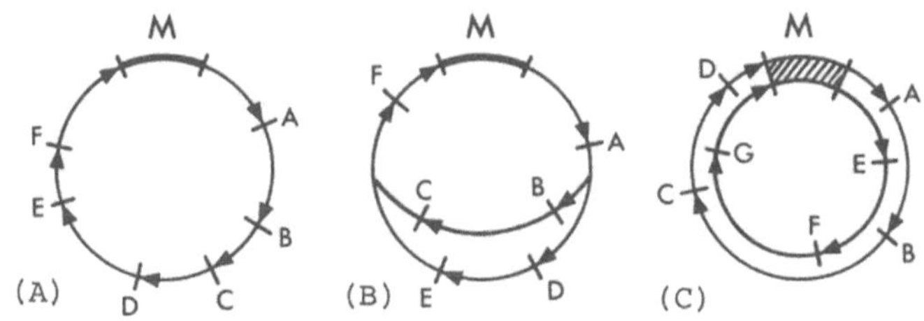

Fig. 2 A-C. (A) Models of dependent, or (B and C) independent causal sequences of events

Figs. 1 and 2. Models of the cell division cycle

lack of cytokinin requirement of bulk plant tissue or dense cell suspension cultures only means that enough endogenous cytokinin is present (8, 34).

It is not in the scope of this report to discuss the papers describing effects of plant hormones at the molecular level in actively dividing cells. We shall only comment on the central question of whether the cell division cycle is affected by the gross modifications of nucleic acid and protein metabolism or whether a more subtle effect of a given hormone controls one or several specific steps of the cell cycle, leading in turn to mass changes of the cell metabolism.

It seems adequate to examine first what is meant by the concept "cell division cycle" then to analyze experiments whose aim is to identify some link between the physiologic effect of a hormone and a molecular step of the cell cycle. We shall conclude by discussing the possibility, despite experimental gaps, of postulating some model(s) of hormone impact on the cell division cycle.

B. The Cell Division Cycle

Before it became possible to locate biochemical markers, the only known events available for timing the mitotic cycle were the step-wise classical mitotic phases. Between one mitosis and the next, a black period of variable duration was called interphase. The important techniques of labeling DNA with ^{3}H-thymidine and subsequently counting the labeled mitoses by autoradiography were then discovered (19). Together with microspectrophotometric measurements of the amount of DNA in nuclei during interphase, these techniques led to the introduction of the well-known description of G_1, S, and G_2 phases, as a consequence of discontinuous DNA synthesis, observed only during a given period of interphase, the S phase (36) (Fig.1). The accurate determination of these phases may be difficult, but the relative position and duration of each phase are now well known for meristematic cells or exponentially growing cell cultures (11, 15). To be complete, this model should also include an additional fundamental state, generally called G_0, of indefinite duration, for cells that are not committed to nuclear division (15, 17).

The previous description refers to "normal" cell division cycles in which no experimental device was used to alter the timetable of the cycle. Experimentally, the various phases of mitosis and the S phase, by its start and end, could only be observed, whereas G_1 and G_2 phases could be defined by their boundaries between M and S phases. However, extranuclear phenomena were also observed during the nuclear division cycle, such as enlargement of the cell wall, cytokinesis, formation of the phragmoplasma, and separation of the daughter cells. These microscopic changes at the cytoplasmic level are correlated with the timetable of the nuclear division cycle, whether or not they have any causal relationship. With microscopic and radiolabeling techniques, it was shown that the mitotic cycle and the cytoplasmic sequence of events may be naturally altered in a variety of ways (1). Some steps are omitted, curtailed, or amplified during S phase or mitosis itself (see 37 for a review). Hormone activity may be connected to such altered mitotic cycles: for instance, endopolyploidy occurs through endomitotic or endoreplicative cycles in the pea root cortical cells (39) and the presence of a cytokinin in the culture medium increases this status. This phenomenon is followed by normal mitotic cycles and differentiation of daughter cells. This example introduces the additional problem of the connection between division and differentiation of the cells (see reviews 33, 37). In the pea root cortical explants, cytokinins favor the formation of high ploidy cells. This observation is restricted by experiments in which cell enlargement and cytokinesis are promoted by an optimum auxin supply. In this situation, an increase in cytokinin concentration favors the nuclear division rate and decreases the average ploidy (4).

The previous model of the cell division cycle (Fig.1) implies a series of step-wise events occurring in a given order along the cycle (36). Obviously, such a fixed order requires a direct causal connection between one step and the next. As an alternative to the dependent pathway model (Fig.2A), other models may be proposed in which a defined number of unconnected pathways are triggered by a starting event A (Fig.2B), after which events B and C occur along their own timetable, independent of D and E. Marker events of the interphase, e.g., S phase, should be experimentally located on one of these pathways. In the model of Figure 2B, event F located shortly before mitosis would be double-controlled by the end events of pathways B, C, and D, E. From this model, a more elaborate scheme can be derived as an array of interacting dependent pathways with positive or negative controls. A third model postulates several independent sequences that contribute to the organization of the structure and functions of the cell division cycle (Fig.2C).

In all models, the timetable of events will be fixed by the physiologic situation. Nevertheless, both the timing and the sequence of observed events may be altered experimentally according to models B and C (Fig. 2) whereas, according to model 2A, only the timing can be modified. To probe the adequacy of any model, we must find marker events of any detectable kind, especially markers of the G_1 or G_2 black phases. Such markers have already been described, either with the help of a genetic approach to the problem or from the physiologic and biochemical analysis of synchronized cell populations.

In yeast, temperature-sensitive mutants have been selected with a block in the cell division cycle at the restrictive temperature (17). The result was to show that some of these mutants could by-pass specific events normally occurring after the step blocked at the restrictive temperature and that at least two independent sequences were controlled by a common start signal.

If biochemical markers are sought, the cell divisions should be synchronized: this has been achieved in many ways, such as deprivation of an essential nutrient, respiratory restriction, step-wise exposure to light and dark, temperature jumps, and arrest of DNA synthesis by base analogs (53). It is noteworthy that the temporary omission of cytokinin from the culture medium was a means to obtain partially synchronized plant cells (22, 27, 40). These techniques allowed the identification of markers in animal cell cultures grown in vitro: G_1 marker-proteins include phosphorylated F_1 histone (3, 18) and a cytoplasmic protein synthesized shortly before the onset of the S phase (32). Specific chromatin proteins are synthesized during the S phase (5), and DNA synthesis itself was divided in stepwise periods, the significance of which is not yet fully understood (30). A large number of marker proteins appear in G_2, including a chromatin-specific protein (6) and a cytoplasmic colchicin-binding protein identified as tubulin (29).

Few markers of the cell division cycle have been discovered in plant cells. However, a G_1 protein, which seems to be an initiation factor of DNA synthesis, has been studied in *Allium cepa* root meristems (13). As in animal cells, G_2 proteins exist, as shown by the sensitivity of particular microscopic events of mitosis to protein synthesis inhibitors (13, 23). It may be rewarding to use synchronized plant cells for histologic and biochemical studies of the mitotic apparatus itself in conjunction with the effect of plant hormones: with animal cells, elucidation of the stepwise organization of the nature and functions of the spindle fibers has already progressed significantly (35).

C. Plant Hormones and Cell Division

As already mentioned, most recognized plant hormones regulate cell division directly indirectly. This statement does not lead to the Arian thread of the cell cycle in the labyrinth of physiologic reactions. A good illustration of this difficulty arises from the experiments bearing on the hormonal control of sex in plants. Application of cytokinin to the dioic plant *Mercurialis annua* generates female flowers on a genetically male individual; reciprocally, auxins masculinize the genetic *Mercurialis* females (10). Similarly, gibberellins favor the expression of male flowers in cucurbits (20). Clearly, both cell multiplication and differentiation are under the control of hormones: in such processes, it escapes molecular analysis. Therefore, we seek simpler biological systems, hoping to gain enough information to tackle the problem with whole plants later. At the moment, this enquiry will be restricted to clear and specific effects of defined hormone factors on the cell cycle of experimentally simplified biological material.

This will enable us to skip comments on the effects of ethylene, gibberellins, and abscisic acid, simply because nothing certain correlates these factors with the cell division cycle. However, simple experiments could alleviate our ignorance: for instance, the usual, although not simple, anticytokinin effect of abscisic acid would be useful to check whether the impact of cytokinins on the cell cycle is counteracted by abscissic acid, and in what way.

D. Auxins and the Cell Division Cycle

Despite the amount of work recently undertaken to unravel the mode of action of auxin at the subcellular level (see review, 50 and discussion, 49), very little has elucidated the relationship between auxin and the cell cycle. One may ask first if the plasticizing effect of auxin on the cell wall and the promotion of cell enlargement are connected to cell division. In many cases the mother cells may repeatedly divide without wall enlargement, the daughter cells remaining clustered in the mother cell wall (25, 38). It is therefore doubtful that cell enlargement is directly connected to the nuclear division cycle.

The generally accepted activation by auxin of the transcription of DNA will certainly have a bearing on protein synthesis machinery, and the activation of this process would be the basis for any other physiologic event, including cell division, although no evidence has been found of a specific connection. More promising are the results showing that 2,4-D induced DNA synthesis and promoted the synchronization of cell divisions in slices of Jerusalem artichoke tubers (41, 53). The synchrony might mean that the auxin addition started a unique pathway, eventually connected to the initiation of DNA replication. The artichoke tuber system is not beyond criticism before the previous conclusion is confirmed: coconut milk and other growth factors were added simultaneously with auxin, and the whole experiment was started by slicing the tubers, therefore by a wound effect. Other authors (26, 43) also observed that addition of auxin was an absolute requirement for DNA synthesis in tobacco pith tissues or cell suspensions, whether a cytokinin was present or not. In contrast to cytokinins (22), auxins were not observed to have a synchronizing effect on the cell divisions in these systems. The suggestion (52) that initiation of the S phase requires activation by auxin of RNA and bulk protein biosynthesis is therefore the best explanation at hand. More experiments should be undertaken to see whether auxin specifically elicits the biosynthesis of initiation or elongation factors of DNA replication.

These findings should not obscure the fact that, apart from promoting DNA synthesis in undifferentiated cells, auxin also promotes specific organogenesis like rooting (45). Initiation of organs requires a large number of cell divisions: the molecular mechanism of this organ initiation is an open field.

E. Cytokinins and the Cell Division Cycle

It is relevant to underline the fact that nothing supports the statement "one hormone, one function." The variety of physiologic responses to cytokinins may well not correspond to a basic unitary mechanism. As an example, the effect of cytokinins in promoting the differentiation of chloroplasts has no obvious connection with the effect of this hormone on cell division (14, 42).

Recent progress in the study of the effects of cytokinins at the molecular level has been reviewed (7, 16, 28, 46, 47); we shall discuss only those results that deal with the involvement of this hormone in cell division. Many model biological systems were used to accumulate information in this field. Since the tobacco pith tissue culture was described, callus cultures and later, suspended cell cultures seem to be the most promising material.

When a cytokinin-sensitive cell suspension is filtered off its medium and transferred into a new medium enriched with 2,4-D, cells will start to divide slowly (25). In freshly excised pith tissue, no cell division is observed unless a cytokinin is added (9). In clonal subculture suspensions, the mitotic index (M.I.) is usually one fifth or less of the M.I. observed when an optimal amount of cytokinin is present (48). A selection of cell clones from a primary callus culture showed that the stimulation of cell division by cytokinins was quantitative and not qualitative: every cell clone behaved in a specific hereditary way, i.e., absolutely or moderately sensitive or autonomous, with a continuous scale of sensitivities (48).

If the cytokinin is added immediately to the transfer medium of sensitive cells, the cell divisions are asynchronous, with a M.I. of about 2% and a complete cell cycle of 32-35 h at 26°C; if the cytokinin is added after an approx. 30 h delay to the new medium, mitoses are partially synchronized, with an optimum M.I. of 5 to 7% and a minimum of 1% or less. Second and third mitotic waves are observed before randomization (22, 27). The delay in the auxin addition is not a synchronizing factor, although it increases the yield of dividing cells and determines to a certain extent the onset of the division waves (22). The following observations are relevant to understanding this synchronizing effect of cytokinins:

1. The first mitotic cycle after the transfer of the cells may be considered as "special," since the cells have to condition their medium before they divide; nevertheless synchronization was observed, regardless of the status of the mother culture, i.e., in stationary or log phase (22).

2. If the cells are kept for 18 h in complete medium and transferred into a cytokinin-free medium, one and only one division of each cell is observed in this medium. After this round, cells return to the minimum M.I. observed in the cytokinin-starved control culture. Further division waves require a new addition of the hormone (26).

3. Inhibition of protein synthesis by 5-methyl-tryptophan during the 18-h period of experiment 2 completely abolished the division wave, but the cells could resume their divisions in the presence of tryptophan and cytokinin (23).

4. DNA synthesis was unaffected by the omission of the cytokinin during at least one generation (26).

5. Inhibition of DNA synthesis by fluoro-deoxyuridine during the 18-h period of experiment 2 did not prevent the next mitotic cycle if cells were transferred to a thymidine-enriched medium where DNA synthesis was resumed (26).

From these experiments, it can be concluded that lack of cytokinin brings the mitotic cycle of tobacco cells to an arrest in a specific position connected with a small number of close events of the cell division cycle. This block, once released, leads to one complete cycle, including mitosis, and only one. Functional protein synthesis machinery is essential for the efficiency of the cytokinin effect, which occurs during a period covering the second half of the cell cycle period, before mitosis itself. DNA synthesis is apparently not involved in the hormone activity, but a normal DNA replication period as well as the independent cytokinin effect is essential to the function of the mitotic apparatus. The relative independence of the cytokinin effect in relation to the S phase has also been observed in other laboratories (12, 43); the argument that the divided cells might have been in G_2 does not apply to the tobacco cell suspension system since an S phase is a requisite for mitosis, whether the S phase occurs before or after the period of hormone application to the cells.

Other molecular events occurring in connection with the cytokinin effect have been investigated using cell suspensions or tissue explants. Protein synthesis is quantitatively unaffected during at least one cell generation (12, 25). The turnover of proteins is also unaffected during the same period, as measured by tracer amino acid incorporation and chase. Some differences in protein turnover and methylation processing were noticed in the radish cotyledon system (14). It is difficult to correlate this result with the previous findings, since cell enlargement is the primary response to the hormone treatment in the cotyledon system.

That the quantitative rate of protein synthesis is unaffected by cytokinins in the tobacco cells does not mean that qualitative changes are impossible: by acrylamide gel electrophoresis of the soluble proteins of these cells, it was shown that the patterns of proteins are almost identical for cells supplemented or not with the hormone. Nevertheless, a few specific bands appeared on the electrophoretic patterns of the hormone-treated cells after a few hours in the complete medium (21).

Markers of the mitosis itself were investigated with synchronized tobacco cells. It is noteworthy that when cytokinin-deprived cells underwent mitosis at a slow rate, metaphase was considerably longer than in cytokinin-supplemented cells, while the prophase duration was unaffected (26). It should be recalled that, in animal cell cultures, proteins specific for the functions of the mitotic apparatus were specifically affected and cells blocked in metaphase by phenylalanine analogs (44, 51). In root meristems, mitosis proteins have been shown to be synthesized in late G_2 (13). Such proteins may be involved in the pathway controlled by cytokinins.

F. Discussion

Present knowledge of auxin control at the molecular level supports the suggestion that auxin will enhance the activity of RNA polymerase(s) on specific sites of the template DNA whereas cytokinins modify the functions of ribosomes at the translational level (see reviews and this Conference). Nothing in these findings contradicts the available evidence of the indirect impact of auxin on the cell division cycle. However, there are no grounds for an useful model of activation of S phase by auxin. Such a pathway remains a matter of speculation.

Knowledge concerning the interactions of cytokinins with the cell division cycle is more advanced. As stated previously, plant hormones and particularly cytokinins may initiate several unconnected metabolic pathways. Whatever the case, the sequence of early events after addition of the hormone to sensitive cells suggests several hypotheses.

As a whole, the machinery of protein synthesis is unaffected either in the soybean or in the tobacco system. This implies that, if cytokinins actually trigger protein synthesis, they do so only for some specific proteins. It is not yet possible to reconcile this statement with the evidence that ribosomes are submitted to cytokinin control for the normal translation of mRNA.

As previously underlined, cytokinins act independently of DNA synthesis. Experimentally, the S phase can be shifted before or after the activation of the next cell division by a temporary cytokinin exposure of the cells (26). These findings together with the observation of the synchronizing effect of this hormone lead to the postulation of a model of the cell cycle similar to the diagram of Figure 2B in which independent pathways operate. Since a functional protein synthesis mechanism is required for the expression of hormone activity, and as hormone-deprived cells are arrested in metaphase, we suggest the working hypothesis that cytokinins order the synthesis of specific proteins of the mitotic apparatus. The mere logic of the suspended cell system suggests that the hormone might trigger at the translation level the reading of specific mRNA; some insight into this mechanism may soon evolve from the recent data showing that benzyladenine is incorporated in RNA, probably rRNA (2), without modification of the structure of the supplied free base. In experiments with cell suspensions, we also found that undegraded benzyladenine was incorporated in RNA at a rate of 1 molecule for 10^5 nucleotides during an 18-h incubation period (24). It is still unknown whether this incorporation is due to misreading at the transcription level or to a specific post-transcriptional polymerase and what bearing this incorporation has on the reading of specific messengers.

References

1. d'Amato, F.: In: Dynamics of Meristem Cell Populations. Miller, M.W., Kuehnert, C.C. (eds.). New York: Plenum Publ. Co., 1972, p. 149
2. Armstrong, D.J., Murai, N., Taller, B.J., Skoog, F.: Plant Physiol. 57, 15 (1976)
3. Balhorn, R., Jackson, V., Granner, D., Chalkley, R.: Biochemistry 14, 2504 (1975)
4. Bennici, A., Buiatti, M., d'Amato, F., Pagliai, M.: Colloque international du CNRS No. 193, Les cultures de tissus de plantes. Paris: CNRS, (ed.). 1971, p. 245
5. Borun, T.W.: In: Cell Cycle and Cell Differentiation. Reinert, J., Holtzer, H. (eds.). Berlin: Springer, 1975, p. 249
6. Bradbury, E.M., Inglis, R.J., Matthews, H.R.: Nature (London) 247, 257 (1974)
7. Burrows, W.J.: Current Adv. Plant Sci. 7, 837 (1975)
8. Chen, C.M., Hall, R.H.: Phytochemistry 8, 1687 (1969)
9. Das, N.K., Patau, K., Skoog, F.: Physiol. Plantarum 9, 640 (1956)
10. Durand, B., Louis, J.P.: In: Fertilization in Higher Plants. Linskens, H.F. (ed.). Amsterdam: North Holland, 1974, p. 327
11. Eriksson, T.: Physiol. Plantarum 20, 348 (1967)
12. Fosket, D.E., Short, K.C.: Physiol. Plantarum 28, 14 (1973)
13. Gonzales-Fernandez, A., Gimenez-Martin, G., Fernandez-Gomez, M.E., de la Torre, C.: Exptl. Cell Res. 88, 163 (1974)
14. Gordon, M.E., Letham, D.S.: Australian J. Plant Physiol. 2, 129 (1975)
15. Gould, A.R., Bayliss, M.W., Street, H.E.: J. Exp. Botany 25, 468 (1974)
16. Hall, R.H.: Ann. Rev. Plant Physiol. 24, 415 (1973)
17. Hartwell, L.H., Culotti, J., Pringle, J.R., Reid, B.J.: Science 183, 46 (1974)
18. Hohman, P., Tobey, R.A., Gurley, L.R.: Biochem. Biophys. Res. Commun. 63, 126 (1975)
19. Howard, A., Pelc, S.R.: Exptl. Cell Res. 26, 395 (1951)
20. Jones, R.L.: Ann. Rev. Plant Physiol. 24, 571 (1973)
21. Jouanneau, J.P.: Physiol. Plantarum 23, 232 (1970)
22. Jouanneau, J.P.: Exptl. Cell Res. 67, 329 (1971)
23. Jouanneau, J.P.: Exptl. Cell Res. 91, 184 (1975)
24. Jouanneau, J.P., Gandar, J.C., Peaud-Lenoel, C.: PD. réf. 77 p. 174

25. Jouanneau, J.P., Peaud-Lenoel, C.: Physiol. Plantarum 20, 834 (1967)
26. Jouanneau, J.P., Tandeau de Marsac, N.: Exptl. Cell Res. 77, 167 (1973)
27. Jouanneau, J.P., Teyssendier de la Serve, B.: C. R. Acad. Sci. (D) 270, 320 (1970)
28. Kende, H.: Intern. Rev. Cytol. 31, 301 (1971)
29. Klevecz, R.R., Forrest, G.L.: Ann. N. Y. Acad. Sci. 253, 292 (1975)
30. Klevecz, R.R., Keniston, B.A., Deaven, L.L.: Cell 5, 195 (1975)
31. Letham, D.S., Shannon, J.S., McDonald, J.R.: Proc. Chem. Soc. p. 230 (1964)
32. Ley, K.D.: J. Cell Biol. 66, 95 (1975)
33. Meins, Jr., F.: In: Cell Cycle and Cell Differentiation. Reinert, J., Holtzer, H. (eds.). Berlin: Springer 1975, p. 151
34. Mackenzie, I.A., Street, H.E.: New Phytologist 71, 621 (1972)
35. McIntosh, J.R., Cande, W.Z., Snyder, J.A.: In: Molecules and Cell Movement. Inoue, S., Stephens, R.E. (eds.). New York: Raven Press, 1975, p. 31
36. Mitchison, J.M.: In: The Cell Cycle, Gene-Enzyme Interactions. Padilla, G.M., Whitson, G.L., Cameron, I.L. (eds.). New York-London: Academic Press, 1969, p. 361
37. Nagl, W.: In: Tissue Cultures and Plant Science. Street, H.E. (ed.). London-New York: Academic Press, 1974, p. 19
38. Phillips, R., Torrey, J.G.: Develop. Biol. 31, 336 (1973)
39. Phillips, R., Torrey, J.G.: Develop. Biol. 39, 322 (1974)
40. Roberts, K., Northcote, D.H.: J. Cell Sci. 6, 299 (1970)
41. Setterfield, G.: Symp. Soc. Exp. Biol. 17, 98 (1963)
42. Seyer, P., Marty, D., Lescure, A.M., Peaud-Lenoel, C.: Cell Different. 4, 187 (1975)
43. Simard, A.: Can. J. Botany 49, 1541 (1971)
44. Sisken, J.E., Iwasaki, T.: Exptl. Cell Res. 55, 161 (1969)
45. Skoog, F.: In: Colloque international du CNRS N° 193, Les cultures de tissus de plantes. Paris: CNRS, (ed.). 1971, p. 115
46. Skoog, F., Armstrong, D.J.: Ann. Rev. Plant Physiol. 21, 359 (1970)
47. Szweykovska, A.: In: Tissue Culture and Plant Science. Street, H.E. (ed.). New York-London: Academic Press, 1974, p. 461
48. Tandeau de Marsac, N., Jouanneau, J.P.: Physiol. Vég. 10, 369 (1972)
49. Teissere, M., Penon, P., van Huystee, R.B., Azou, Y., Ricard, J.: Biochim. Biophys. Acta 402, 391 (1975)
50. Venis, M.A.: Current Adv. Plant Sci. 3, 21 (1973)
51. Wheatley, D.N., Henderson, J.Y.: Nature (New Biol.) 247, 281 (1974)
52. Yasuda, T., Yagima, Y., Yamada, Y.: Plant Cell Physiol. (Tokyo) 15, 321 (1974)
53. Yeoman, M.M.: In: Tissue Culture and Plant Science. Street, H.E. (ed.). New York-London: Academic Press, 1974, p. 1

Paper Demonstration References (see Section IX):

16 - 48 - 68 - 77 - 95 - 126 - 154 - 163 - 168 - 187

27. Hormonal Controls of Abscission: Biochemical and Ultrastructural Aspects

F.T. ADDICOTT and S.M. WIATR

Abstract

Abscission is controlled by auxin (the pricipal natural retardant), and by abscisic acid and ethylene (the principal natural accelerants). The hormones gibberellin and cytokinin also influence abscission, but in secondary ways. Anatomically, abscission includes the development and activity of a *separation layer* and one or more *protective layers*. Cells of the separation layer secrete hydrolytic enzymes that degrade the cell wall, especially the pectic substances of the middle lamella. Present evidence suggests that these enzymes are moved via an endoplasmic reticulum-dictyosome-vesicle carrier system to the plasmalemma and released into the paramural region.

A. Functions of Hormones in Abscission Control

Abscission is a correlation phenomenon. The onset and rate of activity in the abscission zone are largely determined by events elsewhere in the plant. Each organ has a major influence on its own abscission (zone); but events at some distance, especially an event that may put an organ at a competitive disadvantage, can function to accelerate abscission. Hormones are the principal agency in the expression of this correlative behavior of plants. Reports at previous Conferences have included some aspects of the action of hormones in abscission; however, a comprehensive review has not been presented. There is no space for such a review in this report (see Ref.5), but it will be helpful to the understanding of our subject if we introduce this report with a brief overview of the roles of the plant hormones in the control of abscission.

Understanding of the action of hormones in the control of abscission has come slowly because of the complexity of hormonal actions and interactions within the plant. For example, experiments with the application of hormones have, from time to time, given results that were difficult to understand because the response varies with many factors, in particular with the site of application. Table 1 summarizes the differences in abscission responses to three sites of application. Primarily, we are interested in the action of the hormone on the separation layer that develops within the abscission zone, and in general, the levels of hormones in the abscission zone are determined by the hormone production of the subtending organ. The hormones also affect the vigor of the subtending organ; but usually, applications to either the subtending organ or the abscission zone have similar effects on abscission. Gibberellin (GA) is an exception. Applied to the abscission zone it induces a moderate acceleration of abscission; but applied to an organ such as a leaf or a fruit, it tends to promote the vigor of the organ and thus indirectly inhibit abscission. Ethylene (ETH) presents a different pattern; applications to either the abscission zone or the subtending organ promote abscission. However, ETH does not appear to be translocated in connection with abscission, and the promotion of abscission by exposure of an organ to ETH appears to result from changes in the amounts of translocatable hormones, particularly indole-3-acetic

Table 1. Abscission responses to site of application of hormones

Hormone	Organ	Abscission zone	Sink (distant)
Auxin	---	---	++
Gibberellin	---	++	++
Abscisic acid	+++	+++	--(?)
Ethylene	+++	+++	--(?)
Cytokinin	---	---	++

+: promotes abscission; -: retards abscission

acid (IAA) and abscisic acid (ABA), exported from the organ. When hormones are applied to a sink distant from a particular abscission zone, they modify the effectiveness of the sink relative to the subtending organ and hence modify abscission responses. Therefore, in attempting to understand abscission responses, one must keep in mind the complexity of the correlation of plant behavior.

After Laibach's (24) discovery that auxin inhibits abscission, there was a steady expansion of research on the role of hormones in abscission. Now the five major plant hormones (IAA, GA, ABA, ETH, and CK) are all known to have one or more important influences on abscission, and the literature contains evidence of several other endogenous substances that affect abscission (5).

The action of IAA is strongly inhibitory; indeed it appears to be universally involved with the normal inhibition of abscission. However, auxin, being one of the major factors in the competition among the sinks of the plant, can be a factor tending to promote abscission of disadvantaged organs. Also, under certain experimental conditions, auxin can have a promotive influence.

The function of ABA in abscission is primarily promotive. The expanding body of evidence on its occurrence in correlation with abscission and on its induction of abscission responses suggests that ABA may be universally associated with abscission. The relatively few cases where applications of ABA to plants have failed to promote abscission are now understood as having been due to relatively low dosages applied. Those results were similar to some of the early results with IAA, which often failed to promote growth when applied to intact plants!

The effects of ETH on abscission are essentially promotive, but because of its highly volatile nature, its hormonal role is less clear than that of the nonvolatile hormones (3, 38). Indeed, as indicated previously, the action of endogenous ETH in the promotion of abscission may come more from its influence on the levels of other hormones than from direct effects of ETH itself.

While GA has a moderately promotive effect when applied to abscission zones, its major action appears to be to modify the relative effectiveness of the sinks of the plants, as indicated in Table 1.

In a similar pattern, cytokinin (CK) has a moderately retarding effect when applied to the abscission zone, but its major influence on abscission appears to be through modification of the relative effectiveness of the competing sinks on the plant.

Evidence has been presented from time to time that suggests there may be a number of other substances that could have a hormonal influence on abscission. In most cases, either the chemical or physiologic knowledge of the substance is too meager to enable an estimate of its potential. However, one of these, "senescence factor," (39) appears to be an in vivo regulator of ETH production, thereby tending to initiate abscission. The chemical nature of senescence factor is still obscure.

B. Anatomical and Histologic Foundations

With the development of the microscope, early botanists gave some attention to the anatomical nature of abscission. These early investigations culminated in the extensive work of von Mohl (32), who made the important distinction that abscission consisted of two processes, separation and protection, each with appropriate anatomical manifestations. His work was greatly extended by Tison (47) and by Lee (25) who described the cell wall breakdown of separation and the development of protective layers in the leaf abscission zones of a number of woody species. Sampson (43) deepened our knowledge with a comprehensive histochemical investigation. These findings, together with the findings of many subsequent workers, indicate that von Mohl's original concepts of separation and protection were quite valid. Within the abscission zone, a separation layer can be readily distinguished. Its development may or may not parallel the development of the protective layers. It is within the separation layer that cell walls are weakened, enabling the separation of the subtending organ to occur. The main subject of this report is the ultrastructure and activity of the cells of the separation layer; we will not consider further the development of the protective layers.

The early microscopic observations indicated that the disintegration of the cell walls of the separation layer was often accompanied by swelling, gelatinization, and solubilization. Pectic compounds of the middle lamella disintegrate first, and usually cellulosic compounds disintegrate as well. When carefully handled, preparations from many species showed that cells remained intact on each side of the separation. However, the hydrolytic activities are often extensive and sometimes appear to affect entire cell walls and cell contents, as viewed by the light microscope. Almost certainly, other polysaccharides are also hydrolyzed, including xyloglucans and hemicelluloses.

With the light microscope, it was noted early that cells of the separation layer and indeed of the entire abscission zone have more dense cytoplasm than cells of adjacent regions. We can now interpret this observation as indicating that a major portion of the cell contents includes organelles and only a small portion, vacuoles. A deposition of starch regularly appears in and near the separation layer as abscission approaches. Similarly, nucleoli enlarge and become conspicuous in the separation layer. The starch deposition is interpreted as providing a source of substrate for the intensive respiration of the separation layer and for the development of the protective layers generally. The enlarged nucleoli are indicative of increased synthesis of RNA correlated with the synthesis of the hydrolytic enzymes that degrade the cell wall. In some instances, increased deposits of calcium in druses in cells of the abscission zone have been observed as abscission progresses. The calcium was interpreted as having come from calcium pectate in the walls of the separation layer (19). For summaries of the light microscope observations of the separation layer, see Webster (52) and Bornman et al. (8, 9).

In recent years there have been some observations of the effects of hormones on the anatomical and histologic changes in the separation layer. In general, the hormones either retarded or accelerated the changes in a manner parallel to the action of the hormones on the overall process. With some materials, ABA and ETH accelerate the cell wall changes to a very rapid degree, at the same time apparently having little effect on the rate of cell divisions that commonly accompany separation layer activity (9, 54). These hormones also can induce hydrolytic activity and cell wall breakdown in adjacent tissues (54), and ABA appears to induce considerable cytoplasmic breakdown in the separation layer (9).

C. Biochemical Aspects of Abscission

Close to one hundred years ago, Wiesner, and later Molish (33) suggested that abscission was probably caused by enzymes ("ferments"); but it was many years before their suggestions were given serious attention. Further work awaited the development of plant biochemistry to the point that the possibility of the localized synthesis of specific enzymes was an acceptable idea, and the possibility that such enzymes could be excreted into the paramural region and even into the cell wall itself could be accepted.

A number of enzymes are now recognized as increasing in activity in the separation layer in correlation with abscission. These include pectinases (34, 42) and cellulases (21), which are synthesized during the process of abscission (26, 27, 34, 41). It has been known for some time that the pattern of production of such enzymes can be influenced by IAA, ABA, and ETH (2, 3, 13, 15, 28, 29).

Acid phosphatase has been demonstrated, both in association with organelles and in the paramural region of the separation layer (19, 20). Sutcliffe et al. (46) demonstrated increased activity of peroxidase, succinic dehydrogenase, malic dehydrogenase, acid phosphatase, and esterase in and about the separation layer at the time of abscission. Duygu (18) extended these investigations and detected increases in RNAase, protease, and IAA-oxidase, as well. Osborne (37) demonstrated reduced activity of pectin-methyl-esterase (PME) correlated with abscission, a change that would lead to a reduction of the calcium bridges of pectins, thus favoring separation. Thus, there is now significant evidence of the participation of both hydrolytic and respiratory enzymes in the process of abscission, although there is still much to be learned about this aspect of abscission.

That oxygen could be limiting to abscission has been known for some time (43), but the fact that oxygen was an absolute requirement for abscission did not become apparent until the work of Carns et al. (12) in 1951. Carns (11) further showed that inhibitors of respiratory enzymes were also inhibitory to abscission. Biggs and Leopold (6) demonstrated that a minimum supply of carbohydrate substrate must be present for abscission to take place. Collectively the foregoing, and related works of others, emphasized that abscission requires the energy of respiration. The recent confirmation of the acceleration of abscission by levels of oxygen above 21% (Marynick and Addicott, 30) indicates the existence of a highly oxidative reaction that is promotive to abscission. This could well be the reaction catalyzed by IAA-oxidase.

As knowledge of the involvement of enzymes in abscission became apparent, it was of interest to see if RNA synthesis and protein synthesis

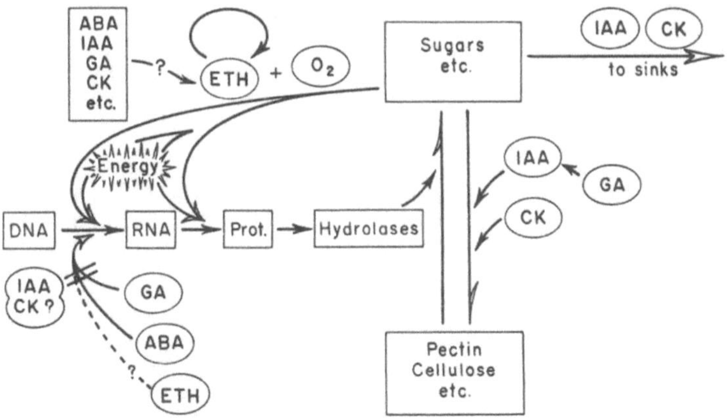

Fig. 1. Schematic representation of influence of plant hormones on activities associated with abscission. Three major biochemical aspects are illustrated: (1) synthesis of hydrolases via RNA; (2) provision of energy for syntheses; (3) hydrolysis of pectins and celluloses to soluble sugars (after Addicott, 5)

in the separation layer were correlated with the development of hydrolytic enzyme activity. That was indeed the case (1, 4, 38). Such evidence gives further support to the interpretation that the hydrolytic enzymes are synthesized de novo. A schematic representation of the influence of the plant hormones on the biochemical activities of abscission is shown in Figure 1.

D. Ultrastructure of Abscission

During the past ten years several ultrastructural investigations of the cytological changes associated with abscission have appeared. While a vast amount of work on this subject remains to be done, the present publications give provocative indications of the cytologic activities associated with abscission.

I. Observations of the Cell Wall

Investigation of cell wall changes associated with abscission is hampered by lack of histochemical methods suitable to electron microscopy. However, the electron microscope has supported, and to a limited degree extended, the earlier observations with the light microscope.

The middle lamella is invariably affected early in the process of abscission. Evidence of swelling of the middle lamella followed by the development of cavities, sometimes in the vicinity of the plasmodesmata (40, 48, 51, 53), and eventual complete disappearance of the middle lamella is a common sequence of events. Chemical identification of the components of the middle lamella has been limited to (1) ruthenium red, which has a high affinity for polymerized acidic polysaccharides, and (2) alkaline hydroxylamine-ferric chloride, which is specific for methylated polygalacturonic acid. Use of these reagents in electron microscopy has confirmed that the middle lamella is largely, if not entirely, composed of pectic substances and is the first cell wall component to be hydrolyzed in the course of abscission. Few differences have been found among the various separation layers studied, including those of the leaves of *Coleus*, *Gossypium*, *Impatiens*, *Phaseolus*, and *Sambucus*; and the pedicels of *Lycopersicum*, *Nicotiana*, and *Phaseolus*; and petals of *Linum* (7, 16, 34, 40, 44, 45, 48, 51, 52, 53, 55). Most of the studies have been restricted to cortical parenchyma, but dissolution of the middle lamella has been reported in epidermis, vascular parenchyma, sieve elements, and companion cells (44).

Knowledge of the steps in the disintegration of the primary cell wall is meager, largely because ultrastructural methods for identifying components of the primary wall (other than certain pectic substances) are still inadequate. However there is some evidence of dispersion of microfibrillar material associated with swelling of the primary wall. In micrographs of late stages of separation the microfibrils appear as tufts of frayed material, suggesting breakage accompanied by swelling or dissolution of matrix materials. Despite the reports of increased cellulase activity during abscission, little evidence of disappearance of microfibrils has been reported. This lack of evidence is supported by Morré's chemical analyses of abscission zones that indicated a general solubilization of pectic fractions during abscission, but little change in cellulose and hemicellulose (34).

II. Observations of Cytoplasmic Organelles

Although some early observations had suggested that breakdown of the plasmalemma preceded the release of hydrolytic enzymes, there is now substantial evidence that the plasmalemma and internal cytoplasmic membranes do not become altered until abscission is well advanced. Neutral red staining, plasmolysis experiments, measurement of electrolyte loss, and recent ultrastructural studies, collectively indicate that the protoplasts of the separation layer are intact as separation commences (19, 32, 44).

There is now little question but that the separation layer secretes hydrolytic enzymes that degrade the cell wall and permit separation. Two major questions remain. One is the mechanism of secretion, and the other is the mechanism of the control of the secretion by the plant hormones. Other reports in the Conference present recent developments in our knowledge of the mechanisms whereby hormones control plant processes. In the balance of this report we will summarize the ultrastructural changes in the cytoplasm during abscission and comment on their significance.

At the onset of abscission the parenchyma cells of the separation layer are normal, undifferentiated plant cells containing a complement of the usual organelles (7, 22, 45). The cytoplasm is relatively dense and contains numerous small vacuoles. One of the earliest changes observed, as noted in the light microscopy comments, is an increase in the size and number of nucleoli. As cytoplasmic changes progress, there may be an increase in the number of mitochondria (19) or alterations in the conformation of cristae and matrix density, which relate to the metabolic state of mitochondria (53). Usually there is little change in the number and appearance of chloroplasts (23, 48), but they may show some signs of degenerative changes such as expanded membranes and partially disrupted thylakoids (53). An increase in the amount of rough endoplasmic reticulum is found frequently prior to cell wall degradation and during abscission (19, 23, 45, 49, 51). Very substantial increases in the number of dictyosomes, the size of dictyosome cisternae, and number of dictyosomal vesicles appear to be significant aspects of abscission at the subcellular level (40, 45, 50; see Fig.2, Table 2). Small vesicles found near or continuous with the plasmalemma are in-

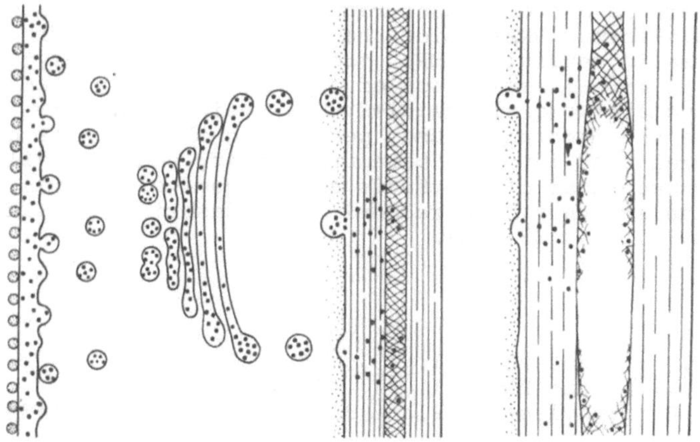

Fig. 2 A-E. Scheme for synthesis and secretion of hydrolytic wall enzymes during abscission. Hydrolases synthesized on rough ER are sequestered in ER cisternae (A). Vesicles carrying the enzymes, perhaps in latent form, are released from ER and then differentiate into flattened dictyosome cisternae (B). At maturing face of dictyosomes, vesicles are formed that fuse with plasmalemma (C) and release enzymes into paramural region of cell. Enzymes, primarily pectinases, diffuse through wall (D), loosen it, and break down pectinaceous middle lamella to effect cell separation (E). This scheme parallels membrane flow and differentiation as described by Morré et al. (36) and is consistent with postulated mechanism for synthesis and secretion of macromolecules (Chrispeels, 14)

Table 2. Percent increase in area occupied by organelles in separation layer cells after abscission induction (from Sexton and Hall, 45)

Endoplasmic reticulum	102
Dictyosomes	488
Mitochondria	1.2
Chloroplasts	26.4
Microtubules	++

terpreted by Gilliland et al. (19) as being derived from the dictyosomes and destined for release into the paramural region. This suggestion was supported by the localization of acid phosphatase in paramural cavities, dictyosomes, and small vesicles (of possible dictyosomal origin) in the cytoplasm (19, 20). Several workers have reported invaginations of the plasmalemma that could well be the result of fusion of dictyosomal vesicles with the plasmalemma.

While our knowledge of the significance of these organelle changes to the secretion of the enzymes of abscission must be considered preliminary, nevertheless we can put together the framework of a hypothesis, drawing on knowledge of the function of the organelles that has been gained in other investigations. When the hormonal signal for the commencement of abscission is received, there is an increased synthesis of RNA as indicated by the increased amounts of nucleolar material.

Presumably there is a concomitant reduction in the synthesis of other RNAs. The newly-formed RNAs guide the synthesis of specific proteins in the endoplasmic reticulum (ER), proteins of the hydrolytic enzymes. These appear to move in some kind of a membrane flow system to the dictyosomes and dictyosomal vesicles. The vesicles migrate to the plasmalemma, where the membranes fuse, permitting the contents of the vesicles to be released into the paramural cavity and the cell wall (Fig.2). This mechanism of synthesis and release of enzymes is similar to other mechanisms that have recently been described in plant and animal cells (14, 17, 31, 35, 36).

References

1. Abeles, F.B.: Plant Physiol. 43, 1577-1586 (1968)
2. Abeles, F.B.: Plant Physiol. 44, 447-452 (1969)
3. Abeles, F.B.: Ethylene in Plant Biology. New York: Academic Press 1970
4. Abeles, F.B., Holm, R.E.: Ann. N. Y. Acad. Sci. 144, 367-373 (1967)
5. Addicott, F.T.: Biol. Rev. 45, 485-524 (1970)
6. Biggs, R.H., Leopold, A.C.: Plant Physiol. 32, 626-632 (1957)
7. Bornman, C.H.: South Afr. J. Sci. 63, 325-331 (1967)
8. Bornman, C.H., Addicott, F.T., Spurr, A.R.: Plant Physiol. 41, 871-876 (1966)
9. Bornman, C.H., Spurr, A.R., Addicott, F.T.: Am. J. Botany 54, 125-135 (1967)
10. Bornman, C.H., Spurr, A.R., Addicott, F.T.: J. South Afr. Botany 35, 253-264 (1969)
11. Carns, H.R.: Ph. D. Thesis. Los Angeles: Univ. Calif. 1951
12. Carns, H.R., Addicott, F.T., Lynch, R.S.: Plant Physiol. 26, 629-630 (1951)
13. Cracker, L.E., Abeles, F.B.: Plant Physiol. 44, 1144-1149 (1969)
14. Chrispeels, M.J.: Ann. Rev. Plant Physiol. 27, 19-38 (1976)
15. Datko, A.H., MacLachlan, G.A.: Plant Physiol. 43, 735-742 (1968)
16. Davenport, T.I., Marinos, N.G.: Australian J. Biol. Sci. 24, 709-715 (1971)
17. DeDuve, C.: In: Subcellular Particles. Hayashi, T. (ed.). New York: Ronald Press, 1959, p. 128
18. Duygu, E.: Bitki 1 (3), 351-366 (1974)
19. Gilliland, M.G., Bornman, C.H., Addicott, F.T.: Am. J. Botany 63, 925-935 (1976)
20. Hall, J.L., Sexton, R.: Ann. Botany 38, 855-858 (1974)
21. Horton, R.F., Osborne, D.J.: Nature (London) 214, 1086-1088 (1967)
22. Jensen, T.E., Valdovinos, J.G.: Planta 77, 298-318 (1967)
23. Jensen, T.E., Valdovinos, J.G.: Planta 83, 303-313 (1968)
24. Laibach, F.: Ber. Deut. Botan. Ges. 51, 386-392 (1933)
25. Lee, E.: Ann. Botany 25, 51-107 (1911)
26. Lewis, L.N., Lew, F.T., Reid, P.D., Barnes, J.E.: In: Plant Growth Substances 1970. Carr, D.J. (ed.). Berlin: Springer, 1972, p. 234
27. Lewis, L.N., Varner, J.E.: Plant Physiol. 46, 194-199 (1970)
28. Linkins, A.E., Lewis, L.N., Palmer, R.L.: Plant Physiol. 52, 554-560 (1973)
29. MacLachlan, G.A., Davies, E., Fan, D.E.: In: Biochemistry and Physiology of Plant Growth Substances. Wightman, F., Setterfield, G. (eds.). Ottawa: Runge Press, 1968, p. 443
30. Marynick, M.C., Addicott, F.T.: Nature (London) 264, 668-669 (1976)
31. Matile, P.: The Lytic Compartment of Plant Cells. New York: Springer 1975
32. Mohl, H. von: Botan. Z. 18, 1-17 (1860)
33. Molisch, H.: Botan. Zentr. 25, 393-394 (1886)
34. Morré, D.J.: Plant Physiol. 43, 1545-1559 (1968)

35. Morré, D.J.: Ann. Rev. Plant Physiol. 26, 441-481 (1975)
36. Morré, D.J., Mollenhauer, H.H., Bracker, C.E.: In: Origin and Continuity of Cell Organelles. Reinert, J., Ursprung, H. (eds.). Berlin: Springer, 1971, p. 82
37. Osborne, D.J.: J. Expl. Botany 9, 446-457 (1958)
38. Osborne, D.J.: In: Shedding of Plant Parts. Kozlowski, T.T. (ed.). New York: Academic Press, 1973, p. 125
39. Osborne, D.J., Jackson, M.B., Milborrow, B.V.: Nature (London) 240, 98-101 (1972)
40. Osborne, D.J., Sargent, J.A.: Planta 130, 203-210 (1976)
41. Reid, P.D., Strong, P.G., Lew, F., Lewis, L.N.: Plant Physiol. 53, 732-737 (1974)
42. Riov, J.: Plant Physiol. 53, 312-316 (1974)
43. Sampson, H.C.: Botan. Gaz. 66, 32-53 (1918)
44. Sexton, R.: Planta 128, 49-58 (1976)
45. Sexton, R., Hall, J.L.: Ann. Botany 38, 849-854 (1974)
46. Sutcliffe, J.F., Arch, P.D., Legett, P.A., Phillips, B.J., Sexton, R.: Abstracts, p. 213 XI Int. Botan. Congr., (Seattle, U.S.A. 1969)
47. Tison, A.: Mem. Soc. Linn. de Normandie 20, 121-137 (1900)
48. Valdovinos, J.G., Jensen, T.E.: Planta 83, 295-302 (1968)
49. Valdovinos, J.G., Jensen, T.E., Sicko, L.M.: Plant Physiol. 47, 162-163 (1971)
50. Valdovinos, J.G., Jensen, T.E., Sicko, L.M.: Planta 102, 324-333 (1972)
51. Webster, B.D.: Am. J. Botany 60, 436-447 (1973)
52. Webster, B.D.: In: Shedding of Plant Parts. Kozlowski, T.T. (ed.). New York: Academic Press, 1973, p. 45
53. Webster, B.D., Chiu, H.W.: J. Am. Soc. Hort. Sci. 100, 613-618 (1975)
54. Webster, B.D., Leopold, A.C.: Botan. Gaz. 133, 292-298 (1972)
55. Wiatr, S.M., Addicott, F.T.: In preparation

Paper Demonstration References (see Section IX):

33 - 50 - 69 - 98 - 111 - 119 - 132

28. Hormonal Regulators of Plant Flowering

M. K. Chaylakhyan

A. Introduction

The concept of hormonal flowering factors was first proposed by us forty years ago. This was the florigen hypothesis in which a complex of flowering hormones of an unknown nature was postulated (7, 8, 9). In development of the hypothesis, it was suggested twenty years later that florigen is bicomponent in nature, consisting of two complementary flowering hormone groups, the gibberellins, which are involved in the formation and growth of flower stems; and the anthesins, which initiate flower formation (12).

Seventeen years later a new approach to the mechanisms of gibberellin and anthesin formation was suggested: that there exist two mechanisms - one autonomous and one inductive (induced by external factors) (15).

Within this period various investigators have proposed various ideas pertaining to the possible role of other regulators in plant flowering (phytohormones and inhibitors), the importance of the relative amounts of these substances, and the order of their action.

The validity of any idea regarding hormonal flowering regulators can be checked best by employing various methods, including chemical ones, for extracting from the plant and identifying the substances that affect plant flowering and also by applying various biological methods. Prominent among the latter are grafting of vegetating and flowering plants of different varieties, species, or genera. The results of such experiments may be crucial for an understanding of the nature of hormonal plant regulators.

The present paper is a review of the main facts underlying our own concepts and those of other authors; a comparative analysis of graft experiments is presented, based on the hypotheses of florigen as a total complex of flowering hormones and as a bicomponent complementary complex of flowering hormones.

B. The Hypothesis of Florigen as a Total Complex of Flowering Hormones

The florigen hypothesis originated from the results of many experiments performed with plants from different photoperiodic groups, and it was always considered to be a hypothesis of a complex of flowering hormones.

The main facts underlying the hypothesis were:

1. The leaf is the organ susceptible to photoperiodic induction of flowering; the photoperiodic stimulus that arises in the leaf moves to the stem buds and induces the morphogenetic effect in them (7, 32, 45, 53).

2. The peculiarities of the flowering stimulus movement from the leaves to the stem buds in various directions and distances were elucidated (9, 46, 47).

3. From graft experiments it was learned that the photoperiodic stimulus moves from leafy shoots of flowering plants to vegetating plants and induces flowering in them. The stimulus is the same for plants of various photoperiodic groups (8, 9, 33, 41, 46, 47).

Many of the biological characteristics of the photoperiodic stimulus that moved from the leaves to the stem buds and induced flowering of the shoots closely resembled the properties of hormones; hence it was suggested that the stimulus is a complex of flowering hormones, i.e., florigen.

The foregoing facts were only an indirect proof of the hormonal nature of the flowering stimulus and of the presence of flowering hormones in plants, since the only direct proof would be the extraction of the substances from the plants and their chemical identification.

This occurred, partially, many years later and in a peculiar way when the gibberellins were discovered (59), and the extraordinarily strong effect of gibberellins on plant flowering was first discovered by Lang (1956) and then confirmed by other investigators (5, 10, 23, 38).

Gibberellic acid and other gibberellins were found to possess the remarkable ability to stimulate growth strongly in the overwhelming majority of plants from various photoperiodic groups. Concurrently it was clearly demonstrated that gibberellins could induce and stimulate the flowering of long-day but not short-day plants. It became clear that the gibberellins are not identical with florigen.

Subsequent investigations showed that the gibberellins do not affect flower formation directly and are only one of the two groups of hormones that comprise the florigen complex. They are a component of the first group since they are essential for the formation and growth of flower stems and for the first phase of plant flowering.

The second group of hormones consists of substances that have conditionally been called anthesins and that induce flower formation and ensure the passage of the second flowering phase. In this way, the hypothesis of florigen as a bicomponent complementary complex of flowering hormones originated (11, 12).

C. The Hypothesis of Florigen as a Bicomponent Complementary Flowering Hormone Complex

According to the florigen hypothesis, gibberellins are essential for flowering of long-day species, and anthesins for flowering of short-day plants. The basic facts are:

1. Flowering of vegetating long-day rudbeckia plants (*Rudbeckia bicolor*) growing under short-day conditions could be induced by an extract of natural gibberellins derived from the leaves of vegetating short-day plants, the tobacco (*Nicotiana tabacum*) var. Mammoth, growing under long-day conditions. Flowering of rudbeckia is the result of the combined action of the Mammoth tobacco gibberellins and of the endogenous rudbeckia anthesins (17).

2. Gibberellins and anthesins are localized in different zones of the apices of the long-day rudbeckia and short-day perilla plants: the gibberellins in the medullary zone, which is responsible for stem formation and growth, and the anthesins in the central zone, which is responsible for flower formation (44).

3. A difference exists in the polarity and paths of translocation of gibberellins and anthesins in whole perilla plants; gibberellins move along the phloem and xylem acropetally, whereas the anthesins move along the phloem in both directions (20).

Our knowledge of the two groups of flowering hormones comprising the florigen complex is quite different. The gibberellins have been extracted from plants, their chemical nature and structure are known, and there are direct and unequivocal proofs of their functioning as regulators or hormones of flowering. Gibberellins induce flowering in many long-day, long-short-day, and short-long-day plants, and they accelerate flowering of conifer seedlings (36, 50, 57).

A comparative study of the effect of exogenous gibberellins and of the variation of content of endogenous gibberellins under various photoperiodic conditions yields consistent results and indicates a close connection between the morphogenetic effect and level of gibberellin content.

Anthesins have not been extracted from plants, their chemical nature is unknown, and their existence is inferred from data based on biological tests. Of special importance is the floral initiation of vegetating red perilla plants under long-day conditions as the result of grafting the leaves induced by a short day in the isolated state onto these plants (38). Of importance also is the elucidation of the pathways and rate of translocation of labeled assimilates that affect the flowering of perilla shoots (16).

According to the hypothesis of florigen as a bicomponent complementary flowering hormone complex, the induction of plant flowering depends on a certain relation or balance in the amounts of gibberellins and anthesins. It is assumed that together with hormones that induce or stimulate flowering, there are substances in plants that retard flowering and that are produced under day lengths that are unfavorable for flowering (9, 18, 19, 28, 46, 47). In long-day plants, the flowering inhibitors are probably substances that are antagonistic to the gibberellins, whereas in the short-day plants, the inhibitors are anthesin antagonists (14). Thus regulation of plant flowering may be regarded as the balance of the action of plant flowering inhibitors and hormones.

D. Theories Concerning the Role of Other Phytohormones in Plant Flowering

Of the various theories concerning hormonal regulation of plant flowering, the first to be mentioned is the idea that the transition to flowering is regulated not by specific flowering hormones but by familiar growth hormones. This hypothesis was originated in 1939 by Kholodny, who saw no need to assume the existence of specific flowering hormones, since the ordinary growth hormones, such as the auxins, are polyvalent, and the spectrum of their action is broad enough to regulate both the growth and morphogenesis of vegetative organs and the formation of flowers.

For the last thirty years new phytohormones have been discovered, such as the gibberellins and the cytokinins, phenolic and terpenoid inhibitors, and other physiologically active substances. Numerous cases of acceleration or retardation of plant flowering by these substances have been recorded. A natural consequence of this was the recent revival of the idea of the complex action of various phytohormones, inhibitors, and other physiologically active substances upon plant flowering (6, 25, 26).

Detailed data on the effect of these substances on plant flowering is presented in the books by Evans (25) and Vince-Prue (57). From the data presented, it can be seen that in the overwhelming majority of cases the auxins, cytokinins, abscisic acid, ethylene, and other active substances exert an action on plant flowering that is much less definite and pronounced than that exerted by the gibberellins. As a rule, these substances are not capable of inducing flowering under noninductive day-length conditions and evoke only a quantitative change in the stimulation or retardation of flowering. In certain cases, however, an appreciable effect on plant flowering has been observed. Thus, α-naphthyl-acetic acid has been found to accelerate flowering and fruit-bearing of pineapple plants (49); ethylene stimulates flowering in leadwort (*Plumbago indica*) (48), and abscisic acid, in morning-glory (*Pharbitis Nil*) and black currant (24). Flowering has been found to be induced in chrysanthemum (*Chrysanthemum morifolium*) by ^{6}N-benzyl adenine (51) and in *Wollfia microscopia* by zeathin (40).

These and many other facts demonstrating the acceleration of flowering by phytohormones and other physiologically active compounds are not sufficient to propose any definite ideas regarding the role of these substances as specific hormonal factors in the regulation of plant flowering.

Other trends in investigations pertaining to hormonal regulators of flowering are based on experimental studies involving the extraction of endogenous flowering stimulators from plants. Thus extracts have been obtained from flowering cocklebur plants (29), and "anthogens" in lipid extracts have been obtained from the leaves of cocklebur and other plants (54). In later experiments steroids were extracted from chrysanthemum buds (4), and extracts containing "florigenic acid" were obtained from cocklebur and sunflower leaves (37). As a whole, these investigations resulted in the separation of substances of a lipid and steroid nature that had a mild stimulatory effect on the flowering of short-day plants under conditions that were near the flowering threshold. However, no one has been able to extract substances that we would call anthesins, i.e., substances capable of inducing flower formation in short-day plants growing on a long day.

E. Autonomous and Induced Mechanisms of Formation of Flowering Hormones

Further work on the hypothesis of florigen as a bicomponent complementary complex of flowering hormones led us to the idea of the presence of two mechanisms of formation of flowering hormones in plants - autonomous and inductive mechanisms. In both cases, the formation is regulated by an intrinsic hereditary program. However, the autonomous mechanism consists of a genetic regulation of formation of flowering hormones that is independent of day length and other external factors and is operative in all plant organs including the stems, stem buds, and leaves. The inductive mechanism also involves genetic control of the formation of the flowering hormones but depends on external factors and, in the case of photoperiodism, is dependent on the leaves as receptive organs.

In photoperiodically neutral species, the autonomous mechanism of hormone formation is found to be the most pronounced; in short-day and long-day plants, the autonomous and inductive mechanisms are of approximately equal importance. In plants with a two-step photoperiodic reaction, i.e., long-short-day and short-long-day plants, the inductive mechanism is the most pronounced (14, 15).

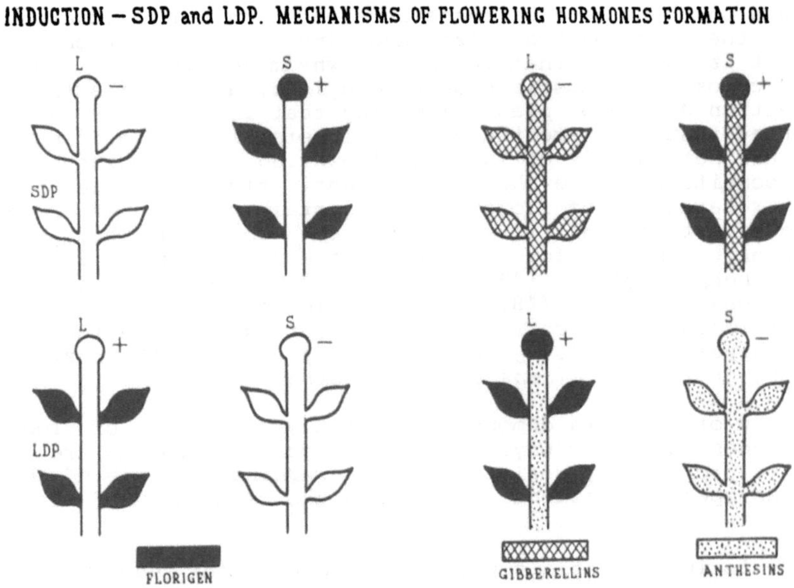

Fig. 1. Photoperiodic induction of flowering in short-day (*SDP*) and long-day (*LDP*) plants. Left, (according to hypothesis of florigen as total complex of flowering hormones), induction in SDP on short day (*S*) and in LDP on long day (*L*), and absence of induction in SDP on L and in LDP on S. Right, (according to hypothesis of florigen as bicomponent complementary complex of flowering hormones), in SDP, gibberellin formation in leaves, stems, and stem buds (autonomous mechanism) is observed under L and S conditions, and anthesin formation, in leaves on S (inductive mechanism). In LDP, anthesins are formed in leaves, stems, and stem buds under L and S conditions (autonomous mechanism) and gibberellins are formed in leaves on L (inductive mechanism). Plant flowering: (+); vegetative growth: (-)

Different hormones are synthesized in the leaves with different clear-cut photoperiodic reactions, such as short-day perilla and long-day rudbeckia. In the first case, anthesins are formed on a short day, and in the latter, gibberellins are formed under long-day conditions. Since in both cases the effect is the same, i.e., formation of flower organs is initiated, it should be apparent that hormones of the other group are permanently present in the plants (gibberellins in perilla and anthesins in rudbeckia), and hence their formation is independent of day length and depends on the autonomous mechanism.

A diagram of formation of flowering hormones that includes the autonomous and inductive mechanisms is depicted in Figure 1. The formation of florigen as a total complex of flowering hormones in shown on the left-hand side of the diagram for short-day plants (SDP) under short-day (S) conditions and for long-day plants (LDP) under long-day (L) conditions. The formation of the components (gibberellins and anthesins) and of the complete florigen complex in the same plant under different day-length conditions is shown at the right-hand side. Gibberellin formation in the stems and leaves of SDP on a long day and a short day is due to the autonomous mechanism; on a short day, anthesin formation in leaves occurs in addition and is conditioned by the inductive mechanism. As a result, the complete florigen complex appears in the leaves and stem buds since anthesins move from the leaves to the stem buds.

Anthesin formation in the stems and leaves of LDP on a short day and a long day is due to the autonomous mechanism; on a long day, gibberellin formation in leaves occurs additionally and is conditioned by the inductive mechanism. The complete florigen complex now appears in the leaves and stem buds since gibberellins move from the leaves to the stem buds.

The hypothesis of florigen as a bicomponent complementary complex of flowering hormones offers some new possibilities in interpreting the results of many graft experiments and may serve as a criterion in assessing various concepts regarding hormonal regulators of plant flowering.

F. Comparative Analysis of Grafting Experiments Based on Both Hypotheses of Florigen

The results of grafting experiments are considered in terms of the hypotheses of florigen both as a total complex of flowering hormones (Fig.1, left side) and as a bicomponent complex of flowering hormones (Fig.1, right side).

In this comparative analysis, it is necessary to consider the following:

1. The grafts may be based on the translocation of flowering hormones that are produced in the donor plants as a result of either the inductive or the autonomous mechanisms or of both.

2. The formation of hormones by the inductive mechanism in leaves of photoperiodically neutral plants, as well as subsequent translocation of the hormones to the stem buds, is very rapid under favorable day-length conditions; in contrast, the formation of hormones by the autonomous mechanism is relatively slow because the hormones are produced dispersively in all plant organs.

3. The translocation of separately obtained florigen components (gibberellins and anthesins) should be considered in view of the fact that their modes of translocation in plants are different.

4. In assessing the results of grafting experiments based on the formation of flowering hormones in the donor plants, it should be kept in mind that flowering inhibitors that are antagonistic to hormones may be present in leaves under unfavorable day-length conditions.

I. Graftings of Short-day and Long-day Plants: Inductive Mechanism

It has been demonstrated by many experiments that flowering of long-day plants under short-day conditions may be induced by grafting them onto short-day plants and, vice versa, flowering of short-day plants under long-day conditions may be induced by grafting them onto flowering long-day plants. Moreover, not only receptor plants of the same species as the donor plant can be induced to flower but also plants of other species. A review of the results of such grafting experiments has been written by Lang (36).

In some intergeneric grafting experiments, short-day Kalanchoe plants grafted onto long-short-day *Bryophyllum* plants could be induced to flower, as could the short-day *Xanthium* plant grafted onto long-day *Rudbeckia*, long-day Silene grafted onto short-day *Perilla* (52), and long-day *Blitum* plants grafted onto short-day *Chenopodium* plants (30).

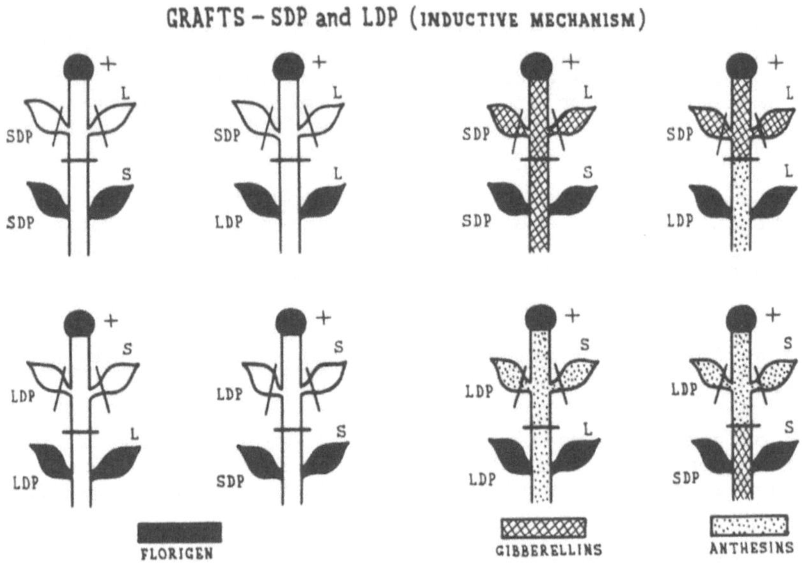

Fig. 2. Graftings of short-day (*SDP*) and long-day (*LDP*) plants (induced). Left (according to hypothesis of florigen as total complex of flowering hormones) - flowering of SDP scions on long day (L) as result of florigen translocation from induced leaves of SDP stocks on short day (S) and of LDP stocks on L (*upper row*); flowering of LDP scions on S as result of florigen translocation from induced leaves of LDP stocks on L or SDP stocks on S (*lower row*). Right (according to hypothesis of florigen as bicomponent complementary complex of flowering hormones) - flowering of SDP scions possessing gibberellins (autonomous mechanism) on L as result of combined anthesin and gibberellin translocation from induced leaves of SDP stocks on S and of LDP stocks on L; flowering of LDP scions possessing anthesins (autonomous mechanism) on S as result of combined anthesin and gibberellin translocation from induced leaves of LDP stocks on L or SDP stocks on S. Plant flowering: (+); vegetative growth: (-)

The grafting scheme is shown in Figure 2. On the left, flowering of defoliated SDP scions on a long day is due to florigen from induced stocks of SDP on a short day and from stocks of LDP on a long day (above); Flowering of scions of LDP on a short day is due to florigen from induced stocks of LDP on a long day and of SDP on a short day (below). The same explanation is also possible for the right-hand part of the diagram. However if the scions of SDP on a long day contain gibberellings, their flowering is probably due to the stock anthesins. Likewise, if the scions of LDP on a short day contain anthesins, their flowering should be due mainly to the stock gibberellins.

II. Grafting of Neutral (NDP), Short-day (SDP), and Long-day (LDP) Plants: Autonomous mechanism

In the early grafting experiments involving photoperiodically neutral plants, it was found that under long-day conditions, short-day plants grafted onto neutral plant stocks would flower. Examples are Jerusalem artichoke (*Helianthus tuberosus*) on sunflower (*Helianthus annuus*), tobacco (*Nicotiana tabacum*) var. Mammoth on var. Samsun, etc. (9, 43, 46, 47). On the other hand, a contradiction appeared when it was de-

1 2 3 4

A B

Fig. 3 A,B. Flowering of photoperiodically sensitive tobacco plants (*Nicotiana sylvestris* and *Nicotiana tabacum*) grafted on plants of neutral species. (A), flowering of scions of L species, var. Sylvestris, on stocks of neutral species, var. Trapesond, under S conditions (*2*), control nongrafted plants, var. Sylvestris, vegetate on short day (*1*). (B), flowering of scions of SDP, var. Mammoth, on stocks of neutral species, var. Trapesond, on L (*4*), control Mammoth plants grafted on Mammoth (*3*) vegetate on L (Photo 31.III and 13.II. 1976)

monstrated later that flowering of the short-day Mammoth variety is not induced by the stock of the photoperiodically neutral Delecrest tobacco variety. However flowering of the scion of the long-day Sylvestris tobacco variety on a short day can be evoked (60).

Recent experiments carried out in our laboratory (22) have shown that scions of the short-day Mammoth plants flower under long-day conditions when grafted onto flowering, photoperiodically neutral Trapesond stocks. If grafted onto Mammoth stock, the scions remain in the vegetative state. In a similar manner, scions of the long-day Sylvestris variety flower under short-day conditions when grafted onto Trapesond stocks, whereas the control nongrafted plants remain in the rosette state (Fig. 3).

The scheme of these grafting experiments is presented in Figure 4. On the left it can be seen that flowering of defoliated scions of SDP on a long day and of LDP scions on a short day proceeds at the expense of florigen, which enters them from the stocks of NDP in which it is produced by the autonomous mechanism under L as well as S conditions. The same explanation applies to the right-hand side of the diagram with the distinction that flowering of scions of SDP containing gibberellins is probably due, predominantly, to anthesins of the NDP stocks. On the other hand, flowering of scions of LDP with anthesins is due mainly to gibberellins of the NDP stocks.

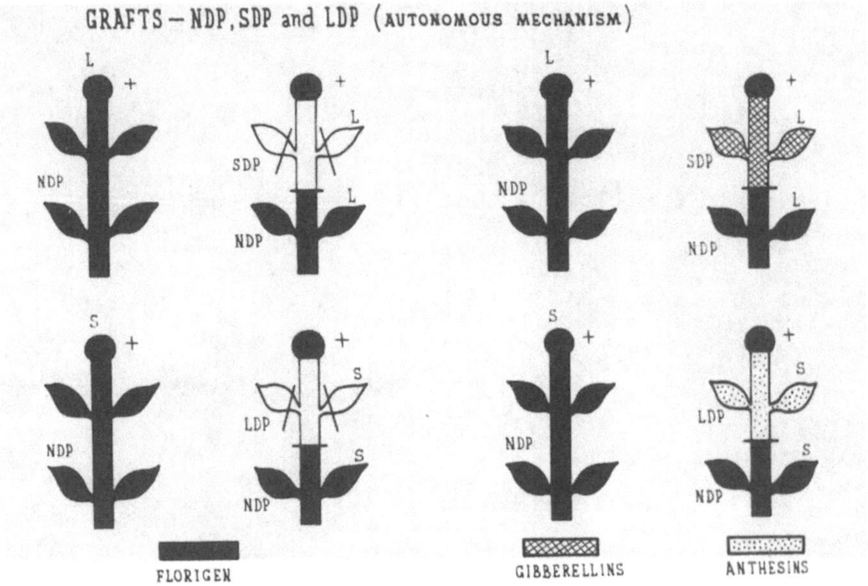

Fig. 4. Graftings of NDP, SDP, and LDP. Left - flowering of SDP scions on long day (L) (*upper row*) and LDP scions on short day (S) (*lower row*) as result of translocation of florigen formed under L or S conditions from NDP stocks (autonomous mechanism). Right - flowering of SDP scions possessing gibberellins on L as result of combined anthesin and gibberellin translocation from NDP, flowering of LDP scions possessing anthesins on S as result of combined gibberellin and anthesin translocation from NDP stocks

III. Grafting of Short-day and Long-day Plants: Autonomous Mechanism

The overwhelming majority of grafting experiments between long-day and short-day species and varieties, carried out for the purpose of initiating flowering, were performed in such a way that the donors were induced, i.e., were in the flowering state. In a few cases, however, the donors were not induced and were in the vegetative state, and nevertheless flowering of the receptors was initiated under unfavorable day-length conditions. It was found that vegetating short-day Mammoth tobacco plants under long-day conditions can induce flowering in defoliated long-day Sylvestris scions grafted onto them, whereas non-grafted defoliated Sylvestris plants on a long day do not flower (36, 60). It has also been demonstrated in grafting experiments that donor plants of the long-day species *Sedum*, with their leaves moved under short-day conditions, induced flowering in defoliated receptor plants of the short-day Kalanchoe plant but did not induce flowering in defoliated *Sedum* receptors (6).

The results of these grafting experiments cannot be explained on the basis of the hypothesis of florigen as a total complex of flowering hormones but can be explained from the viewpoint of the hypothesis of florigen as a bicomponent complementary complex of flowering hormones (Fig.5).

From the left-hand side of Figure 5 it can be seen that flowering of defoliated SDP scions on a long day cannot occur since the stocks of

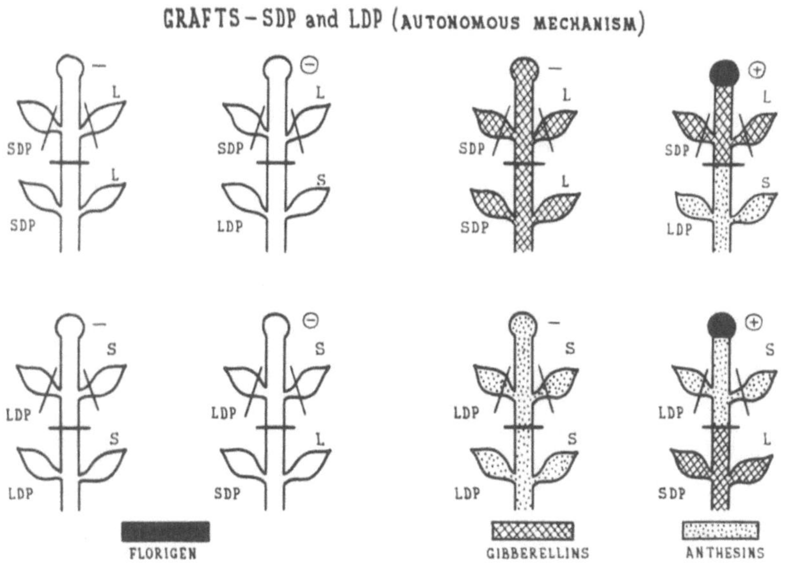

Fig. 5. Graftings of SDP and LDP (noninduced). Left - SDP scions on long day (L) (*upper row*) and LDP scions on short day (S) (*lower row*) do not flower because there is no florigen in LDP stocks on S and in SDP stocks on L. Right - flowering of SDP scions possessing gibberellins on L as result of anthesin translocation from LDP stocks on S (autonomous mechanism); SDP scions grafted on SDP that are on L do not flower on L (*upper row*). Flowering of LDP scions possessing anthesins on S as result of gibberellin translocation from SDP stocks on L (autonomous mechanism); LDP scions grafted on LDP on S conditions do not flower on S

the SDP on a long day and of LDP on a short day are not induced (*above*). Similarly, scions of LDP on a short day cannot flower since the stocks of LDP on a short day and of SDP on a long day are not induced (*below*). On the right, it can be seen that flowering of defoliated scions of SDP on a long day does not occur if the stocks are SDP on a long day. But it occurs if the stocks are LDP on a short day since in the latter case the flowering of the scions (SDP on L) with gibberellins is due to anthesins entering them from the stocks of LDP on a short day (*above*). In a similar way, defoliated scions of LDP on short day do not flower if the stocks are LDP on a short day but do flower if the stocks are SDP on a long day (*below*).

It appears to be important to carry out grafting experiments in which both graft components are not induced and are in the vegetative state.

IV. Grafting of Biennial Long-day Plants (2-LDP) and Short-day Plants (SDP): Autonomous Mechanism

In one of the earliest grafting experiments in which flowering was initiated, grafting of the stems of short-day Mammoth tobacco plants growing under long-day conditions onto rosette biennial henbane (*Hyoscyamus niger*) plants in their first year of growth resulted in stem growth and flowering of the henbane plants (42). Subsequently converse experiments were carried out in our laboratory (13) in which rosette biennial henbane plants were grafted onto stem plants of Mammoth to-

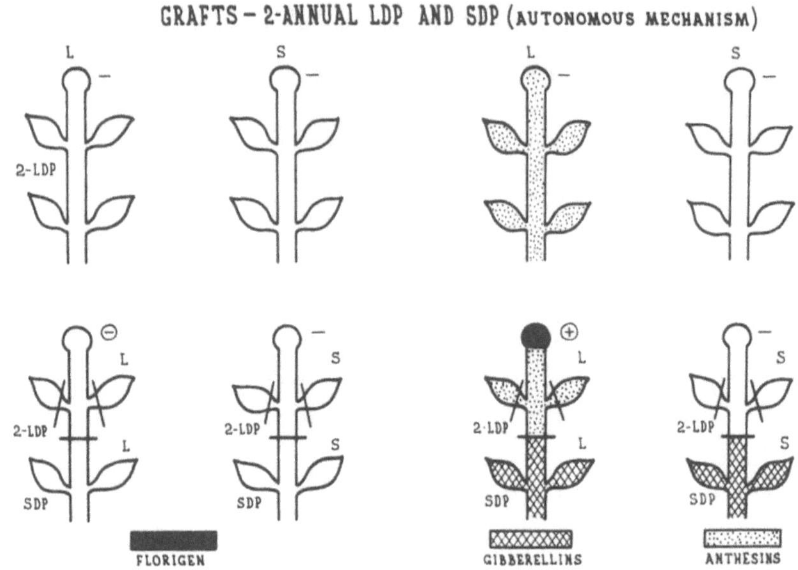

Fig. 6. Graftings of biennial long-day (2-LDP) and SDP. Left - biennial plants, both nongrafted and grafted on SDP stocks, do not flower under L and S conditions because there is no florigen either in the scions or in the stocks. Right - flowering of 2-LDP scions possessing anthesins on L as result of gibberellin translocation from SDP stocks on L (autonomous mechanism); 2-LDP scions grafted on SDP under L conditions do not flower on S because of absence of anthesins

bacco under long-day conditions. As a result, the scions produced stems and flowered the action of the stocks.

The scheme of these grafting experiments is shown in Figure 6. Nongrafted biennial henbane plants on a long or short day (above) and grafting onto SDP Mammoth plants on a long day are depicted in the left-hand part of Figure 6. According to the hypothesis of florigen as a total complex of flowering hormones, the scions should not flower. The same grafts are shown on the right-hand side, and it can be seen that scions of biennial henbane, which possess anthesins on a long day, flower, thanks to the gibberellins that are present in the stocks of the Mammoth tobacco on a long day.

The presence of anthesins in rosette biennial henbane plants on a long day may be inferred from the results of previous experiments in which plant flowering was induced by gibberellin treatment under L but not S conditions (34). These experiments also demonstrate the possibility of inducing flowering of shoots by combining the two vegetating components.

V. Grafting of Long-short-day (LSDP) Plants: Inductive Mechanism

Bryophyllum daigremontianum, in which flowering can be induced only by replacing a long day by a short day, is a representative of long-short-day species. From the viewpoint of the hypothesis of florigen as a bicomponent complementary complex of flowering hormones, this signifies that, initially, gibberellin formation proceeds on a long day and an-

Fig. 7. Flowering of long-short-day grafted plants *Bryophyllum daigremontianum* when combining L and S components. Flowering of *Bryophyllum* middle shoots situated between two stock leaves that were on L and two grafted leaves that were on S (*2*); same plant without covering of S leaves (*3*); control middle shoots situated between two stock leaves that were on L and two grafted leaves that were on L (*1*) do not flower (Photo 15.XI. 1971)

thesin formation, on the short days that follow. With respect to the gibberellins, it has been shown that treatment of plants with dilute gibberellin solutions may replace the long-day exposure, i.e., it may evoke the flowering of plants that had been growing continuously under short-day conditions. But if this is the case, it may be expected that in a grafting experiment in which the stock with leaves in on a long day and the scion with leaves is on a short day, the middle shoots located between them should flower as a result of the gibberellins from the stock leaves and of the anthesins from the scion leaves. Such grafts were performed in our laboratory (21) and yielded positive results (Fig.7). These grafting experiments differ from previous ones in which the donors were stocks induced by replacing long-day by short-day conditions or by gibberellin treatment under short-day conditions (62). The scheme of these experiments is shown in Figure 8. Nongrafted *Bryophyllum* LSDP plants on a long or short day are shown on the left (*above*), as are grafts: short-day scions on long-day stocks and, vice versa, long-day scions on short-day stocks, with removal of the apices and retainment of the scion and stock leaves and the middle shoots (*below*). According to the hypothesis of florigen as a total complex of flowering hormones, the middle shoots cannot flower because of the absence of florigen in the stocks and scions. In the right-hand part of the scheme, it is shown that the middle shoots flower due to anthesins entering them from the scion leaves on a short day and to gibberellins from the leaves of stocks on a long day. One might expect that the shoots would flower if the grafting components were arranged in the opposite order, but in reality this does not happen because the translocation of gibberellins from scions on a long day downward to the stock

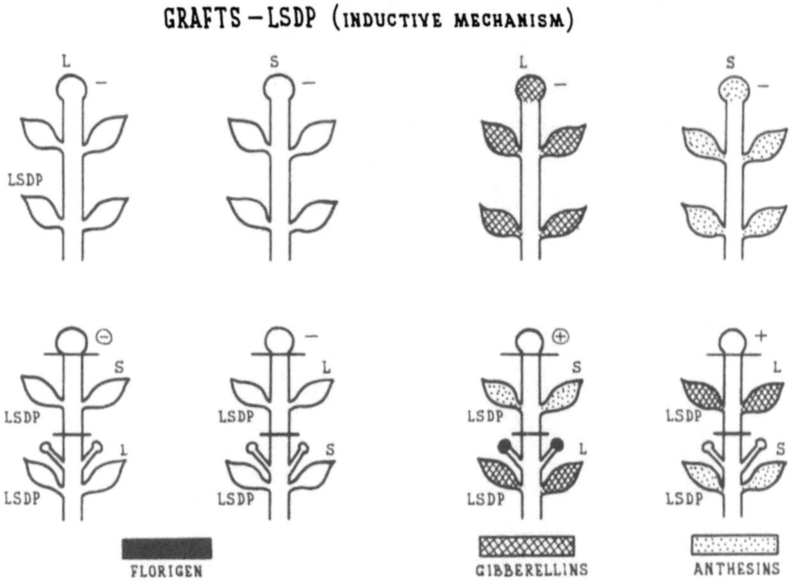

Fig. 8. Graftings of long-short-day (LSDP) plants (inductive mechanism). Left: in LSDP both nongrafted and middle shoots situated between stock leaves on L and grafted leaves on S do not flower, because there is no florigen either in the scions or in the stocks. Right: in nongrafted LSDP, gibberellin formation is induced on L and anthesin formation, on S. Flowering of middle shoots that obtain gibberellins from lower leaves of stocks and anthesins from upper grafted leaves; when leaves are arranged in reverse order, the shoots do not flower because of difficulty of gibberellin translocation from scions downward to stocks

on a short day is impeded, in agreement with our data (20) on the slow and hampered gibberellin translocation in whole plants in the basipetal direction.

The results of these grafting experiments are a clear-cut argument in favor of the hypothesis of the presence in plants of two different groups of flowering hormones, comprising the florigen complex.

G. Conclusion

In recent years there has been an increase in the interest of many investigators in problems of hormonal regulation of plant flowering and in the various relevant concepts and, in particular, in the florigen hypothesis (1-3, 6, 25, 27, 36, 55-58, 61). The different original approaches employed by the various authors have prompted us in the present paper to review briefly the principal experimental facts, including plant transplantation experiments, which form the basis of the hypothesis of florigen, and the notions of other authors regarding the role of familiar phytohormones and inhibitors as flowering regulators.

In its general form the hormonal concept of plant flowering has received wide recognition. However, the precise nature of the flowering hormones remains unknown. In particular it is not clear whether they are specific substances, or whether the flowering function can be fulfilled by well-known hormones with a broad spectrum of action. Most investigators seem to agree that the gibberellins are hormonal substances that differ radically from others by their high activity and ability to induce and stimulate plant flowering. There is, however, much less unanimity regarding the substances directly responsible for flower formation, i.e., the anthesins as we understand them. In the chain of our considerations, a very important link is lacking; there is no direct proof of the presence of such substances, nor are there data on their chemical structure, modes of extraction, and methods of chemical identification.

Naturally the discovery of this link would be a major scientific achievement, and many scientists are striving to accomplish this. The first twenty years (1937-1958) of research were rewarded by the discovery of gibberellins; now, at the end of another twenty years of work (1958-1978), we may only express the hope that the hormones that induce the formation of flower organs will finally be found.

References

1. Aksenova, N.P., Bavrina, T.V., Konstantinova, T.N.: Nauka-Moscow, 1973
2. Audus, L.J.: Plant Growth Substances, London: Leonard Hill Books 1972
3. Bernier, G., Bodson, M., Kinet, J.M., Jacquemard, A., Havelange, A.: In: Proc. 8th Int. Conf. Plant Growth Substances, Tokyo, 1973, Tokyo: Hirokawa Publ. Co., 1974, p. 980
4. Biswas, P.K., Paul, K.B., Henderson, J.H.M.: Physiol. Plantarum 19, 875 (1966)
5. Bunsow, R., Harder, R.: Naturwissenschaften 43, 479 (1956)
6. Carr, D.J.: Ann. N. Y. Acad. Sci. 144, 305 (1967)
7. Chailakhyan, M.Kh.: Dokl. Acad. Nauk USSR 12, 443 (1936a)
8. Chailakhyan, M.Kh.: Dokl. Acad. Nauk USSR 13, 79 (1936b)
9. Chailakhyan, M.Kh.: Acad. Nauk USSR-Moscow-Leningrad, 1937
10. Chailakhyan, M.Kh.: Dokl. Acad. Nauk USSR 117, 1077 (1957)
11. Chailakhyan, M.Kh.: 25th Timiriazev Lecture, Nauka-Moscow, 1964
12. Chailakhyan, M.Kh.: Biol. Zentr. 77, 641 (1958)
13. Chailakhyan, M.Kh.: In: Biochemistry and Physiology of Plant Growth Substances, Wightman, F.W., Setterfield, G. (eds.). Ottawa: Runge Press, 1968, p. 1317
14. Chailakhyan, M.Kh.: In: Plant Growth Substances, Carr, D.J. (ed.). Berlin: Springer-Verlag, 1972, p. 745
15. Chailakhyan, M.Kh.: Fiziol. Rast. 22, 1265 (1975)
16. Chailakhyan, M.Kh., Butenko, R.G.: Fiziol. Rast. 4, 450 (1957)
17. Chailakhyan, M.Kh., Loshnikova, V.N.: Fiziol. Rast. 7, 521 (1960)
18. Chailakhyan, M.Kh., Kochankov, V.G.: Fiziol. Rast. 14, 773 (1967)
19. Chailakhyan, M.Kh., Podolny, V.Z.: Fiziol. Rast. 17, 502 (1970)
20. Chailakhyan, M.Kh., Khlopenkova, L.P.: Fiziol. Rast. 19, 1002 (1972)
21. Chailakhyan, M.Kh., Yanina, L.I.: Dokl. Acad. Nauk USSR 208, 749 (1973)
22. Chailakhyan, M.Kh., Khazhahyan, Kh.K., Ahamyan, L.B.: Dokl. Acad. Nauk USSR, 229, 1340 (1976)
23. Chouard, P.: C. R. Acad. Sci. Paris 245, 5520 (1957)
24. El-Antably, H.M.M., Wareing, P.F.: Nature (London) 210, 328 (1966)
25. Evans, L.T.: The Induction of Flowering. Some Case Histories. Ithaca, New York: Cornell Univ. Press 1969
26. Evans, L.T.: Ann. Rev. Plant Physiol. 22, 365 (1971)
27. Evans, L.T.: Daylength and Flowering of Plants, Benjamin, W.A. (ed.). Menlo Park, California: Reading 1975

28. Gibby, D.D., Salisburry, F.B.: Physiol. Plantarum 47, 784 (1971)
29. Hamner, K.C., Bonner, J.: Botan. Gaz. 100, 388 (1938)
30. Jacques, M.: C. R. Acad. Sci. Paris 276, 1705 (1973)
31. Kholodny, N.G.: Phytohormones, 1939, Selected Works, vol.2, Acad. Nauk USSR-Kiev, 1957, p. 159
32. Knott, J.E.: Proc. Am. Soc. Hort. Sci. 31, 152 (1934)
33. Kuijper, J., Wiersum, L.K.: Proc. Kon. Needrl. Acad. Wet. 39, 1114 (1946)
34. Lang, A.: Naturwissenschaften 43, 284 (1956)
35. Lang, A.: Ann. Rev. Plant Physiol. 3, 265 (1952)
36. Lang, A.: In: Encyclopedia of Plant Physiology, Ruhland, W. (ed.), Vol. XV, pt.I, 1952, Berlin: Springer 1965, p. 1380, 1952
37. Lincoln, R.G., Cunningham, A., Hamner, K.C.: Nature (London) 202, 559 (1964)
38. Lona, F.: L'anteneo Parmense 27, 867 (1956)
39. Lona, F.: Bull. Soc. Ital. Biol. Sper. 25, 761 (1949)
40. Maheshwari, S.C., Venkataraman, R.: Planta 70, 304 (1966)
41. Melchers, G.: Biol. Zentr. 57, 568 (1937)
42. Melchers, G.: Ber. Deut. Botan. Ges. 57, 29 (1939)
43. Melchers, G., Lang, A.: Biol. Zentr. 61, 16 (1941)
44. Milyaeva, E.L., Chailakhyan, M.Kh.: Izvest. Acad. Nauk USSR, Ser. Biol. 3, 342 (1974)
45. Moshkov, B.S.: Ser. A. Sotsialist. Rastenievodstvo 17, 25 (1936)
46. Moshkov, B.S.: Tr. Prikl. Bot. Genet. i selek. Ser. A Sotsialist. Rastenievodstvo 19, 107 (1937a)
47. Moshkov, B.S.: Tr. prikl. Bot. Genet. i selek. Ser. A. Sotsialist. Rastenievodstvo 21, 145 (1937b)
48. Nitsch, C., Nitsch, J.P.: Planta 72, 371 (1967)
49. Overbeek, J. van: Botan. Gaz. 108, 64 (1946)
50. Pharis, R.P., Morf, W.: Biosci. 19, 719 (1969)
51. Pharis, R.P.: Planta 105, 205 (1972)
52. Pol, P.A. van de: Meded. Landb. Hoogesch. Wageningen 72, 1 (1972)
53. Psarev, G.M.: Sovetsk. Botanica. 3, 88 (1936)
54. Roberts, R.H., Struckmeyer, B.E.: J. Agr. Res. 59, 633 (1939)
55. Salisbury, F.B.: Ann. N. Y. Acad. Sci. 144, 295 (1967)
56. Steward, F.C., Krikorian, A.D.: Plants Chemicals and Growth. New York-London: Academic Press 1971
57. Vince-Prue, D.: Photoperiodism in Plants. New York: McGraw Hill 1975
58. Wareing, P.F., Phillips, I.D.J.: The Control of Growth and Differentiation in Plants, Pergamon Press, 1970
59. Yabuta, T., Sumiki, J.: Agr. Chem. Soc. Japan 14, 1526 (1938)
60. Zeevaart, J.A.D.: Meded. Landb. Hoogesch. Wageningen 58, 1 (1958)
61. Zeevaart, J.A.D.: Ann. N. Y. Acad. Sci. 137, 723 (1962)
62. Zeevaart, J.A.D., Lang, A.: Planta 58, 531 (1962)

Paper Demonstration References (see Section IX):

34 - 74 - 107 - 121 - 136 - 147 - 153 - 188 - 189 - 192

29. Hormonal Relationships in Whole Plant Senescence

P. J. DAVIES, W. M. PROBESTING, and T. J. GIANFAGNA

Abstract

The hormonal control of whole plant senescence, as determined by the senescence of the apex, is being investigated using a genetic line of peas (*Pisum sativum* L.). This line, designated G2, has dominant genes at two different loci, both of which function in short days to delay apical senescence and thus greatly extend the reproductive phase. GA_3 and GA_{20} applied to the apex in long days prevented or delayed senescence, respectively. GA_9, IAA, benzyladenine, or abscisic acid were without effect. Five different inhibitors of gibberellin synthesis failed to induce senescence under short days. Grafting experiments showed that in short days G2 stock plants could delay senescence in scions from plants possessing one of the dominant alleles for the photoperiodic control of senescence, but not if the plant lacked both dominant alleles. Senescence required flower and fruit development as well as long days. A senescence-promoting stimulus was produced throughout fruit development. It is proposed that senescence is a result of a change in gibberellin production by the leaves coupled with the production of a senescence factor, or the diversion of substances required for apical growth, by the developing fruits.

A. Introduction

Relatively little is known about the cause or internal control of senescence in whole plants. One of the principal problems in such work is that in monocarpic annuals senescence usually follows flowering and fruiting, so that the cause of any changes during plant senescence is uncertain. As a result, senescence has often been attributed to a diversion of nutrients from the growing apex of the plant to the developing fruits (12, 19). Such a nutritional relationship was disproved by Leopold et al. (8) who showed that fruits exerted their greatest promotion of leaf senescence not at the time of fruit growth, as would be expected if nutrient exhaustion was a prime factor, but at fruit maturation. Similar results have been noted with regard to apical senescence, which is a better indicator of the demise of the whole plant than leaf senescence (10, 15). (Apical senescence is herein defined as those processes preceding the death of the apex, which, in monocarpic annuals, is inevitably followed by the death of the whole plant.) Lockhart and Gottschall (9) found that removal of young fruits from Alaska peas enabled the stem apex to grow for a longer period than when the fruits matured on the plants. In Greenfeast peas only fruit maturation was found to alter the physiology of stock plants to the extent that grafted young buds failed to grow (10). These experiments suggest that the production of a "senescence factor" by the fruits may be involved.

We have been able to separate the effects of flowering and fruiting on senescence by the use of an early flowering genetic line of peas, designated G2, which is photoperiod-dependent with regard to senescence (11). In long days, the plants fruit and subsequently senesce after a short reproductive phase. In short days, however, despite entering the reproductive phase and developing mature fruits, apical growth continues for a protracted period of time and apical senescence does not occur. This behavior is dependent on the presence of dominant genes at two different loci. Recent evidence indicates that the two loci in question are *Sn* and *Hr* (13). Plants with recessive alleles at either one or both of the loci flower and senesce regardless of photoperiod. Use of the G2 line enables plants to develop mature fruits in short days without senescing. The plants are then transferred to long days, under which senescence takes place. Since the control of senescence appears to involve translocatable factors, we have begun an investigation of the hormonal control of senescence in G2 peas.

B. Materials and Methods

Pea (*Pisum sativum* L.) plants, line G2, were grown singly in 15-cm clay pots filled with a mixture of sand, soil, and peat. Plants were fertilized every two weeks with a complete nutrient solution, and lateral branches were routinely trimmed.

Experiments were conducted in a growth cabinet under fluorescent and incandescent lights, or in a glasshouse with supplemental fluorescent lighting to extend the daylength if needed. Standard short and long days were 9 and 18 h, respectively, and day and night temperatures were 20°C and 15°C (or as close as possible in the glasshouse).

In all cases the criteria for senescence of the shoot apex were a decrease in the size of the apical bud (enclosed by the oldest pair of unexpanded stipules), rate of growth, and a loss of green coloration. Once the apex has died further growth cannot occur, yet a resumption of growth may occur at any phase of senescence prior to the death of the apex. Node counts were used to measure responses. The first reproductive node was the node at which the first flower bud was initiated, even if aborted.

C. Results and Discussion

I. Environmental Control of Senescence in G2 Peas

Under long days with high light intensity in summer, the first flower appeared after about five weeks, the first symptoms of senescence after about six weeks, and death of the apex after about eight weeks. Development was slightly slower under lower light intensities. Under short days the first few flower buds usually aborted, and apical growth continued indefinitely. It was found that apical death occurred after a minimum of 25 long days following transfer of plants with fruits from short to long days. The effect of long days could be completely reversed by the restoration of short days within 3-4 days before the appearance of the final node. The response to daylength was quantitative. Under 9-h photoperiods, growth was indeterminate, while with increasing daylengths from 12 to 18 h the number of reproductive nodes declined gradually with a minimum number under 18-h to 24-h photoperiods. The senescence response of G2 was based on the total duration of illumination (or darkness) received per 24 h, rather than by the length of uninterrupted darkness. Brief interruptions of the night period by white light were without effect. Plants senesced when a short day was extended with low intensity blue, red, or far-red light. These results are reported more fully elsewhere (15).

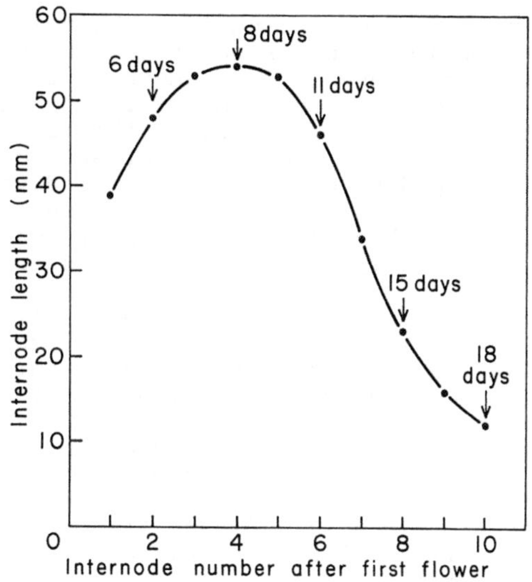

Fig. 1. Change in internode length of G2 plants in long days following production of first flower. Number of days listed are number of days from anthesis of first flower

II. The Involvement of Hormones in the Photoperiodic Control of Senescence

Grafting experiments, described below, indicated that the leaves were the photoperiodic receptors. Deflorated G2 stock plants transmitted a senescence-delaying stimulus to photoperiodically insensitive scions. Covering the apical bud and subtending stipules of ungrafted G2 plants in long days did not delay senescence, indicating that the apex was not the sole photoperiod receptor. As the response to photoperiod occurs in the apex, some signal must be transferred from the leaves to the apex and a hormone would seem probable. To test possible candidates, 0.1 ml of 10 µg/ml naphthalene acetic acid, benzyladenine, or gibberellic acid (GA_3) was applied to shoot apices of G2 peas in long days three times weekly. The auxin and cytokinin applications had no effect with respect to the controls, the shoot apices senescing after 13 reproductive nodes. GA_3, however, extended the reproductive phase indefinitely (in excess of 40 reproductive nodes) in all cases. Termination of growth was averted when GA_3 was applied to nearly senescent plants, and growth continued until the plants had produced 30-35 reproductive nodes, at which point the experiment was terminated. Thus the effect of GA_3 resembled the effect of short days on the senescence of G2 peas. In addition GA_3 increased the mean internode length from 2.7 cm to 11 cm, increased the rate of flower abortion (which also occurs in short days), and reduced leaf size and chlorophyll content. GA_3 has previously been shown to delay senescence of Alaska pea (3, 9). This was attributed to a delay in fruit set.

Because GA_3 does not occur in peas (4, 5) and because it induced stem elongation highly abnormal for this line, we tested gibberellins native to pea. The three principal gibberellins in peas are thought to be GA_{20}, GA_{29}, and GA_9 (5, 6) (in decreasing order of amount), with biosynthesis proceeding in the order of increasing polarity $GA_9 \rightarrow GA_{20} \rightarrow GA_{29}$ (5, 6, 16, 17). GA_9 was without effect when applied to stem apices as above, while GA_{20} produced a delay in senescence (from 11.0 to 27.3 reproductive nodes) and also caused a slight increase in internode length (from 2.9 to 3.5 cm). Insufficient GA_{29} was available for testing. It is thus clear that both apex senescence and stem elongation are sensitive to the application of one native gibberellin (GA_{20}) but not another (GA_9). One of the phenomena associated with the induction of senescence in peas is a decrease in the length of the internodes produced as senescence of the apex proceeds (Fig.1), which could be the result of a declining GA supply or a rise in the level of an elongation inhibitor. Thus a decrease in the amount of GA_{20} or other similar GA reaching the apex in long days compared to short days would be one possibility. This decrease could be due to a drop in the total gibberellin content or to a change in gibberellin interconversions, compartmentation in the leaves, or transport from the leaves to the apex.

In most long-day plants in which GA affects flowering, GA substitutes for long days. In peas, however, GA_3 has been shown to produce effects on flowering similar to short days. Barber et al. (1) showed previously that GA_3 may modify the magnitude of the *Sn*-controlled photoperiodic effect on flowering under certain conditions and concluded that *Sn* is involved with gibberellin metabolism. The gene *Sn* has also been shown to produce a graft-transmissible substance inhibitory to flowering (14). GA metabolism in several species including peas has been shown to be affected by light (7, 16) and has been associated with the chloroplast (17, 20). It is clearly possible that in G2 plants changes in daylength may alter the content and/or the export of gibberellin by the leaves. Qualitative and quantitative changes in the endogenous gibberellins, interconversions of applied radiolabeled GAs, and the synthesis of GAs from precursors are now being investigated in order to attempt to answer these questions.

Gibberellin physiology was also investigated using inhibitors of gibberellin biosynthesis. Various compounds have been used as growth retardants in many plants, with the effect in some cases shown to be caused by a reduction in gibberellin biosynthesis (2). Plants not responding to one growth retardant often respond to another, so five different inhibitors were applied as a soil drench in 200 ml of water in amounts noted below: AMO 1618 [2'-isopropyl-4'-(trimethylammonium chloride)-5'-methylphenyl piperidene carboxylate] (10 mg); CCC (β-chloroethyltrimethyl ammonium chloride) (400 mg); phosphon (tributyl-2,4-dichlorobenzyl phosphonium chloride) (100 mg); B-9 (N-dimethylaminosuccinamic acid) (100 mg); and ancymidol [α-cyclopropyl-α-(4-methoxyphenyl)-5-pyrimidinemethanol] (2 mg) (18). The compounds were applied to plants in short days, both pre- and postfruiting, and in long days with transfer to short days before the final death of the apex to measure the capacity for recovery. In no case did the inhibitors induce senescence. Except for ancymidol, which reduced internode length by 37%, the inhibitors had no significant effect on growth. While these results are not conclusive, they indicate that gibberellin biosynthesis may not control the delay of senescence. Instead some form of GA interconversion or release may be the controlling factor.

III. Grafting Studies

Further evidence of the hormonal nature of the control of whole plant senescence was provided by grafting plants that did not possess the genetic capacity for continued growth to G2 plants in short days. These plants fall into three categories: I1 (*sn, Hr*); I2 (*Sn, hr*); I3 ('Alaska') (*sn, hr*). G2 stocks were utilized just prior to flowering when four weeks old with 10 nodes. I-type scions were obtained from 10-day-old seedlings with the entire epicotyl grafted laterally onto the stock at node 8. It was necessary to defruit the G2 stocks to enable growth of the scions, which were allowed to develop fruit. Senescence was not affected in grafted plants grown in long days. In short days very dif-

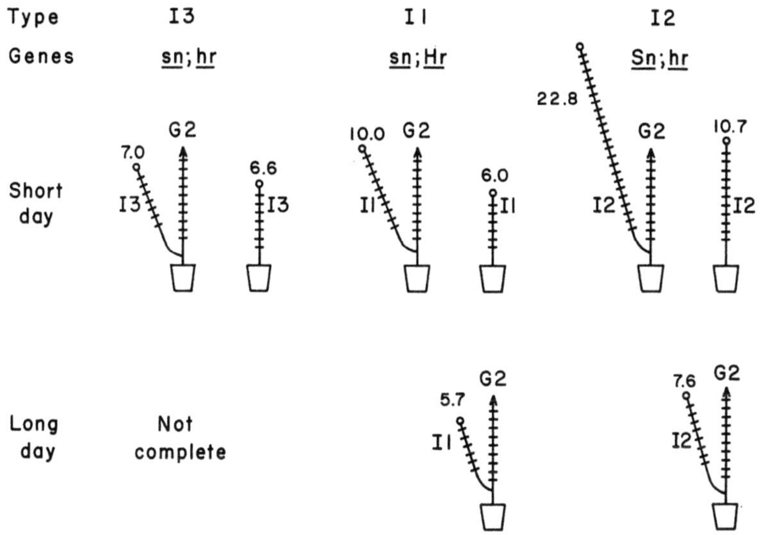

Fig. 2. Influence of G2 stock plants on grafted I-type scions under short (9 h) daylength. Fruits developed on I-type plants while G2 stocks were defruited. Growth of G2 stocks was indeterminate; that of I-type determinate in all cases. Numbers represent number of internodes produced by scions prior to apical death

ferent results were obtained, depending on the genetic make-up of the I-type scions (Fig.2). I3 scions with fruit exhibited no response (i.e., no increase in reproductive nodes) when grafted onto G2, despite the fact that the entire plant was in short days. Deleafing the I3 scion (in order to ensure the movement of organic compounds from stock to scion) made no difference. When I2 scions were grafted onto G2 in short days, senescence was considerably delayed, with scions producing over twice the number of reproductive nodes produced by the control plants prior to senescence. The combination I1/G2 produced 66% more reproductive nodes in short days than in long days. Reciprocal grafts were not attempted due to the rather weak growth of the I type plants in short days compared to the very robust nature of the G2 plants.

These grafting experiments clearly show that in short days, but not in long days, G2 plants produce some transmissible substance that is capable of delaying senescence of nonphotoperiodically sensitive plants, provided that they possess at least one of the two dominant alleles required for the delay of senescence in short days. Thus it seems that each allele might provide some additive change that retards senescence under short days. This low level, plus the influence of the G2 stock, is sufficient to delay senescence. Without any contribution from the I-type scion, the G2 produces an insufficient effect to delay senescence of the scion. The transmissible factor seems to delay senescence in short days rather than to induce senescence, because I-type plants did not affect the G2 stocks. The requirement for defruiting the stocks does not, however, enable a definitive conclusion on this point. Grafts of I1 and I2 in short days are in progress to determine whether the complementary lines delay senescence when combined.

Table 1. The stage of fruit development at the first indication of apical senescence (A) and at the death of the apex (B)

Fruit number	Length (% of maximum) (± SE)	Width (% of maximum) (± SE)
(A) 1	100	51 ± 1.2
2	100	36 ± 1.9
3	89 ± 2.0	-
4	62 ± 4.2	-
5	37 ± 1.8	-
6	-	-
(B) 1	100	100
2	100	77 ± 2.4
3	100	75 ± 2.2
4	100	68 ± 2.6
5	100	62 ± 2.1
6	100	52 ± 3.6
7	100	40 ± 4.0
8	55 ± 9.1	-

Measurements were taken: (A) at time of full expansion of 6th internode[a] after first flower (11 days after anthesis of first flower), and (B) 18 days after anthesis of first flower. Experiment was conducted in growth cabinet in long days

[a] Because reproductive internode 6 is the first clearly shorter internode, senescence stimulus production actually commenced prior to this time

IV. Senescence Factor Production by Fruits

The presence of fruits is required by G2 plants for senescence in long days. The first indication of senescence is decreasing internode length (Fig.1). Internodes maintain an approximately constant length in the presence of fruits in short days. Under continuous long days, senescence started with the development of the first fruit (Table 1A), increased to a maximum rate of development coincident with the filling of the first fruits, and the death of the apex occurred when most fruits had reached their full length but only the oldest fruit had reached its full size (Table 1B). Senescence was not coincident with fruit maturation, as has been reported for Alaska peas (9, 10) and leaf senescence of soybeans (8). Removal of fruits prior to maturity caused, however, a resumption of growth. Defloration of G2 plants in long days prevented senescence indefinitely. Under these conditions the newly formed internodes were shorter and the leaves were convoluted and darker green in color. Up to 90 reproductive nodes were produced before the experiment was terminated. The number of fruits required for senescence varied with daylength, light quality, temperature, and nutrition. In the glasshouse conditions described previously, when various combinations of flowers were removed, an average of five fruits were required to cause senescence (Table 2). The nondeflorated controls produced more fruits than the minimum number required for senescence because some fruits were very young when apical death occurred, but these fruits continued to grow to maturity. With near threshold numbers of fruit, apical growth resumed, in some cases, as the oldest fruits matured. It appeared that mature fruits did not continue to produce the senescence stimulus, and the influence of the remaining fruits was insufficient to prevent growth.

Table 2. Number of fruits required to induce apical death in G2 plants in long days

Treatment	% of Plants senescing	No. of fruits at apex death (± SE)	% of Senesced apices that re-started growth	% Restarts that died
No flower removal	100	8.2 ± 0.5	27.3	100
Alternative flowers removed[a]	72.7	5.0 ± 0.4	37.5	33.3

[a] Removal of more flowers resulted in no apical death. With 2/3 of flowers removed, 25% showed symptoms of senescence, but all recovered and continued growing

The nature of the senescence factor is uncertain. Nutrition is not a controlling factor, because the symptoms of senescence appear very early in fruit development and, in short days, G2 plants produce a full fruit complement with no senescence. Growing fruits do appear to export a small quantity of material. When radioactive sucrose solution was injected into growing pods, radioactivity was detected in other pods and in some cases in the apices. These experiments are currently being repeated and quantified. If a substance is exported from the fruits to the apex, it may be responsible for the senescence of the apex. Eeuwens and Schwabe (4) have reported the appearance of abscisic acid (ABA) in Alaska pea seeds during the final stages of their development, coincident with the time at which Alaska fruits exert their greatest senescence-promoting effect (9, 10), indicating that ABA may be the senescence factor. Applications of 0.1 ml of a 500 μM ABA solution to apical buds of G2 peas 3 times weekly produced no enhancement of apex senescence when the plants were grown in long days with either a full complement of fruits, or when alternate fruits were removed to produce a near threshold fruit number. Thus ABA does not seem to be involved in apical senescence of G2, though uptake of the compound might have been a factor, and additional uptake methods for ABA are being investigated.

D. Conclusions

I. The Hormonal Balance Controlling Apical Senescence in G2 Peas

The grafting experiments clearly show that a translocatable substance inhibits senescence in short days. The photoperiodic stimulus is perceived by the leaves and prevents senescence of the apex. Applications of GA_{20}, a native gibberellin of pea, delays senescence concomitant with causing a small increase in internode elongation, so that the progressive decrease in length that is part of the senescence process does not occur. In short days the gibberellin supply is adequate for growth even in the presence of fruits. In long days the gibberellin supply becomes deficient and, under the influence of the fruits, apical senescence occurs.

The presence of fruits is required for senescence. The fruits could either be producing a senescence factor that kills the apex in the absence of GA, or they could act as hormonal sinks leaving the apex deficient in GA needed for growth (and possibly other hormones needed with the GA). GA transport is being studied to check this possibility. It was noted in the grafting experiments that fruits on the G2 stocks often caused premature senescence of the I-type scions. This phenomenon could be explained either in terms of an excess of senescence factor production or of the fruits providing a sink such that there was insufficient GA to provide for both the stock and scion. If the fruits do indeed produce a senescence factor, then possibly the GA either prevents the synthesis of the senescence factor in short days or interferes with the action of the senescence compound in the apex.

While the exact nature of the hormonal control of senescence in G2 peas is currently undetermined, it seems clear that whether apical senescence does or does not take place is dependent on a hormonal balance in the plants.

Acknowledgements

We thank Dr. N. Murofushi for kindly supplying us with a sample of GA_9 and GA_{20}. This work was supported by Grant PCM76-05616 from the National Science Foundation.

References

1. Barber, H.N., Jackson, W.D., Sprent, J.I.: Nature (London) 182, 1321 (1958)
2. Corcoran, M.R.: In: Gibberellins and Plant Growth. Krishnamoorthy, H.N. (ed.) Wiley: New York, 1975, p. 239
3. Ecklund, P.R., Moore, T.G.: Am. J. Botany 55, 494 (1968)
4. Eeuwens, C.J., Schwabe, W.W.: J. Exp. Botany 26, 1 (1975)
5. Frydman, V.M., Gaskin, P., MacMillan, J.: Planta 118, 123 (1974)
6. Frydman, V.M., MacMillan, J.: Planta 125, 181 (1975)
7. Jones, R.L.: Ann. Rev. Plant. Physiol. 24, 571 (1973)
8. Leopold, A.C., Niedergang-Kamien, E., Janick, J.: Plant Physiol. 34, 570 (1959)
9. Lockhart, J.A., Gottschall, V.: Plant Physiol. 36, 389 (1961)
10. Malik, N.S.A., Berrie, A.M.M.: Planta 124, 169 (1975)
11. Marx, G.A.: Bioscience 18, 505 (1968)
12. Molisch, H.: The Longevity of Plants. New York: Translated and published by E.H. Fulling 1938
13. Murfet, I.C., Marx, G.A.: Pisum Newsletter 8, 46 (1976)
14. Murfet, I.C., Reid, J.B.: Australian J. Biol. Sci. 26, 675 (1973)
15. Proebsting, W.M., Davies, P.J.: Plant Physiol. 58, 800 (1976)
16. Railton, I.D.: Plant Sci. Lett. 3, 207 (1974)
17. Railton, I.D., Reid, D.M.: Plant Sci. Lett. 3, 303 (1974)
18. Shive, J.B., Sisler, H.D.: Plant Physiol. 57, 640 (1976)
19. Sinclair, T.R., deWitt, C.T.: Science 189, 565 (1975)
20. Stoddart, J.L.: Planta 81, 106 (1968)

Paper Demonstration References (see Section IX):

12 - 27 - 34 - 40 - 51 - 101 - 128 - 181 - 195 - 196

30. Growth Substances and Protoplasts

E. C. COCKING

Abstract

The removal of the cell wall of higher plant cells enables protoplasts to be isolated. Current work involves attempts to detect any specific effect of growth substances on the plasma membrane, including reactions leading to the regeneration of a cell wall. The subsequent growth and development of protoplasts vary with the species, but recent work is providing evidence for a rather tight genetic control of growth and development, which may be correlated largely with growth substance effects.

A. Introduction

In recent years studies on the effects of growth substances on isolated protoplasts have centred upon several areas of investigation. The first of these has involved considerations of the influence of various growth substances, in particular, auxins, on the osmotic stability of isolated protoplasts. This interest stemmed from the early observations of the effects of auxins, and of other growth substances, on isolated root protoplasts (2). The other major area of investigation has centred on the effects of growth substances, supplied in various culture media, on the growth and development of isolated protoplasts. The early work was concerned mainly with the establishment of the conditions under which protoplasts would regenerate a new cell wall. From these studies (1) it was clear that such isolated protoplasts first regenerated some form of cellulose wall and then assumed in culture the behaviour of cultured cells. Conditions were soon established for the regeneration of the new cell wall, and this has been the subject of many detailed fine-structural and biochemical studies (18). Even though detailed results have not been reported, all evidence suggests that cell wall regeneration is not directly influenced by exogenously supplied growth substances. Isolated protoplasts are known to be leaky, and, as will be discussed later, this may make such assessments difficult. To some extent studies of the action of growth substances on isolated protoplasts have been overshadowed by assessments of the consequences of fusion of protoplasts isolated from different species, and the production, by this means, of interspecific somatic hybrids. The need for readily applicable, general methods for the selection of interspecies somatic hybrids has, however, recently prompted the search for naturally occurring differences between the sensitivities of various protoplast systems to drugs, including growth substances (3). Moreover, the interactive effects between growth substances and culture media on the growth and division, and plant regeneration capability, of different species and their hybrids are now being intensively investigated. This is because of the ready application of such differences to the selection of interspecies somatic hybrids following the culture of fused protoplasts (14).

B. Rapid Effects of Auxins on Isolated Protoplasts

Interest in the action of growth substances on isolated protoplasts was stimulated by studies on isolated tomato protoplasts (1). Protoplasts were liberated from root tips of tomato seedlings by treatment with *Myrothecium* cellulase. Additions of exogenous levels of 10^{-5}, 10^{-7}, 10^{-9}, and 10^{-11} g/ml of indole-3-acetic acid (IAA) induced various changes in the protoplasts. At 10^{-5} g/ml there was not only a rapid increase in the size of existing vacuoles present in the protoplasts, but also formation of new vacuoles, both in already vacuolated protoplasts and in the densely cytoplasmic protoplasts. At the level of 10^{-11} g/ml of IAA, these effects took place very rapidly, and within one minute the protoplasts had burst. The bursting was probably due to a very rapid increase in the volume of the vacuoles within the protoplast, resulting in a very sudden expansion of the plasma membrane and its sudden bursting. A similar response has also more recently been observed with isolated *Allium cepa* root protoplasts (11). Several studies on isolated protoplasts from roots, fruit, and leaves have indicated a specific effect of auxins that frequently results in rapid enlargement and bursting. This protoplast system therefore, at least superficially, would appear to be particularly suitable for investigations of the effects of growth substances on the plasma membrane, e.g., enabling the detection of induced changes in the enzyme complement associated with the plasma membrane, or ionic changes. Analysis at this level is, however, rendered difficult by the fact that protoplasts are usually very leaky, and, as will be discussed later, many substances including growth substances leak through the plasma membrane. Thus physiologically and biochemically the isolated protoplast cannot be regarded merely as a cell from which the cell wall has been removed. Even *Avena* coleoptile protoplasts have now been shown to respond somewhat similarly to IAA (7), and the earlier report that this protoplast system was insensitive to added IAA has been discounted. These earlier negative results (16) were almost certainly due to these workers having given their protoplasts a pre-treatment with IAA in 0.5 M mannitol for 50 min, which would almost certainly obviate any effect of IAA at a lower molarity of mannitol, similar to that reported by Hall and Cocking (7).

Although the explanation of these rapid effects of IAA almost certainly involves the plasma membrane, nothing is known as to how they are mediated. What is known, however, is that this effect on protoplasts is a growth response specifically induced by auxin, since there is an observed marked inhibition of the IAA bursting response of *Avena* coleoptile protoplasts by 4-Cl-phenoxyisobutyric acid, which is known to be a specific anti-auxin for IAA.

C. Effects of Growth Substances on the Growth and Development of Isolated Protoplasts

Protoplasts can now be readily isolated from a wide range of different plant species and from their cultured cells, by digestion of the cell wall with suitable cell wall-degrading enzymes (4). These protoplasts are naked cells, and it can be shown by freeze-etching (deep etching) studies that essentially all the wall has been removed. The usual cytochemical methods utilising fluorescent detection establishes that the wall has been removed (15), but very occasionally a few microfibrils of cellulose can be detected on the plasma membrane by the more refined freeze-etching methods. Apparently there is no readily detectable effect of growth substances on wall synthesis by isolated protoplasts. The early work on wall synthesis by isolated protoplasts clearly showed that wall synthesis could take place merely in a simple plasmolyticum, containing sucrose, mannitol, $CaCl_2$, KCl, and White's salts (12). Most workers have ignored any possible specific effects of growth substances on wall regeneration and have concentrated their attention on the ultrastructure of the isolated protoplasts, and the wall that is synthesised de novo on the surface of the plasma membrane (6). One of the complicating factors that has probably made interpretation of experiments, designed to show any effects of exogenous growth substances on such isolated protoplasts, difficult is the extensive leakage of metabolites and growth substances from freshly isolated protoplasts. Evidence for such a leakage was provided by the work on the culture of protoplasts isolated from cell suspensions of *Parthenocissus tricuspidata* crown gall (17). In these studies it was observed that, although cultured crown gall tissues are independent of an exogenous supply of growth regulators, protoplasts isolated from such tissues are, at least initially, dependent on their presence in the medium, for the initiation of division. Cell suspensions could be readily grown on Heller's salts medium with the addition of 2% sucrose and 0.4 mg/l thiamine HCl; but protoplast suspensions required a medium such as that of Nagata and Takebe with 0.1 mg/l NAA and 0.2 mg/l kinetin. The most likely explanation of these differences in growth substance requirements is that removal of the cell wall during protoplast isolation results in leakiness of the plasma membrane of the isolated protoplasts. As a result of this general leakage, the internal pool of growth regulators is diminished, and, as a result, external supplementation by added NAA and kinetin is necessary for the initiation and maintenance of division. Cell wall regeneration and the initiation of division result in reduced leakiness, and the resultant cell suspensions can be grown in the absence of added growth regulators. Additional indirect evidence for such leakiness is provided by the finding that very sophisticated culture media containing various amino acids, sugars, and other metabolites enable protoplasts to undergo division at very low cell densities (9). Growth of the protoplasts at higher cell densities, on less sophisticated media, is probably aided by the leakage of metabolites from the protoplasts.

Protoplasts isolated from different species differ in their sensitivity to exogenously supplied growth substances including anti-auxins (3). The most comprehensive study so far has been on *Petunia hybrida* and *Nicotiana tabacum*. *P. hybrida* was found to be much more sensitive than *N. tabacum* to the anti-auxin 4-Cl-phenoxy-isobutyric acid (Table 1). Abscisic acid showed the converse situation, with *P. hybrida* being substantially less sensitive than *N. tabacum* (Table 1). The reasons for these differences are not at present known. They may be related to differences in the internal pool concentrations of the various growth regulators, and a detailed analysis of the levels of growth regulators in these two species would be useful.

Following the regeneration of the cell wall and the onset of division, the ability of protoplast cultures for sustained division is dependent on the growth substance-media milieu. *Petunia parodii* protoplasts, isolated from leaves, grow with a high plating efficiency in MSP1 medium [MS = salts and vitamins medium of Murashige and Skoog (10) with 2% sucrose and 10% mannitol; P1 = growth substances NAA = 2.0 mg/l and BAP = 0.5 mg/l) to a small colony stage (approx. 50 cells). Growth ceases at this stage, and the colonies fall apart followed by eventual death (13). Protoplasts of *P. parodii* will, however, produce callus on F1 medium (5), and eventually whole plant regeneration is possible (8). This medium contains the growth substances NAA (1.5 mg/l), 6BAP (1.0 mg/l), and 2,4-D (0.5 mg/l). Protoplasts isolated from leaves of *P. hybrida* grow on MSP1 and produce callus. *P. hybrida* can be crossed sexu-

Table 1.

	4-Cl-phenoxyisobutyric acid (µg/ml)								
	0	0.1	0.5	1.0	2.5	5.0	10.0	25.0	50.0
TOB	2	2	2	2	2	2	2	2	5
PET	2	2	3	5	5	5	5	5	5

	Cis-trans abscisic acid (µg/ml)					
	0	0.005	0.025	0.05	0.1	0.5
TOB	2	2	3	5	5	5
PET	2	2	2	2	2	2

In the case of 4-Cl-phenoxyisobutyric acid, petunia is more sensitive than tobacco. Abscisic acid shows converse situation, with petunia being substantially les sensitive than tobacco. For details of categories of protoplast growth (*2*, *3*, *5*) see reference (3); *2* and *3* indicate adequate growth; *5*, majority of protoplasts dead

ally with *P. parodii*, and the cross *P. hybrida* × *P. parodii* (and the reciprocal cross) can be readily obtained. Protoplasts from these crosses will grow and produce callus on MSP1. Thus in simple genetic terms the specific growth response of *P. parodii* to the MSP1 growth substance -media milieu is not dominant, or, conversely, positive growth, leading to callus formation, is dominant. The effects are not cytoplasmically controlled. We have also found that *P. parodii* protoplasts will also produce callus on KM medium (9), i.e., the 8p medium described in this paper with NAA = 1.0 mg/l, zeatin = 0.5 mg/l, and 2,4-D = 0.2 mg/l.

The further development of callus normally involves root and shoot development, resulting in the regeneration of whole plants from isolated protoplasts. However, callus proliferation can continue without such organogenesis, and recently we have obtained some evidence that such plant regeneration is under genetic control. *P. parodii* protoplast callus was maintained on MSD3 medium (D3 growth substances, IAA = 2.0 mg/l and BAP = 1.0 mg/l), and root and shoot development was observed and subsequently whole plants were regenerated. *P. hybrida* protoplast callus would not similarly undergo organogenesis on MSD3. However, *P. hybrida* × *P. parodii* protoplasts (F1 hybrid protoplasts) were grown to callus, and when maintained on MSD3, organogenesis took place, and, subsequently whole plant regeneration. The regeneration potential of *P. parodii* on this particular growth substance-media milieu would appear to be genetically dominant.

These results are of particular interest because they indicate that there is a rather precise genetic control by the plant of protoplast and cell division and also of regeneration capability. This rather precise genetic control is not, however, directly correlated with auxins or cytokinins, but with growth regulators and medium combined. With 2,4-D plus 0.5 mg/l 6BAP, or 5.0 mg/l 2,4-D plus levels of 6BAP greater than 0.5 mg/l, no sustained division is observed, contrasting with the sustained division to form callus in F1 medium containing NAA (1.5 mg/l), 6BAP (1.0 mg/l), and 2,4-D (0.5 mg/l). Differential growth effects are therefore not due solely to 2,4-D influences, but there is a medium interaction effect as well. The differential response of *P. hybrida* and *P. parodii* to the growth substance-media milieu has recently been exploited as a component half of the complementation selection of somatic hybrids following the induced fusion of the protoplasts of these two species (14). In addition, differential growth substance responses have also aided the selection of these interspecies somatic hybrids [for a detailed discussion of the use of naturally occurring differential drug sensitivities in the selection of somatic hybrids following the induced fusion of protoplasts, please see (3)]. There is an urgent need for a fuller understanding of the genetic control of the growth and development of cultured plant cell protoplasts and for an assessment of the extent of any correlation with endogenous and exogenous growth substance levels.

References

1. Cocking, E.C.: Nature (London) 191, 780 (1961)
2. Cocking, E.C.: Nature (London) 193, 998 (1962)
3. Cocking, E.C., Power, J.B., Evans, P.K., Safwat, F., Frearson, E.M., Hayward, C., Berry, S.F., George, D.: Plant Sci. Lett. 3, 341 (1974)
4. Evans, P.K., Cocking, E.C.: Isolated Plant Protoplasts in Plant Tissue and Cell Culture, 2nd ed. Street, H.E. (ed.) Oxford: Blackwell Sci. Publ. 1976
5. Frearson, E.M., Power, J.B., Cocking, E.C.: Develop. Biol. 33, 130 (1973)
6. Grout, B.W.W.: Planta 123, 275 (1975)
7. Hall, M.D., Cocking, E.C.: Protoplasma 79, 225 (1974)
8. Hayward, C., Power, J.B.: Plant Sci. Lett. 4, 407 (1975)
9. Kao, K.N., Michayluk, M.R.: Planta 126, 105 (1975)
10. Murashige, T., Skoog, F.: Physiol. Plantarum 15, 473 (1962)
11. Pilet, P.E.: C. R. Acad. Sci. Paris 273, 2253 (1971)
12. Pojnar, E., Willison, J.H.M., Cocking, E.C.: Protoplasma 64, 460 (1967)
13. Power, J.B. *et al.*: 1976, Plant Sci. Lett. 7, 51 (1976)
14. Power, J.B., Frearson, E.M., Hayward, C., George, D., Evans, P.K., Berry, S.F., Cocking, E.C.: Nature (London) 263, 500 (1976)
15. Prat, R.: J. Microscopie 18, 65 (1973)
16. Ruesnik, A.W., Thiman, K.V.: Proc. Natl. Acad. Sci. U.S. 54, 56 (1965)
17. Scowcroft, W.R., Davey, M.R., Power, J.B.: Plant Sci. Lett. 1, 451 (1973)
18. Willison, J.H.M., Cocking, E.C.: Protoplasma 75, 397 (1972)

Paper Demonstration References (see Section IX):

29 - 82 - 90 - 125 - 126 - 144 - 145

IX. Paper Demonstrations

The paper demonstrations are as far as possible quoted in the reports dealing with a nearly similar topic.

1. de Agazio, M. and Toponi, M.A.: Plant Biochemistry and Ecophysiology Laboratory, C. N. R., Rome (Italy) Study of a Soluble Protein Fraction Related to IAA-Treatment in Pea Internodes

2. Aimi, R. and Iwami, S.: Dept. of Biological Sciences, University of Tsukuba, Ibaraki (Japan) Bioelectric Approach on the Hormonal Control of Geotropic Response in the Root of *Vicia faba*

3. Altman, A. and Mozes, R.: Dept. of Horticulture, Faculty of Agriculture, Rehovot (Israel) Root-to-Shoot Cytokinin Transport: its Significance for Hormonal Balance

4. Anderson, P., Brenner, M.L., and Lovrien, R.: University of Minnesota, St. Paul, Minnesota (USA) The Use of Microcalorimetry to Detect Initial Reactions Imparted by Plant Hormones

5. Anderson, J.D. and Lieberman, M.: Beltsville Agricultural Research Center, ARS, USDA, BELTSVILLE, Maryland (USA) Relationship Between IAA-Stimulated Ethylene Evolution and Protein Synthesis in Pea Subhook Sections

6. Antoszewski, R.: Research Institute of Pomology, Skierniewice (Poland) Influence of Abscisic Acid on the Transpiration Rate of Strawberry Leaves ("Physiological Memory" of Stomata)

7. Bagni, N., Audisio, S., and Serafini Fracassini, D.: Institute of Botany, University of Bologna, Bologna (Italy) Polyamine Interactions with Nucleic Acids in Normal and Tumor Tissues of *Nicotiana*

8. Barendse, G.W.M.: Dept. of Botany, University of Nijmegen, Nijmegen (The Netherlands) The Metabolism of Gibberellins in *Pharbitis nil*

9. Bausher, M.G. and Cooper, W.C.: U.S. Dept. of Agriculture, Orlando, Florida (USA) HPLC Separation of Endogenous Growth Regulators in Citrus

10. Bednar, T.W., Linsmaier-Bednar, E.M., and King, C.M.: Div. of Cancer Research, Michael Reese Medical Center, Chicago, Illinois (USA) Peroxidase Oxidation of Indole-3-Acetic Acid: Biological Inactivation and Activation?

11. Bejaoui, M. and Pilet, P.E.: Institute of Plant Biology and Physiology, University of Lausanne, Lausanne (Switzerland) Oxygen Uptake of Growing and Georeactive Roots

12. Benzioni, A. and Itai, C.: R&D Authority and Biology Dept., Ben-Gurion University, Beer-Sheva (Israel) The Effect of Senescence of Preconditioning of Tobacco Leaves to Heat Shock

13. Best, P.H.: Limnological Institute, Nieuwersluis (The Netherlands) The Influence of Abscisic Acid on the Sugar, Starch and Protein Contents of the Submersed Aquatic Macrophyte *Ceratophyllum demersum*

14. Beyer, E.M., Jr.: CR&D dept., Experimental Station, E.I. du Pont de Nemours & Co., Wilmington, Delaware (USA) Silver ion: a Potent Inhibitor of Ethylene Action in Plants

15. Blommaert, K.L.J.: Fruit and Fruit Technology Research Institute, Stellenbosch (Rep. of South Africa) Metabolic Effects of Ethephon (2-chloro-ethyl-phosphonic acid) on the Ripening of Barlinka Grapes

16. Bocchi, A.: Faculty of Veterinary Medecine, University of Parma, Parma (Italy) The Estimation of Animal Virus Multiplication in Cultured Cell Monolayers as a Method for Bioassay of Plant Hormone Action

17. Bocion, P.F. and de Silva, W.H.: Biological Laboratories, Dr. R. Maag Ltd., Dielsdorf (Switzerland) The Interaction of Dikegulac-Sodium and Environmental Factors on Apical Dominance of *Rhododendron simsii* (Planch)

18. Bopp, M., Overlach, U., Erichsen, I., and Knoop, B.: Institute of Botany, University of Heidelberg, Heidelberg (Germany BRD) Receptor Proteins in Cytokinin Competent Cells of the Moss Protonema

19. Bown, A.W., Dymock, I.J., and Hill, B.: Dept. of Biological Sciences, Brock University, St. Catharines, Ontario (Canada) Dark Carbon Dioxide Fixation and the Regulation of Growth by Ethylene and IAA

20. Brenner, M.L., Andersen, C.R., Ciha, A.J., Mondal, M., and Brun, W.: Horticulture Dept., University of Minnesota, St. Paul, Minnesota (USA) Relationship of Endogenous Plant Hormone Content to Changes in Source-Sink Relationships

21. Buchen, B. and Sievers, A.: Botanical Institute, University of Bonn, Bonn (Germany BRD) Secretion and Growth of the Cell Wall During Megasporogenesis of *Selaginella*

22. Buschmann, C. and Lichtenthaler, H.K.: Botanical Institute, University of Karlsruhe, Karlsruhe (Germany BRD) Influence of IAA and Kinetin on Composition and Photosynthetic Activity of Chloroplasts

23. Chang, W.C., Chang, D.Y., and Chiu, P.L.: Institute of Botany, Academia Sinica, Taipei (Taiwan) Effects of DPX-1840 on the Gravimorphism and the Transport of IAA-2-^{14}C in Mungbean Hypocotyl Segments

24. Cocucci, M., Ballarin, A., and Scacchi, S.: Institute of Botanical Sciences, C.N.R., University of Milano, Milano (Italy) Characteristics of FC-Induced Changes of Transmembrane Potential in Maize Roots

25. Cocucci, S. and Cocucci, M.: Institute of Botanical Sciences, C.N.R., University of Milano, Milano (Italy) Effects of FC, GA_3, and ABA on the Development of K^+ Uptake in Germinating Radish Seeds

26. Collet, G.F.: Federal Station of Agronomic Research of Changins, Nyon (Switzerland) Activity on the Growth of Plant Growth Regulators and Their Glucosides

27. Coombe, B.G.: Waite Agricultural Research Institute, University of Adelaide, Glen Osmond (Australia) Abscisic Acid and Sugar Accumulation in the Grape Berry

28. Coombe, B.G., Mander, L.N., and Turner, J.V.: Waite Agricultural Research Institute, University of Adelaide, Glen Osmond (Australia) Molecular Structure and Gibberellin Activity: a Systematic Study Using Synthetic Analogues

29. Cuendet, P. and Pilet, P.E.: Institute of Plant Biology and Physiology, University of Lausanne, Lausanne (Switzerland) Kinetics of Auxin Uptake by Protoplast Suspensions

30. Curtis, R.W.: Botany and Plant Pathology, Purdue University, W. Lafayette, Indiana (USA) Phytochrome Involvement in Malformin Action

31. Digby, J., Firn, R.D., and Pinsent, C.: Biology Dept., University of York, York (United Kingdom) The Role of Gibberellins in Inducing the Differential Growth associated with Tropic Curvature

32. Dobozy, O.K.: United Chemical Works, Budapest (Hungary) Cationic Surfactants as Plant Growth Regulators

33. Doerffling, K. and Boettger, M.: University of Hamburg, Hamburg (Germany BRD) Physiology and Chemistry of Abscission Accelerators in *Coleus* Petioles

34. Dunberg, A.: The Royal College of Forestry, Umea (Sweden) Metabolism of Gibberellins in Norway Spruce (*Picea abies*)

35. Durley, R.C. and Kannangara, T.: Crop Science Dept., University of Saskatchewan, Saskatoon, Saskatchewan (Canada) Use of High Pressure Liquid Chromatography for the Analysis of Plant Growth Regulators in *Sorghum bicolor* (L.) Moench

36. Eliasson, L. and Tillberg, E.: Dept. of Plant Physiology, University of Umea, Umea (Sweden) Interaction between Abscisic Acid and Auxins in Regulation of the Growth of *Lemna gibba* L.

37. Elliott, D.C.: School of Biological Sciences, Flinders University of South Australia, Bedford Park (Australia) Cytokinine Action in Enzyme Induction is a Membrane-Related Phenomenon

38. Esquerre-Tugaye, M.T.: Center of Plant Physiology, University Paul Sabatier, Toulouse (France) Cell Wall Glycoproteins and Enzymes as Related to the Physiological Age of the Plant

39. Evans, M.: Dept. of Botany, Ohio State University, Columbus, Ohio (USA) Effects of Auxin, Acidic pH and 3-Methyleneoxindole on Intact Root Elongation

40. Even Chen, Z. and Itai, C.: Ben Gurion University of the Negev, Beer Sheva (Israel) The Role of Ethylene in Senescence

41. Firn, R.D. and Digby, J.: Dept. of Biology, University of York, York (England) Plant Shoot Geotropism - a New Hypothesis

42. Fletcher, R.A. and Poku, J.A.G.: Dept. of Environmental Biology, University of Guelph, Guelph, Ontario (USA) Cytokinins in Relation to Chlorophyll Production in Aging Cucumber Cotyledons

43. Foy, C.L. and Chang, I.K.: Dept. of Plant Pathology and Physiology, Virginia Polytechnic Institute and State University, Blacksburg, Virginia (USA) Studies on the Mode of Action of Picloram, an Auxin

44. Ganguly, S.N., Saha, P.K., and Sircar, S.M.: Bose Institute, Calcutta (India) Studies on the Seed of *Shorea robusta* Gaertn. in Relation to Growth Regulators

45. Gardner, G., Sussman, M.R., and Kende, H.: MSU/ERDA Plant Research Laboratory, Michigan State University, East Lansing, Michigan (USA) Binding of Cytokinins to Particulate Fractions from Protonema of the Moss *Funaria hygrometrica*

46. Gaspar, Th., Coumans, M., and Ceulemans, E.: Institute Van Beneden, University of Liège, Liège (Belgium) Abscisic Acid in the Sugarbeet Fruit. Role in Stabilized Dormancy

47. Geuns, J.M.C.: University of Louvain, Louvain (Belgium) Physiological Activity of Corticosteroids in Higher Plants

48. Gogala, N., Stegnar, P., and Pohleven, F.: Institute of Biology and "Josef Stefan" Institute, University of Ljubljana, Ljubljana (Yugoslavia) Cytokinins in the Root Exudate and the Ion Uptake in Mycorrhiza

49. Goldschmidt, E.E., Rubin, A., and Zamski, E.E.: Dept. of Horticulture, The Hebrew University of Jerusalem, Rehovot (Israel) Are Chloroplast-Membrane Components Actually Degraded During Senescence?

50. Goren, R.: Fac. of Agriculture, The Hebrew University of Jerusalem, Rehovot (Israel) Anatomical, Hormonal, and Enzymic Aspects of Abscission in *Citrus*

51. Göring, H.: Dept. of Biology, Humboldt-University, Berlin (Germany DDR) Relations between Cytokinine Effect and K/Ca-Ratio in Tissues of Higher Plants

52. Gressel, J., Kadouri, A., Atsmon, D., and Cohen, N.: Dept. of Plant Genetics, Weizmann Institute of Sciences, Rehovot (Israel) Effects of Dikegulac-Sodium on Nucleic Acid Synthesis

53. Guignard, R. and Pilet, P.E.: Institute of Plant Biology and Physiology, University of Lausanne, Lausanne (Switzerland) Relation between Cellulase Activity and the Degree of Polymerisation of its Substrate

54. Günther, G., Tewes, A., and Heidemann, M.: Dept. of Botany, Teachers College Potsdam, Potsdam (Germany DDR) Growth Intensity and Regulation in Cell and Tissue Cultures form Weeds

55. Hahn, H., Heitmann, I., and Blumbach, M.: Botanical Institute, University of Bonn, Bonn (Germany BRD) The Biogenesis of ^{6}N-(Δ^{2}-Isopentenyl)Adenine in *Agrobacterium tumefaciens* B6

56. Hall, M.A., Darvill, A.G., and Smith, C.J.: Dept. of Botany and Microbiology, University College of Wales, Aberystwyth (United

Kingdom) Cell Wall Structure in *Zea mays* in Relation to Auxin-Induced Growth

57. Hashimoto, T. and Wakimori, H.: Institute for Physical and Chemical Research, Saitama-Ken (Japan) Effect of Batatasins on Auxin Transport in Oat Coleoptiles

58. Hashizome, T., Sakai, S., and Sugiyama, T.: Laboratory of Bioorganic Chemistry, Tokyo University of Agriculture and Technology, Tokyo (Japan) Cytokinin Activity of ^{6}O-Substituted Guanine and Hypoxanthine Derivatives

59. Hattori, H. and Marumo, S.: Dept. of Agricultural Chemistry, Nagoya University, Nagoya (Japan) Biological Activities of a Plant Growth Promotor, (-)-cis-Sativenediol, Produced by Fungi

60. Hedden, P. and Phinney, B.O.: Dept. of Biology, University of California, Los Angeles, California (USA) The Dwarf-5 Mutant of *Zea mays*: a Genetic Lesion Controlling the Cyclization Step (B Activity) in Kaurene Biosynthesis

61. Heimann-Matile, J. and Pilet, P.E.: Institute of Plant Biology and Physiology, University of Lausanne, Lausanne (Switzerland) Lipase Activity on Growing Roots

62. Hensel, W. and Sievers, A.: Botanical Institute, University of Bonn, Bonn (Germany BRD) Influence of the Slowly Rotating Clinostat on the Polarity of Statocytes and on the Gravitropic Reaction of Roots

63. Hertel, R., Dohrmann, U., Jesaitis, A.J., and Petersen, W.: Institute of Biology III, University of Freiburg, Freiburg i. Br. (Germany BRD) Various Receptor Sites in Membrane Fractions from Corn Coleoptiles: in vitro Binding of Auxins, Riboflavin, and a Naturally Occurring Analog of Morphactin

64. Hofer, R.-M., Schlienger, C., and Pilet, P.E.: Institute of Plant Biology and Physiology, University of Lausanne, Lausanne (Switzerland) Analysis of the Surface of Growing and Geostimulated Roots

65. Holtz, J.: Botanical Institute, University of Bonn, Bonn (Germany BRD) Δ^2-Isopentenylation of Oligo A-Nucleotides: an Additional Pathway of Cytokinine Biosynthesis?

66. Homes, J.: University of Brussels, Brussels (Belgium) Induction of Root Hairs from Maize Roots and Abscisic Acid

67. Homes, J.: University of Brussels, Brussels (Belgium) Auxin Effect and the Formation of Bud Cells

68. Horgan, R.: Dept. of Botany and Microbiology, University College of Wales, Aberystwyth (United Kingdom) The Metabolism of Cytokinins in Callus Cultures and Cell Suspension Cultures of Soybean

69. Horton, R.F.: Dept. of Botany and Genetics, University of Guelph, Guelph, Ontario (Canada) The Regulation of Stem Abscission in Succulents

70. Ihl, M.: Institute of Food Sciences and of Food Technology, University of Chile, Santiago (Chile) Primary Action of IAA on Cotyledon Tissues

71. Imaseki, H., Kondo, K., Odawara, S., and Watanabe, A.: Research Institute for Biochemical Regulation, Faculty of Agriculture, Nagoya University, Nagoya (Japan) Action of Plant Hormones in Ethylene Production

72. Iversen, T.-H. and Hestnes, A.: Dept. of Botany, The University of Trondheim, Trondheim (Norway) Movements of Cell Organelles and the Redistribution of Gibberellins in Geotropically Stimulated Plant Root Cells

73. Jaccard, M. and Pilet, P.E.: Institute of Plant Biology and Physiology, University of Lausanne, Lausanne (Switzerland) Creep-Relaxation of Collenchyma Cells: Change of Rheological Parameters During Growth and Differentiation

74. Jindal, K.K.: Botanical Institute, University of Stockholm, Stockholm (Sweden) Endogenous Gibberellins in Male and Female Papaya Plants During Floral Ontogeny

75. Johri, M.M. and Avtar Krishan Handa: Molecular Biology Unit, Tata Institute of Fundamental Research, Bombay (India) Hormonal Balance and Cell Differentiation in the Protonema of the Moss *Funaria hygrometrica*

76. Jotterand-Dolivo, M.C. and Pilet, P.E.: Institute of Plant Biology and Physiology, University of Lausanne, Lausanne (Switzerland) Wall Hydroxyproline in the Bending Zone of Georeactive Maize Roots

77. Jouanneau, J.-P., Gandar, J.-C., and Peaud-Lenoel, C.: Functional Plant Biochemistry, University Center of Luminy, Marseille (France) Covalent Linkage of Benzyladenine Nucleotide in Tobacco RNA

78. Kamisaka, S. and Shibata, K.: Dept. of Biology, Faculty of Sciences, Osaka City University, Osaka (Japan) Regulatory Role of Phenylpropanoids in Gibberellic Acid-Stimulated Lettuce Hypocotyl Elongation

79. Kang, B.G.: Dept. of Biology, Yonsei University, Seoul (Korea) Differential Distributions of Amyloplasts and Golgi Bodies in Geotropically Stimulated Pea Stems with Regard to Ethylene Action

80. Karssen, C.M.: Dept. of Plant Physiology, Agricultural University, Wageningen (The Netherlands) Two Sites of Hormonal Action During Germination of *Chenopodium album* Seeds

81. Kasamo, K.: Institute for Plant Virus Research, Chiba (Japan) Cancellation of Kinetin-Induced Retardation of Chlorophyll Degradation in Tobacco Leaf Discs During Senescence by Removal of the Lower Epidermis

82. Katsumi, M.: Dept. of Biology, International Christian University, Tokyo (Japan) Inhibition by DCCD of Endogenous Growth, IAA-Induced Elongation and ATPase Activity of Cucumber Hypocotyl Sections and its Prevention by GA_3 and some Adenine Nucleotides

83. Keller, E. and Pilet, P.E.: Institute of Plant Biology and Physiology, University of Lausanne, Lausanne (Switzerland) Action of a Morphactin on the Contents of Total Nitrogen and Some Amino Acids of Young *Lens* Roots

84. King, P.J.: Friedrich Miescher Institute, Basel (Switzerland) Utilisation of 2.4-dichlorophenoxyacetic Acid (2.4-D) by Steady-

85. Knypl, J.S.: Laboratory of Plant Growth Regulators, University of Lodz, Lodz (Poland) Induction of Phosphatase and Ribonuclease Isoenzymes by N,N-Dimethyl-morpholinium Chloride in *Spirodela oligorrhiza*

86. Konings, H.: Botanical Laboratory, University of Utrecht, Utrecht (The Netherlands) Differences in the Auxin-Ethylene Interrelationship between Roots of Mono- and Dicotyledonous Species

87. Koshimizu, K., Hayashi, H., and Asahira, T.: Dept. of Food Science and Technology, Faculty of Agriculture, Kyoto University, Kyoto (Japan) Substances with Cytokinin Activity in Tomato Fruits

88. Kuraishi, S.: Dept. of Biology, College of General Education, University of Tokyo, Tokyo (Japan) Synthesis of Indole-3-Acetic Acid from Indole in *Avena* Coleoptiles

89. Kutacek, M., Kefeli, V., Vackova, K., and Machackova, I.: Institute of Experimental Botany, CSAV, Prague (Czechoslowakia) Natural Sinapic Acid Derivatives in Kohlrabi-Sinapine and 1-O-Sinapoyl-β-D-Glucose, Their Biological Activity

90. Lado, P., Rasi-Caldogno, F., Colombo, R., and de Michelis, M.I.: Institute of Botanical Sciences, University of Milano, Milano (Italy) Effects of Fusicoccin and Auxin on the Active Uptake of Sugars, Amino Acids and Inorganic Anions

91. Laloue, M., Gawer, M., Terrine, C., and Guerin, J.: Laboratory of Plant Physiology, P. and M. Curie University, Paris (France) Significance of Cytokinin 7-Glucoside and Cytokinin Nucleotides to the growth of Tobacco Cell Suspensions

92. Larsen, P. and Iversen, T.-H.: Dept. of Botany, The University of Trondheim, Trondheim (Norway) The Effect of Abscisic Acid (ABA) and 3,5-diiodo-4-hydroxybenzoic acid (DIHB) on Elongation and Geotropic Curvature of *Lactuca* Seedlings

93. Lavee, S. and Epstein, E.: Institute of Horticulture, Volcani Center, Bet Dagan (Israel) Metabolic Differences in Degradation of ^{14}C-IAA in Apple and Olive Tissue Cultures

94. Lee, T.T.: Research Institute, Agriculture Canada, London, Ontario (Canada) Mechanism of Action of Phenolic Inhibitors in Enzymic Degradation of Indole-3-Acetic Acid

95. Leguay, J.J.: P. and M. Curie University, Paris (France) Regulation by 2,4-D of the Number of Cell Generations in *Acer Pseudoplatanus* Cell Population. Influence of 2,3,5 Triiodobenzoic Acid

96. Leshem, Y., Gilboa-Garber, N., and Mizrahi, L.: Dept. of Life Sciences, Bar-Ilan University, Ramat-Gan (Israel) GA and Permeability Effects on Biological Membranes

97. Leshem, Y., Zieslin, N., Haveli, A.H., and Spiegelstein, H.: Dept. of Life Sciences, Bar-Ilan University, Ramat-Gan (Israel) Promotive Effect of GA on Anthocyanidin Synthesis : PAL Activation or Membrane Associated Effects?

98. Lewis, L.N., Koehler, D.E., Leonard, R.T., Vanerwoude, W.J., and Linkins, A.E.: University of California, Riverside (California) Association of Latent Cellulase Activity with Plasma Membranes from Kidney Bean Abscission Zones

99. Longo, C.P., Longo, G.P., Rossi, G., and Servettaz, O.: Institute of Botanical Sciences, University of Milano, Milano (Italy) The Effect of Benzyladenine on Mitochondria and Microbodies in Excised Cotyledons of Sunflower and Watermelon

100. Lörz, H., Harms, C.T., and Potrykus, I.: Max-Planck-Institute for Plant Genetics, Ladenburg/Heidelberg (Germany BRD) Callus Cultures of Maize and Oat with Respect to Plant Regeneration

101. Mapelli, S., Frova, C., Torti, G., and Soressi, G.P.: Laboratory of Plant Biosynthesis, C.N.R., Milano (Italy) Relative Level of Endogenous Plant Hormones in Parthenocarpic Tomato Mutant During Development

102. Marcelle, R.: Laboratory of Plant Physiology, Research Station of Gorsem, Sint-Truiden (Belgium) The Effect of Growth Regulators on the Greening and the Rooting of Bean Plants

103. Margaris, N.S., Karamanos, A., and Adamantiadou, S.: Institute of General Botany, University of Athens, Athens (Greece) Changes of some Soluble Constituents During the Release from Apical Dominance

104. Martinez-Honduvilla, C.J. and Santos-Ruiz, A.: Dept. of Biochemistry, Faculty of Pharmacy, University of Madrid, Madrid (Spain) Pine Seed Inhibitors: Effect of Heat and H_2O_2 Treatments on Seed Coat

105. Mayor, G. and Pilet, P.E.: Institute of Plant Biology and Physiology, University of Lausanne, Lausanne (Switzerland) Longitudinal and Transversal Distribution of Acid Phosphomonoesterase in Maize Roots

106. Menschick, R. and Hild, V.: Botanical Institute, University of Münster, Münster (Germany BRD) Decarboxylation of IAA in Coleoptiles in Relation to Phototropism

107. Miginiac, E., Fontaine, D., and Sotta, B.: Phytotron Laboratory, CNRS, Gif-sur-Yvette (France) The Influence of Roots and Cytokinins on the Suppression of Flowering of Photoperiodic Plants

108. Minocha, S.C.: Dept. of Botany, University of New Hampshire, Durham (USA) Cytokinin Metabolism in Relation to Differentiation of Tracheids in Cultured Tuber Tissue of Jerusalem Artichoke (*Helianthus tuberosus*)

109. Monselise, S.P., Erner, Y., and Goren, R.: Dept. of Horticulture, Hebrew University of Jerusalem, Rehovot (Israel) The Rough Fruit Condition of the Shamouti Orange. An Example of Hormonal Control of Tissue Development

110. Moore, T.C.: Dept. of Botany and Plant Pathology, Oregon State University, Corvallis, Oregon (USA) Compartmentation of Gibberellin (Ent-Kaurene) Biosynthesis in Chloroplasts

111. Morgan, P.W., Davenport, T.L., and Jordan, W.R.: Dept. of Plant Sciences, Texas A&M University, College Station, Texas (USA) Whole Plant Regulation of Water Stress-Induced Leaf Abscission

112. Muir, R.M.: Botany Dept., University of Iowa, Iowa (USA) Ethylene Production by Epicotyl Tissue of the Pea

113. Musatenko, L.I., Boichuk, O.B., and Sytnik, K.M.: Institute of Botany, Academy of Sciences, Kiev (Ukrainian SSR) Phytohormones of Primary *Phaseolus* Leaves

114. Nash, L.J., Reeve, D.R., and Crozier, A.: Dept. of Botany, University of Glasgow, Glasgow (Scotland, U.K.) Gibberellin Metabolism in Seedlings of *Phaseolus coccineus* L

115. Neskovic, M., Grubisic, D., and Konjevic, R.: Institute of Botany, Faculty of Science, Beograd (Yugoslavia) Phototropism of Dwarf Pea Stems in Relation to the Metabolism of Hormones

116. Nevins, D.: Dept. of Botany, Iowa State University, Ames, Iowa (USA) An Evaluation of Nojirimycin as an Inhibitor of IAA-Induced Growth

117. Nicolas, G. and Zarra, I.: Dept. of Plant Physiology, Faculty of Sciences, University of Salamanca, Salamanca (Spain) Induction of Rice Coleoptile Elongation by Mercaptoethanol. SH Groups and Disulphide Reductase Activity

118. Oostrom, H. and van Loopik-Detmers, M.A.: Dept. of Plant Morphogenesis, Botanical Laboratory, Leiden (The Netherlands) Partial Purification of a High Affinity Receptor for Indole-3-Acetic Acid

119. Osborne, D.J., Milborrow, B.V., Wright, M., Mallaby, R., and Roberts, J.A.: Agricultural Research Council Unit of Developmental Botany, Cambridge (England) Differentiation of Abscission Zones and the Role of Ethylene and Senescence Factor in the Separation Process

120. Paton, D.M., Osawa, T., Willing, R.R., Mosedale, J., and Crow, W.D.: Dept. of Botany, Australian National University, Canberra (Australia) Inhibitor-Promotor Balance in *Eucalyptus*

121. Paulet, P.: Dept. of Plant Physiology, University of Orléans, Orléans (France) The Regulation of Flowering on *Cichorium intybus* L. Root Tissues Cultured in vitro

122. Paunovic, A.S. and Jankovic, R.: Fruit Research Institute, Cacak (Yugoslavia) Effect of SADH and CCC on Growth and Flower Bud Formation in Nursery Trees of *Plum* cv. Pozegaca

123. Paunovic, A.S., Ogasanovic, D., Plazinic, R., and Vukmirovic, M.: Fruit Research Institute, Cacak (Yugoslavia) Effects of Off-Shoot-O and Off-Shoot-T on Preventing Lateral Shoot Growth on Maiden Trees of Apples, Peaches and Stolon of Strawberry Plants

124. Pernet, J.J., Cherix-Verrey, M.C., and Pilet, P.E.: Institute of Plant Biology and Physiology, University of Lausanne, Lausanne (Switzerland) Basipetal Movement of Radioactivity from (5-^{3}H)-IAA in Maize Roots: Action of Fusicoccin on Decapped and Intact Apical Segments

125. Poovaiah, B.W. and Leopold, A.C.: Dept. of Horticulture, Washington State University, Pullman, Washington (USA) Effects of Inorganic Solutes on Growth and Auxin Binding

126. Potrykus, I., Harms, C.T., and Lörz, H.: Max-Planck-Institute for Plant Genetics, Ladenburg/Heidelberg (Germany BRD) Methods for Large-Scale Screening of the Hormonal, Nutritional, and Environ-

mental Conditions Which Affect Cell Division and/or Differentiation in Single-Cell Cultures. Combination of Hanging-Droplets and Dilution Series

127. Przybecka, D. and Pilet, P.E.: Institute of Plant Biology and Physiology, University of Lausanne, Lausanne (Switzerland) DNA Distribution in Intact and Decapped Maize Roots

128. Raafat, A. and Sadeghian, E.: Dept. of Agricultural Botany, Faculty of Agriculture, Ain Shams University, Cairo (Egypt) Endogenous IAA and GA_3 Levels in Bean Plant as Influenced by Senescence Retardation of Primary Leaves Created by Benzyladenine or Disbudding Treatment

129. Rajagopal, R.: Dept. of Biology, Carleton University, Ottawa (Canada) Effect of TMV Infection on the Endogenous Cytokinins of Tobacco Leaves in Various Stages of Development

130. Phipps, J. and Wightman, F.: Dept. of Biology, Carleton University, Ottawa (Canada) Influence of Virus Infection on the Levels of Endogenous Growth Regulators in Tobacco Leaves at Various Stages of Development

131. Raschke, K.: MSU/ERDA Plant Research Laboratory, Michigan State University, East Lansing, Michigan (USA) Effects of Stress on Photosynthesis Possibly Mediated by Abscisic and Phaseic Acids

132. Rasmussen, G.K.: U.S. Dept. of Agriculture, Orlando, Florida (USA) Influence of Growth Regulators on Maturation of *Citrus* Fruit

133. Rivier, L., Milon, H., and Pilet, P.E.: Institute of Plant Biology and Physiology, University of Lausanne, Lausanne (Switzerland) Quantitative Determinations of ABA in the Cap and Apex of Maize Root by Gas Chromatography-Mass Spectrometry

134. Ruddat, M. and Bensen, P.L.: The University of Chicago, Chicago, Illinois (USA) Regulation of Enzyme Activity of the Aleurone Grain-Spherosome Complex in Wheat Aleurone Tissue by Gibberellin

135. Rudnicki, R.M., Saniewski, M., and Nowak, J.: Research Institute of Pomology, Skierniewice (Poland) The Growth of Inflorescence Stalk in Hycinth Related to Hormonal Distribution

136. Sachs, R.M. and Hackett, W.P.: Dept. of Environmental Horticulture, University of California, Davis, California (USA) Relating Gibberellin-Induced Inhibition of Flowering to Requirement for Far-Red Radiation and Promotion by Removal of Young Leaves

137. Sakurai, N.J., Nevins, D.J., and Masuda, Y.: Dept. of Biology, Faculty of Science, Osaka City University, Osaka (Japan) Auxin- and Hydrogen Ion-Induced Elongation and Physical and Chemical Changes in the Cell Wall of *Avena* Coleoptiles

138. Sangwan, R.S.: Laboratory of Botany, University of Paris VII, Paris (France) Hormonal Regulation of Totipotency in *Petunia* Cell and Pollen Cultures in vitro

139. Sankhla, N., Sankhla, D., and Huber, W.: University of Jodhpur, India and Technical University, Munich (Germany BRD) Role of Bioregulants in Enzyme Activity with Special Reference to Abscisic Acid and Farnesol

140. Schlienger, C., Hofer, R.-M., and Pilet, P.E.: Institute of Plant Biology and Physiology, University of Lausanne, Lausanne (Switzerland) Auxin Effect on Growing Coleoptiles and Ultrastructural Analysis of Walls

141. Schmid, A. and Meier, H.: University of Fribourg, Fribourg (Switzerland) Studies on the Natural Inhibition of Adventitious Root Formation on Woody Cuttings of Aspen (*Populus tremula* L.)

142. Scott, T.K. and Edwards, K.L.: Dept. of Botany, University of North Carolina, Chapel Hill, North Carolina (USA) Rapid Growth Responses of Corn Root Segments: Effect of Auxin on Elongation

143. Sembdner, G., Knöfel, H.D., Müller, P., and Kramell, R.: Institute of Plant Biochemistry, Research Center for Molecular Biology and Medecine, Academy of Sciences of the GDR, Halle (Saale) (Germany DDR) Gibberellin Affinity Chromatography

144. Senn, A., Pilet, P.E., and Zryd, J.P.: Institute of Plant Biology and Physiology, University of Lausanne, Lausanne (Switzerland) Electrophoretic Mobilities of NAA-Treated *Nicotiana sylvestris* Protoplasts

145. Senn, A. and Zryd, J.P.: Institute of Plant Biology and Physiology, University of Lausanne, Lausanne (Switzerland) Production and Culture of Mesophyll Protoplasts from "in vitro" Cultured Plantlets of *Nicotiana sylvestris*

146. Shen-Miller, J., McNitt, R.E., and Wojciechowski, M.: Argonne National Laboratory, Argonne, Illinois (USA) Light and Electron Microscopy of Georesponding Corn Roots

147. Shinozaki, M.: Applied Botany, Faculty of Agriculture, Kyoto University, Kyoto (Japan) Comparative Studies of Long-Day Flowering by Continuous High-Intesity Light and CCC in *Pharbitis nil*

148. Silk, M.W.K.: Dept. of Biology, University of Pennsylvania, Philadelphia, Pennsylvania (USA) Gibberellin A_1 Metabolism and Hypocotyl Growth

149. Simonin, P.A. and Pilet, P.E.: Tate and Lyle Ltd., Philip Lyle Memorial Research Laboratory, Reading (England) α-Amylase Activity in GA_3-Treated Lentil Roots

150. Simpson, G.M.: Crop Science Dept., University of Saskatchewan, Saskatoon, Saskatchewan (Canada) Hormonal Control of Metabolism in Dormant Seeds of *Avena fatua*

151. Sobick, V., Sievers, A., Volkmann, D., and Briegleb, W.: Botanical Institute, University of Bonn, Bonn (Germany BRD) Gravireaction and Ultrastructure of Roots in Simulation of Weightlessness

152. Spencer, M. and Phillips, D.R.: Dept. of Plant Biochemistry, University of Alberta, Edmonton, Alberta (Canada) Ethylene and Mitochondrial Adenosine Triphosphatase

153. Sponsel-Frydman, V.M. and MacMillan, J.: School of Chemistry, The University of Bristol (United Kingdom) The Metabolism of Gibberellins in Immature Seeds of *Pisum sativum* cv. Progress No. 9

154. Stange, L.: Institute of Plant Physiology, University of Kassel, Kassel (Germany BRD) Auxin and Meristeme Differentiation in *Riella helicophylla*

155. Steck, P. and Pilet, P.E.: Institute of Plant Biology and Physiology, University of Lausanne, Lausanne (Switzerland) ATP, ADP and AMP Changes in Growing Roots

156. Stoddart, J.L. and Foster, C.A.: Welsh Plant Breeding Station, Aberystwyth (United Kingdom) "Bakanae barley" - a New Mutant of *Hordeum vulgare* L. with an Accelerated Growth Rate

157. Stonier, T.: School of Science and Society, University of Bradford, Bradford (England) The Role of Auxin Protectors in Cell Division

158. Stowe, B.B.: Kline Biology Tower, Yale University, New Haven, Connecticut (USA) Molecular Characteristics of Oleanimin Growth Substances

159. Fujii, T. and Shimokoriyama, M.: Dept. of Botany, Faculty of Science, University of Tokyo, Tokyo (Japan) Roles of Hydroxyproline-Protein and of Auxin in Oxygen-Sensitive Growth

160. Takahashi, N., Murofushi, N., and Yamane, H.: Dept. of Agricultural Chemistry, University of Tokyo, Tokyo (Japan) Metabolism of Gibberellins in Maturing and Germinating Bean Seeds

161. Takeuchi, S., Sanai, S., Ota, Y., Okuda, S., and Kawarada, A.: Institute of Physical and Chemical Research, Saitama (Japan) The Favourable Effect of Nicotinamide on the Plant Growth

162. Tanimoto, E. and Igari, M.: Biological Laboratory, College of General Education, Nagoya City University, Nagoya (Japan) Auxin-Induced Enhancement of Glycosidase Activities in Pea Stems and Artichoke Tubers

163. Teissere, M., Penon, P., and Ricard, J.: Laboratory of Plant Biochemistry, Marseille (France) Effect of 2,4-Dichlorophenoxyacetic Acid Treatment upon the Level of DNA-Dependent RNA Polymerase Activities from Wheat Germs

164. Teyssendier de la Serve, B. and Peaud-Lenoel, C.: Laboratory of Plant Biochemistry, Universitary Center of Luminy, Marseille (France) Potential Turn-Over of Tobacco Cell RNA, as a Source of Natural Cytokinins

165. Theimer, R.R.: Botanical Institute, University of München, München (Germany BRD) Interactions of Kinetin and Light During Microbody Development in Cotyledons of Sunflower (*Helianthus annuus* L.) Seedlings

166. Thomas, T.H., Biddington, N.L., and Palevitch, D.: National Vegetable Research Station, Warwick (England) Hormonal Regulation of Celery and Wheat Seed Germination

167. Thorpe, T.A. and Halmer, P.: Dept. of Biology, University of Calgary, Calgary, Alberta (Canada) Growth Regulator-Induced Changes in Cell Wall Polysaccharides in Tobacco Callus

168. Truelsen, T.A. and Wyndaele, R.: Institute of Plant Physiology, University of Arhus, Arhus (Denmark) Cyclic Adenosine 3' : 5'-Monophosphate in *Helianthus callus*

169. Tucker, D.J.: Glasshouse Crops Research Institute, Littlehampton (United Kingdom) Hormonal Control of Side Shoot Growth in the Tomato

170. Vandendriessche, T.: Dept. of Molecular Biology, Free University of Brussels, Brussels (Belgium) Circadian Rhythms and Plant Growth Regulation

171. Vanderhoven, C. and Zryd, J.P.: Institute of Plant Biology and Physiology, University of Lausanne, Lausanne (Switzerland) CO_2 Effects on Sycamore Cell Growth

172. Van Loon, L.C.: Dept. of Plant Physiology, Agricultural University, Wageningen (The Netherlands) Hormone-Mediated Changes in Symptom Expression in Virus-Infected Tobacco Plants During Growth and Senescence

173. Van Onckelen, H.A., Caubergs, R., Van Hoof, R., and De Greef, J.A.: Dept. of Biology, University of Antwerpen, Antwerpen (Belgium) Influence of Light and Endogenous Hormone Balance on the Amylolytic Activity in Cotyledons of *Phaseolus vulgaris*

174. Vardar, Y. and Baltepe, S.: Dept. of Botany, University of Ege, Izmir (Turkey) Some Experimental Results on the Growth and Hormonal Regulation in Lettuce in Relation to Light

175. Parkash, V.: Indian Agricultural Research Institute, New Delhi (India) Effects of Cytokinins and Auxin on the Induction of Nitrate Reductase Activity in Cotyledons and Hypocotyls of Fenugreek (*Trigonella foenum-graecum* L.) in Relation to Their Growth Responses

176. Veen, H. and Frissel, M.J.: Centre for Agrobiological Research, Wageningen (The Netherlands) Simulation Model on Hormone Transport

177. Venketeswaran, S. and Hemphill, J.K.: Dept. of Biology, University of Houston, Houston, Texas (USA) Growth Studies and Pigment Analysis of Three Cultured Soybean Phenotypes, *Glycine max* (L.) Merrill

178. Vieitez, E. and Vieitez, A.: Dept. of Biology, University and Plant Physiology, C.S.I.C., Santiago de Compostela (Spain) Effect of Synthetic Hormones and Etiolation on Rootability in Chestnut

179. Volkmann, D.: Botanical Institute, University of Bonn, Bonn (Germany BRD) The Development of the Polar Arrangement of Organelles in Statocytes of Cress Roots

180. Whittingham, C.P., Lenton, J.R., Milford, G.F.J. and Pocock, T.O.: Botany Dept., Rothamsted Experimental Station, Harpenden (United Kingdom) The Role of Growth Regulators in Sugar Beet Development

181. Wielgat, B. and Kahl, G.: Dept. of Biology, University of Frankfurt, Frankfurt (Germany BRD) In vivo and in vitro RNA Synthesis in Potato Tuber Slices Under Hormonal Control

182. Wilkins, H. and Wilkins, S.M.: A.R.C. Unit, Wye College, University of London, Ashford (England) Growth Promotion of Mechanically Stressed Seedlings

183. Wright, S.T.C.: A.R.C. Unit, Wye College, University of London, Ashford (England) The Relationship Between Leaf Water Potential (Ψ_{leaf}) and the Levels of ABA and Ethylene in Excised Wheat Leaves

184. Wurgler, W.: Federal Station of Agronomic Research of Changins, Nyon (Switzerland) Studies on Absorption and Translocation of Glyphosate

185. Yamaki, T., Nishinari, N., Yoshizawa, K., Ichimura, H., Namekawa, K., and Shimojo, E.: College of General Education, University of Tokyo, Tokyo (Japan) Regulation of Cell Division by IAA in Plants and Animals

186. Yamamoto, R.: Dept. of Biology, Faculty of Science, Osaka City University, Osaka (Japan) Comparison of Sugar Compositions of the Cell Wall of Mono- and Dicotyledonous Plants

187. Yamada, Y., Yajima, Y., and Yasuda, T.: Dept. of Agricultural Chemistry, Kyoto University, Kyoto (Japan) The Relationship Between 2,4-D and DNA Synthesis in Callus Induction from Jerusalem Artichoke Tuber Tissue

188. Zatyko, J.M.: Horticultural Research Station, Fertöd (Hungary) Root-Promoting Effect of Distilled Water

189. Zeevart, J.A.D.: MSU/ERDA Plant Research Laboratory, Michigan State University, East Lansing, Michigan (USA) Photoperiodic Control of Growth and Flower Formation in *Bryophyllum* as Mediated by Gibberellins

190. Zmrahl, Z. and Machackova, I.: Institute of Crop Production, Prague (Czechoslovakia) Regulation of Phenolics Metabolism by Indole-3-Acetic Acid in Wheat

191. Zryd, J.P.: Institute of Plant Biology and Physiology, University of Lausanne, Lausanne (Switzerland) Sugars and Growth of Sycamore Cells at Low Cell Density

192. Cleland, C.F.: Smithsonian Insitution Radiation Biology Laboratory, Rockville, Maryland (USA) Synergistic Interaction of Red Light and Cytokinin in Promotion of Growth and Flowering in *Lemna gibba* G_3

193. Cleland, R.E.: Botany Dept., University of Washington, Seattle, Washington (USA). Hormone-Induced H^+-Excretion and K^+-Uptake

194. Degani, N. and Itai, C.: Nuclear Research Centre Negev, Beer Sheva (Israel) Growth and ABA in Irradiated Wheat Seedlings

195. Fensom, D.: Mount Allison University, Sackville (Canada) Effects of Auxin and Membrane Conditioners on Electro-Osmosis and Transcellular Water Flow in *Nitella*

196. Garg, O.P. and Swaraj, L.: Haryana Agricultural University, Hissar (India) The Role of Ascorbic Acid in Symbiotic Nitrogen Fixation

197. Goldthwaite, J., Manos, P., and Perdrizet, G.: Dept. of Biology, Boston College, Chestnut Hill, Massachusetts (USA) Assessing the Importance of Gibberellins, Cytokinins and Abscisic Acid as Senescence Regulation Hormones

198. Hillman, J.R., Colqhoun, A.J., and McDougall, J.: Botany Dept., University of Glasgow, Glasgow (United Kingdom) A Study of the Role of ABA in the Regulation of Foliar Senescence in *Raphanus*, *Phaseolus* and *Taxus*

199. Hillman, J.R., Math, V.B., and McDougall, J.: Botany Dept., University of Glasgow, Glasgow (United Kingdom) IAA and Apical Dominance in the French Bean

200. Ilahi, I.: Department of Botany, University of Peshawar (Pakistan) The Nature and Balance of Growth-Regulatory Substances in the Reproductive Organs of *Ceratonia siliqua* L.

201. Kulescha, Z.: Botanisch Laboratorium, Leiden (The Netherlands) Absorption and Degradation of IAA by Jerusalem Artichoke

202. Laidman, L.: Dept. of Biochemistry and Soil Science, University College of North Wales, Bangor (United Kingdom) The Probable Lysosomal Basis for Gibberellin Responses in Wheat Aleurone Tissue

203. Rubinstein, B., Mahar, P., and Tattar, T.A.: Botany Dept., University of Massachusetts, Amherst, Massachusetts (USA) Use of Osmotic Shock to Localize and Characterize a Site of Auxin Action

Because the abstracts of paper-demonstrations No 130, 159 and 175 on one hand and No 192-203 on the other hand have not reached us in time, their authors have not been arranged in alphabetical order.

Subject Index

MIX
Papier aus verantwortungsvollen Quellen
Paper from responsible sources
FSC® C105338

If you have any concerns about our products,
you can contact us on
ProductSafety@springernature.com

In case Publisher is established outside the EU,
the EU authorized representative is:
Springer Nature Customer Service Center GmbH
Europaplatz 3, 69115 Heidelberg, Germany

Printed by Libri Plureos GmbH
in Hamburg, Germany